Methods in Molecular Biology™

Series Editor
John M. Walker
School of Life Sciences
University of Hertfordshire
Hatfield, Hertfordshire, AL10 9AB, UK

For further volumes:
http://www.springer.com/series/7651

Phytoplasma

Methods and Protocols

Edited by

Matt Dickinson

School of Biosciences, University of Nottingham, Sutton Bonington, Nottingham, UK

Jennifer Hodgetts

The Food and Environment Research Agency, Sand Hutton, York, UK

Humana Press

Editors
Matt Dickinson
School of Biosciences
University of Nottingham
Sutton Bonington
Nottingham, UK

Jennifer Hodgetts
The Food and Environment Research Agency
Sand Hutton, York, UK

ISSN 1064-3745 ISSN 1940-6029 (electronic)
ISBN 978-1-4939-5965-5 ISBN 978-1-62703-089-2 (eBook)
DOI 10.1007/978-1-62703-089-2
Springer New York Heidelberg Dordrecht London

Printed on acid-free paper

Humana Press is a brand of Springer
Springer is part of Springer Science+Business Media (www.springer.com)

Preface

Phytoplasmas are a group of bacteria that lack a cell wall and can not as yet be grown in axenic culture. They are capable of causing severe diseases in a wide range of plant species, and are vectored between plants by insect vectors (mainly leafhoppers) in which the bacteria can also multiply. The aim of this book is to present a range of important protocols that can form the basis for anyone intending to develop a research programme on phytoplasmas or intending to set up a diagnostics facility for identifying the presence of these pathogens in plants or their insect vectors.

Following an introductory chapter on the importance of phytoplasma diseases, a number of protocols for maintaining collections of plants and insects and for transferring phytoplasmas between plant species by insects, grafting or dodder are presented. This is followed by methods for detection and diagnosis, ranging from microscopy-based methods through PCR and real-time PCR to field-based detection methods. Techniques are also included for separating and classifying the phytoplasmas into their different taxonomic groups and subgroups, as well as methods that have been developed for proteomics analyses. The final chapters cover the methods for separating phytoplasma genomic and plasmid DNA from plant DNA for whole genome sequencing, along with methods for mapping phytoplasma genomes. The target audience is primarily plant pathologists and molecular biologists, including scientists in developing countries where phytoplasmas are often a serious and devastating problem of crop plants.

Nottingham, UK *Matt Dickinson*
York, UK *Jennifer Hodgetts*

Contents

Contributors

MARK T. ANDERSEN • *Plant & Food Research, Auckland, New Zealand*
NANCY M. ANDRADE • *Institute of Production and Plant Protection, Universidad Austral de Chile, Valdivia, Chile*
NOLBERTO L. ARISMENDI • *Institute of Production and Plant Protection, Universidad Austral de Chile, Valdivia, Chile*
ASSUNTA BERTACCINI • *Alma Mater Studiorum—University of Bologna, DiSTA, Plant Pathology, Bologna, Italy*
SABRINA BERTIN • *DIVAPRA—Entomologia e Zoologia applicate all'Ambiente "Carlo Vidano", Università degli Studi di Torino, Grugliasco, Italy*
JANA BOBEN • *National Institute of Biology, Ljubljana, Slovenia*
DOMENICO BOSCO • *DIVAPRA—Entomologia e Zoologia applicate all'Ambiente "Carlo Vidano", Università degli Studi di Torino, Grugliasco, Italy*
PATRICIA CARLE • *INRA, University of Bordeaux, UMR1332 Fruit Biology and Pathology, Villenave d'Ornon, France*
NYNNE MEYN CHRISTENSEN • *Center for Advanced Bioimaging (CAB), University of Copenhagen, Copenhagen, Denmark*
NICOLETTA CONTALDO • *Alma Mater Studiorum—University of Bologna, DiSTA, Plant Pathology, Bologna, Italy*
DAVID CROSSLEY • *The Food and Environment Research Agency, Sand Hutton, York, UK*
JEAN-LUC DANET • *INRA, University of Bordeaux, UMR1332 Fruit Biology and Pathology, Villenave d'Ornon, France*
ROBERT E. DAVIS • *USDA, Beltsville, MD, USA*
MARINA DERMASTIA • *National Institute of Biology, Ljubljana, Slovenia*
B. JEAN DEVONSHIRE • *Bioimaging, Plant Pathogens and Microbiology, Rothamsted Research, Harpenden Herts, UK*
MATT DICKINSON • *School of Biosciences, University of Nottingham, Sutton Bonington, Nottingham, UK*
BOJAN DUDUK • *Institute of Pesticides and Environmental Protection, Laboratory of Applied Phytopathology, Belgrade, Serbia*
XAVIER FOISSAC • *INRA, University of Bordeaux, UMR1332 Fruit Biology and Pathology, Villenave d'Ornon, France*
YINGPING GAI • *State Key Laboratory of Crop Biology, Shandong Agricultural University, Tai'an, Shandong, People's Republic of China*
LUCIANA GALETTO • *Istituto di Virologia Vegetale, CNR, Torino, Italy*
JÜRGEN GROSS • *Federal Research Centre for Cultivated Plants, Institute for Plant Protection in Fruit Crops and Viticulture, Dossenheim, Germany*
KRISTINA GRUDEN • *National Institute of Biology, Ljubljana, Slovenia*

NIGEL A. HARRISON • *Plant Pathology Department, Research & Education Center, University of Florida, Fort Lauderdale, FL, USA*
ERICKA E. HELMICK • *Plant Pathology Department, Research & Education Center, University of Florida, Fort Lauderdale, FL, USA*
JENNIFER HODGETTS • *The Food and Environment Research Agency, Sand Hutton, York, UK*
MATJAŽ HREN • *National Institute of Biology, Ljubljana, Slovenia*
XIANLING JI • *College of Forestry, Shandong Agricultural University, Tai'an, Shandong, People's Republic of China*
HEATHER KINGDOM • *John Innes Centre, Norwich Research Park, Norwich, UK*
ING-MING LEE • *Molecular Plant Pathology Laboratory, USDA, ARS, Beltsville, MD, USA*
LIA W. LIEFTING • *Plant Health and Environment Laboratory, Ministry for Primary Industries, Auckland, New Zealand*
OLGA MAKAROVA • *Department of Agroecology, Aarhus University, Slagelse, Denmark*
SYLVIE MALEMBIC-MAHER • *INRA, University of Bordeaux, UMR1332 Fruit Biology and Pathology, Villenave d'Ornon, France*
CARMINE MARCONE • *Department of Pharmaceutical and Biomedical Sciences, University of Salerno, Salerno, Italy*
PAOLO MARGARIA • *Istituto di Virologia Vegetale, CNR, Torino, Italy*
MARTA MARTINI • *Department of Agriculture and Environmental Sciences, University of Udine, Udine, Italy*
CRISTINA MARZACHÌ • *Istituto di Virologia Vegetale, CNR, Torino, Italy*
NATAŠA MEHLE • *National Institute of Biology, Ljubljana, Slovenia*
RITA MUSETTI • *Department of Agricultural and Environmental Sciences, University of Udine, Udine, Italy*
MILAN NAVRÁTIL • *Department of Cell Biology and Genetics, Palacký University, Olomouc, Czech Republic*
MOGENS NICOLAISEN • *Department of Agroecology, Aarhus University, Slagelse, Denmark*
PETRA NIKOLIĆ • *National Institute of Biology, Ljubljana, Slovenia*
HIROMI NISHIDA • *Graduate School of Agricultural and Life Sciences, The University of Tokyo, Tokyo, Japan*
HENRIETTE NYSKJOLD • *Department of Agroecology, Aarhus University, Slagelse, Denmark*
SABRINA PALMANO • *Istituto di Virologia Vegetale, CNR, Torino, Italy*
SAMANTA PALTRINIERI • *Alma Mater Studiorum—University of Bologna, DiSTA, Plant Pathology, Bologna, Italy*
NINA PREZELJ • *National Institute of Biology, Ljubljana, Slovenia*
JAROSLAVA PŘIBYLOVÁ • *Institute of Plant Molecular Biology, Biology Centre of the Academy of Sciences of the Czech Republic, České Budějovice, Czech Republic*
MAJA RAVNIKAR • *National Institute of Biology, Ljubljana, Slovenia*
MATEVŽ RUPAR • *National Institute of Biology, Ljubljana, Slovenia*
DANA ŠAFÁŘOVÁ • *Department of Cell Biology and Genetics, Palacký University, Olomouc, Czech Republic*

Pascal Salar • *INRA, University of Bordeaux, UMR1332 Fruit Biology and Pathology, Villenave d'Ornon, France*
Bernd Schneider • *Julius Kühn-Institute (JKI), Federal Research Centre for Cultivated Plants, Institute for Plant Protection in Fruit Crops and Viticulture, Dossenheim, Germany*
Martina Šeruga Musić • *Department of Microbiology, Division of Biology, University of Zagreb, Zagreb, Croatia*
Jonathan Shao • *Molecular Plant Pathology Laboratory, USDA-Agricultural Research Service, Beltsville, MD, USA*
Majid Siampour • *Plant Virology Research Center, College of Agriculture, Shiraz University, Shiraz, Iran*
Dijana Škorić • *Department of Microbiology, Division of Biology, University of Zagreb, Zagreb, Croatia*
Josef Špak • *Institute of Plant Molecular Biology, Biology Centre of the Academy of Sciences of the Czech Republic, České Budějovice, Czech Republic*
Xiaobing Suo • *Molecular Plant Pathology Laboratory, USDA-Agricultural Research Service, Beltsville, MD, USA*
Mara Tedeschi • *Alma Mater Studiorum—University of Bologna, DiSTA, Plant Pathology, Bologna, Italy*
Rosemarie Tedeschi • *DIVAPRA—Entomologia e Zoologia applicate all'Ambiente "Carlo Vidano", Università degli Studi di Torino, Grugliasco, Italy*
Jenny Tomlinson • *The Food and Environment Research Agency, Sand Hutton, York, UK*
Lucy T.T. Tran-Nguyen • *Department of Resources, Plant Industries, Darwin, NT, Australia*
Melanie Tuffen • *School of Biosciences, University of Nottingham, Sutton Bonington, Nottingham, UK*
Pavla Válová • *Department of Cell Biology and Genetics, Palacký University, Olomouc, Czech Republic*
Wei Wei • *Institute for Bioscience and Biotechnology Research, University of Maryland, College Park, MD, USA*
Phyllis Weintraub • *Gilat Research Center, D.N. Negev, Israel*
Yan Zhao • *Molecular Plant Pathology Laboratory, USDA-Agricultural Research Service, Beltsville, MD, USA*

Chapter 1

The Phytoplasmas: An Introduction

Matt Dickinson, Melanie Tuffen, and Jennifer Hodgetts

Abstract

This volume of "Methods in Molecular Biology" entitled "Phytoplasmas: Methods and Protocols" aims to provide a broad range of protocols for working with this group of plant pathogens. In this first chapter, we provide some background information about the phytoplasmas to put the protocols into context.

Key words: Detection, Diagnostics, Morphology, Pathogenicity, Phytoplasmas, Taxonomy, Transmission

1. Introduction

Phytoplasmas are phloem-limited bacterial plant pathogens that were first identified in 1967 in ultrathin sections of the plant phloem of plants infected with yellows disease, a syndrome that was previously thought to be caused by a virus. Because their observed morphology was similar to a group of animal pathogenic mollicutes (bacteria that lack cell walls) called the mycoplasmas, they were named mycoplasma-like organisms or MLOs (1). Subsequently, many other diseases were discovered to have MLOs associated with them, and in the early 1990s, phylogenetic analyses of DNA sequences from these organisms showed them to be a large, monophyletic group within the mollicutes, and the term phytoplasma was proposed and adopted to describe them. Other mollicutes include the mycoplasmas, acholeplasmas, and spiroplasmas, and they are most closely related to bacteria such as *Bacillus*, *Clostridium*, and *Streptococcus*.

The phytoplasmas have small genomes ranging from 530 to 1,350 kb (amongst the smallest known for any self-replicating organisms) and a low G+C content in their DNA (23.0–29.5 Mol%), and have now been assigned to a novel genus, '*Candidatus* Phytoplasma' (2, 3). This category of *Candidatus* is used by

Matt Dickinson and Jennifer Hodgetts (eds.), *Phytoplasma: Methods and Protocols*, Methods in Molecular Biology, vol. 938, DOI 10.1007/978-1-62703-089-2_1, © Springer Science+Business Media, LLC 2013

taxonomists to classify organisms such as phytoplasmas that cannot be cultured in vitro, and over 20 *Candidatus* species have been identified within the genus, based primarily on the basis that isolates within a species share at least 97.5% sequence identity within their 16S rRNA gene. In addition to the confirmed *Candidatus* species, there are a number of proposed "species" that have not yet been formally approved, and running in parallel to the *Candidatus* species classification system is the more widely used 16Sr group classification system. This system is based on restriction enzyme digest profiles of a defined region of the 16S rDNA (*see* Chapter 14). The standard procedure when using this classification system is that "universal" primers are used to amplify this specified region of the 16S rDNA from infected plant or insect DNA samples, and the polymerase chain reaction (PCR) product is then digested with specific restriction enzymes and the profiles analyzed, normally using polyacrylamide gel electrophoresis to give good resolution of the digest products. The restriction fragment length polymorphism (RFLP) profiles are then compared, and specific patterns are used to classify phytoplasmas into the 16Sr groups and subgroups (4).

Although phytoplasmas cannot be maintained in axenic culture, it has been possible to maintain reference strains for some phytoplasmas in host plants such as the Madagascar periwinkle (*Catharanthus roseus*), either in glasshouse maintained plants or in micropropagation (see Chapters 2, 3, and 4), and collections of these phytoplasmas are held at a number of institutions around the World. However, there are many other phytoplasma "isolates" that have not been maintained and for which the only information available is a description of symptoms and a sequence at GenBank. This makes it difficult to undertake conventional RFLP analysis and compare patterns by gel electrophoresis for all the different groups and subgroups, and has also made it difficult to incorporate new strains and subgroups into the 16Sr or '*Candidatus* species' classification systems. As a result, Wei et al. (5, 6) have devised a system based on sequences deposited at the National Center for Biotechnology Information's (NCBI) nucleotide sequence database, in which the restriction digestion profiles are predicted for a broad range of enzymes using in silico methods (see Chapter 28). Similarity coefficients for the restriction fragments are then calculated by defined formulae (4, 6, 7) to reflect the number of shared and distinct fragments between any given strains, and based on this system there are now over 28 16Sr groups (see Table 1).

Phytoplasmas have been found associated with numerous diseases of plants worldwide, and the symptoms they elicit vary depending on the host, phytoplasma strain, and environmental factors, and often include yellow/purple discolouration of leaves and shoots, virescence (greening of petals), phyllody (conversion of floral organs into leaf-like structures), proliferation of shoots, witches' broom, stunting, general decline, and in some cases plant death.

Table1
Taxonomic groupings of phytoplasmas

16Sr group[a]	*Candidatus* species[b]	Type strain	GenBank accession number	Geographic distribution of the group[c]
I-A	*Candidatus* Phytoplasma asteris	Aster yellows witches' broom	NC_007716	North America, Europe
I-B	*Ca.* Phytoplasma asteris	Onion yellows	NC_005303	Worldwide
I-C	*Ca.* Phytoplasma asteris	Clover phyllody	AF222065	North America, Europe
I-D	*Ca.* Phytoplasma asteris	Paulownia witches' broom	AY265206	Asia
I-E	*Ca.* Phytoplasma asteris	Blueberry stunt	AY265213	North America
I-F	*Ca.* Phytoplasma asteris	Apricot chlorotic leaf roll	AY265211	Spain
II-A		Peanut witches' broom	L33765	Asia
II-B	*Ca.* Phytoplasma aurantifolia	Lime witches' broom	U15442	Arabian peninsula
II-C		Cactus witches' broom	AJ293216	Asia, Africa
II-D	*Ca.* Phytoplasma australasiae*	Papaya yellow crinkle	Y10097	Australia
III-A	*Ca.* Phytoplasma pruni*	Western X disease	L04682	North America
III-B		Clover yellow edge	AF189288	America, Asia, Europe
IV-A	*Ca.* Phytoplasma palmae*	Coconut lethal yellowing	AF498307	Florida, Caribbean
IV-B	*Ca.* Phytoplasma palmae*	Phytoplasma sp. LfY5(PE65)-Oaxaca	AF500334	Mexico
IV-D	*Ca.* Phytoplasma palmae*	*Carludovica palmata* leaf yellowing	AF237615	Mexico
V-A	*Ca.* Phytoplasma ulmi	Elm yellows	AY197655	North America, Europe
V-B	*Ca.* Phytoplasma ziziphi	Jujube witches' broom	AB052876	Asia
V-C	*Ca.* Phytoplasma vitis*	Alder yellows	AY197642	Europe
V-G		Jujube witches' broom related	AB052879	Asia
VI-A	*Ca.* Phytoplasma trifolii	Clover proliferation	AY390261	North America, Asia

(continued)

Table 1 (continued)

16Sr group[a]	*Candidatus* species[b]	Type strain	GenBank accession number	Geographic distribution of the group[c]
VII-A	*Ca.* Phytoplasma fraxini	Ash yellows	AF092209	North America
VIII-A	*Ca.* Phytoplasma luffae*	Loofah witches' broom	AF353090	Taiwan
IX-A		Pigeon-pea witches' broom	AF248957	America
IX-D	*Ca.* Phytoplasma phoenicium	Almond witches' broom	AF515636	Middle East
X-A	*Ca.* Phytoplasma mali	Apple proliferation	AJ542541	Europe
X-C	*Ca.* Phytoplasma pyri	Pear decline	AJ542543	Europe
X-D	*Ca.* Phytoplasma spartii	Spartium witches'-broom	X92869	Europe
X-F	*Ca.* Phytoplasma prunorum	European stone fruit yellows	AJ542544	Europe
XI-A	*Ca.* Phytoplasma oryzae	Rice yellow dwarf	AB052873	Asia, Africa, Europe
XII-A	*Ca.* Phytoplasma solani*	Stolbur	AJ964960	Europe
XII-B	*Ca.* Phytoplasma australiense	Australian grapevine yellows	L76865	Australasia
XII-C		Strawberry lethal yellows	AJ243045	Australia
XII-D	*Ca.* Phytoplasma japonicum	Japanese hydrangea phyllody	AB010425	Japan
XII-E	*Ca.* Phytoplasma fragariae	Strawberry yellows	DQ086423	Europe
XIII-A		Mexican periwinkle virescence	AF248960	Mexico, Florida
XIV-A	*Ca.* Phytoplasma cynodontis	Bermudagrass whiteleaf	AJ550984	Asia, Africa
XV-A	*Ca.* Phytoplasma brasiliense	Hibiscus witches' broom	AF147708	Brazil
XVI-A	*Ca.* Phytoplasma graminis	Sugar cane yellow leaf	AY725228	Cuba
XVII-A	*Ca.* Phytoplasma caricae	Papaya bunchy top	AY725234	Cuba
XVIII-A	*Ca.* Phytoplasma americanum	Potato purple top wilt	DQ174122	North America
XIX-A	*Ca.* Phytoplasma castanae	Chestnut witches' broom	AB054986	Japan

(continued)

Table 1 (continued)

16Sr group[a]	*Candidatus* species[b]	Type strain	GenBank accession number	Geographic distribution of the group[c]
XX-A	*Ca.* Phytoplasma rhamni	Buckthorn witches' broom	X76431	Germany
XXI-A	*Ca.* Phytoplasma pini	Pine shoot proliferation	AJ632155	Spain
XXII-A	*Ca.* Phytoplasma cocosnigeriae*	Coconut lethal decline	Y14175	West Africa, Mozambique
XXIII-A		Buckland valley grapevine yellows	AY083605	Australia
XXIV-A		Sorghum bunchy shoot	AF509322	Australia
XXV-A		Weeping tea witches' broom	AF521672	Australia
XXVI-A		Sugar cane phytoplasma D3T1	AJ539179	Mauritius
XXVII-A		Sugar cane phytoplasma D3T2	AJ539180	Mauritius
XXVIII-A		Derbid phytoplasma	AY744945	Cuba
XXIX-A*	*Ca.* Phytoplasma allocasuarinae	*Allocasuarina muelleriana* phytoplasma	AY135523	Australia
XXX-A*	*Ca.* Phytoplasma cocostanzaniae*	Tanzanian lethal decline	X80117	Tanzania

[a]The 16Sr classification system is based on the computer simulated RFLP work of Wei et al. (5). The groups marked with an asterisk were not included in the original analysis since the available sequences did not cover the complete F2nR2 region required for simulated RFLP analysis
[b]The '*Candidatus* species' classification is based on that described in the review by Hogenhout et al. (25). The species marked with an asterisk are provisional names that have not been ratified as yet
[c]The geographical locations are updated from those described by Lee et al. (2)

Amongst the more economically important diseases are the aster yellows type which occur on a broad range of crops, such as potatoes, carrots, maize, tomatoes, onions and flowers worldwide, the peach diseases of the United States, the apple proliferation/pear decline diseases of Europe, the grapevine diseases that occur worldwide, the sugarcane and grass diseases such as rice yellow dwarf, sugarcane whiteleaf, and napier grass stunt that occur in East Africa and Asia, the potato witches' broom and maize bushy stunt diseases of Central and South America, and the coconut lethal yellowing and lethal yellowing-like diseases that occur in the Caribbean, Central America, and throughout coconut growing regions of Africa. A look at Journals such as "New Disease Reports" or "Plant

Disease" shows that new phytoplasma diseases are constantly emerging in new hosts and at new locations; and Table 1 details some of these important diseases and the 16Sr group and/or *Candidatus* species to which they belong. This list is not exhaustive, and it is important to note that "isolates" are often given names based on the diseases they are associated with, but as mentioned above, in most cases the only information that is provided for these "isolates" will be descriptions of the symptoms and a portion of the 16S rDNA sequence deposited at GenBank.

This sequence will be enough information to assign the "isolate" to a 16Sr group/subgroup and possibly also a *Candidatus* species, but it will not be enough to ascertain how it relates to other "isolates" within that same 16Sr subgroup. Thus there may be many "isolates" obtained from different plant species and given different names that might be identical organisms, but further sequencing of additional genes would be needed to confirm this, and a number of protocols are provided in this volume for examining additional genes beyond the 16S rRNA and for "barcoding" of phytoplasmas (see Chapters 14, 15, 16, 17, 26, and 27). Additional confusion can be caused by the fact that sometimes a disease is given a common name in different parts of the world because of the symptoms caused (e.g., the coconut lethal type diseases), but when these "isolates" are studied in more detail they are found to belong to different 16Sr groups depending on the country of origin. For example the coconut lethal yellows diseases of the United States and Caribbean belong to 16SrIV, whilst the diseases causing similar symptoms in Ghana and Mozambique belong to 16SrXXII and those in Tanzania to a different and currently unassigned 16Sr grouping (provisionally listed as 16SrXXX in Table 1). It is therefore important to be cautious in assigning names to new diseases that are associated with phytoplasma and also when examining "isolates" for which sequences have been deposited at GenBank, since as yet there has been limited independent validation work for large numbers of these sequences. However, there are also many phytoplasmas for which reference samples are available, and that have been studied extensively, and much pioneering work has been undertaken on these organisms to understand their genomes, pathogenicity mechanisms, and transmission, and to try to control and manage the diseases they cause.

2. Morphology

Phytoplasmas can be visualized in infected material by microscopy and their morphology is often described as being pleiomorphic and rounded or filamentous in shape, with diameters less than 1 μm (2). Like other mollicutes, phytoplasmas lack cell walls, and are bound solely by a triple-layered unit membrane (1).

Because they cannot as yet be cultured in axenic culture, visualization has tended to require skilled microscopists and techniques that involve sectioning and fixing tissue followed by DAPI (diamidino-2-phenylindole) staining and fluorescence microscopy, or transmission electron microscopy (TEM). In addition, bioimaging methods have been developed using vital fluorescence dyes and confocal laser scanning microscopy to visualize phytoplasmas in situ (8) and some of these different protocols are presented in Chapters 9, 10, and 11.

3. Detection and Diagnostics

Whilst visualization by microscopy is very useful for identifying and confirming the presence of phytoplasmas in plants and for localization studies, it is not ideal as a routine rapid diagnostic method. Instead most diagnostics for phytoplasmas involves detection of the bacterial nucleic acid in infected plants/insects. Some attempts have also been made to develop antibody-based detection systems. These have either used enriched phytoplasma preparations from infected tissues for injecting into animals to produce polyclonal antibodies, or more recently specific genes cloned into bacterial expression vectors have been used to produce recombinant proteins to produce more specific polyclonal and monoclonal antibodies (see Chapter 30). Whilst the use of enriched fractions has the advantage of being relatively quick and easy, it is not as yet possible to produce a completely enriched preparation that does not contain host tissues, so any polyclonal antibody preparations produced will have poor specificity and detect plant and phytoplasmas, so care needs to be taken in screening and using such preparations. Using recombinant proteins has the advantage of producing much more specific antibody preparations, but if these are to be developed into a diagnostic system the proteins used probably need to be highly expressed proteins that are exposed on the surface of the bacteria, so that they can be readily detected. Major antigenic proteins have been identified and the genes encoding them sequenced for some phytoplasmas, and there is the potential for developing some of these into detection systems, but these would be specific for a particular phytoplasma rather than generic, since there is a high level of diversity in these proteins between different phytoplasmas.

In this protocols book, we have focussed mainly on nucleic acid-based detection and diagnostics systems, since these can be readily developed and used as both a generic tool (to identify whether any phytoplasma is present in a sample) or specific (to identify which 16Sr group or '*Candidatus* species' is present in the tissue). They can also be used for the phylogenetic characterization of phytoplasmas and to enhance understanding of the diversity that exists in this group of organisms.

The simplest and most commonly used diagnostic method is the PCR and phytoplasma-specific primers have been developed that amplify various regions of the rRNA operon (see Chapter 14). Total DNA is extracted from test plant or insect material, and a few protocols for DNA extraction and enriching for the presence of phytoplasma in the tissue are described in Chapters 12 and 13, and the phytoplasma-specific primers will then confirm whether any phytoplasma is present. Where phytoplasmas are present at low titres, a nested PCR approach is often required, sometimes using 16Sr group-specific primers to aid taxonomic classification. Alternatively, and when group-specific primers are not used, RFLP techniques can be used on the PCR products to assign phytoplasmas to 16Sr groupings. In addition, single-strand conformation assays have been developed to detect minor variability between closely related strains within groups (see Chapter 18), and sequencing of the rRNA genes can also be used for accurate identification of strains for taxonomic purposes, and if necessary the sequences can be put through the iPhy classifier (Chapter 28) to assign them to a 16Sr group.

Whilst basic PCR is quick and easy to use as a diagnostic method, care has to be exercised with the protocols and interpretation of results. Some plant DNA extracts contain inhibitors of *Taq* polymerase, so a negative result in PCR does not necessarily mean the absence of phytoplasma. Normally, samples are also tested with primers for plant/insect genes to confirm that they support PCR before conclusions can be drawn. An alternative strategy that has been developed is the use of Terminal Restriction Fragment Length Polymorphism analysis (T-RFLP) (Chapter 20) which allows the plant and phytoplasma to be identified simultaneously in a sample to confirm that it supports PCR. False positives are also a major problem in standard PCR protocols, especially if nested PCR is used. The use of an open system in which tubes have to opened up and samples run on gels can lead to large quantities of PCR products being released into labs as aerosols, etc., and these can back-contaminate reaction mixes and samples if sufficient care is not taken.

Because of these issues, most diagnostics laboratories have moved away from open systems into closed systems, in which amplification products are detected in real-time rather than on gels. These methods include real-time PCR and also real-time Loop Mediated Isothermal amplification (LAMP) (see Chapters 21, 22, 23, 24, and 25) and protocols have been developed that are either generic or specific, and can detect phytoplasmas with similar sensitivity to nested PCR techniques. The advantages of these systems are the reduction in potential for contamination and false positives, so they are recommended for routine diagnostics purposes. The disadvantages are that the equipment, especially for real-time PCR is expensive, and it is difficult to do downstream analysis on the samples, such as RFLP analysis and 16Sr group identification, so either specific assays have to be used for detection, or samples can be analyzed subsequently by conventional PCR and RFLP/sequencing if this additional information is required.

It is important to note that detecting a phytoplasma in an infected plant is not absolute proof that the organism is responsible for the disease, as Koch's postulates cannot be proved for unculturable organisms. As a result there are many examples where the distinction has to be made that the phytoplasmas have been found "associated" with a particular symptom. The problem is compounded by the fact that phytoplasmas can be found in plants exhibiting no obvious symptoms, and conversely there are occasions where plants presenting symptoms have been tested but only a proportion show the presence of the phytoplasma. This is generally attributed to the uneven distribution of phytoplasmas within infected plants such that samples taken from two different parts of the same plant may give different results in PCR analysis. In addition, there is evidence that phytoplasma titres are affected by seasonal variation and environmental factors (2). However, where the vectors are known and appropriate acquisition and feeding experiments have been undertaken, there are now many cases where it has been unambiguously shown that phytoplasmas are the causal agents of disease.

4. Genomics and Pathogenicity Mechanisms

The symptoms caused by phytoplasmas in plants suggest interference with plant signaling mechanisms and/or hormone metabolism. Physiological studies have shown that levels of starch in the roots of fruit trees infected with phytoplasmas are reduced to only around a half to a third of normal, and that starch and other carbohydrates accumulate in the leaves. Radiolabelling experiments have confirmed that transport of photosynthates from shoots to roots and from mature to young leaves is impaired in phytoplasma-infected plants. Metabolomic studies in the Madagascar periwinkle have shown an accumulation in leaves of metabolites related to the biosynthesis of phenylpropanoid and terpenoid indole alkaloid secondary metabolites along with the accumulation of sucrose, glucose, glycine, polyphenols, and succinic acid (9). Sugar levels are also known to be important in regulating floral transition, and gene expression studies in tomatoes infected with the stolbur phytoplasma (which causes virescence and phyllody symptoms) has indicated that genes controlling the identity of the inflorescence meristem are up-regulated, whilst those involved in meristem development and flower identity are down-regulated (10).

To gain further insights into pathogenicity mechanisms of phytoplasmas, genome sequencing projects have been undertaken and genes associated with pathogenicity identified. These sequencing projects have involved the isolation of pure phytoplasma DNA from infected plants/insects using either CsCl gradients or pulse field gel electrophoresis, and protocols for this as well as for the isolation of phytoplasma plasmids and chromosome mapping are described in Chapters 31, 32, 33, and 34.

The full sequences and annotations of four phytoplasma genomes are publicly available, two of them belonging to the 16SrI, '*Ca.* Phytoplasma asteris' group, onion yellows strain (OY) (11) and aster yellows witches' broom strain (AY-WB) (12); along with 16SrXII '*Ca.* Phytoplasma australiense (13), and 16SrX '*Ca.* Phytoplasma mali' (14). '*Ca.* Phytoplasma mali', unlike other mollicutes studied so far, has a linear chromosome (14), but there are many 16Sr groups for which little sequence information is currently available, so whether the occurrence of linear genomes is widespread is unclear. The two aster yellows phytoplasmas possess circular chromosomes (along with 2–4 plasmids) and encode 754 and 671 genes, respectively, of which almost 60% have assigned functions. The '*Ca.* Phytoplasma australiense' genome is also circular, encoding 839 genes, and has one plasmid, whilst '*Ca.* Phytoplasma mali' has no plasmids and encodes 497 genes.

Analysis of the genome sequences reveals that all four of the sequenced phytoplasma genomes lack genes for key metabolic processes, even those found to be conserved in other minimal genome organisms. This includes pathways for the biosynthesis of many amino acids, fatty acids, and nucleotides, suggesting the phytoplasma must import all of these from their hosts (13, 14). Furthermore, despite living in high sugar environments (1 M sucrose in phloem and 0.9 M trehalose in insect hemolymph), phytoplasmas lack the genes for the phosphotransferase systems that most other bacteria use as their energy-efficient means for importing and phosphorylating sugars such as sucrose, glucose, and fructose. Furthermore, they appear to lack sucrose degrading enzymes, and instead it appears that phytoplasmas probably import the hexoses in their phosphorylated form for direct use in their glycolytic pathways, and it may be that it is the resultant imbalance in sugar distribution in plants that accounts for the disease symptoms (15). Additionally, phytoplasmas have incomplete pathways for carbohydrate metabolism and ATP production. The genome of '*Ca.* Phytoplasma asteris' was the first living organism to be identified that has no genes for ATP synthase (11) or genes that may be able to import ATP, and this has also been established in '*Ca.* Phytoplasma mali', such that it is suggested phytoplasmas may import or generate ATP using an unknown mechanism (14).

The small genome size and apparent tendency towards reductive evolution of phytoplasmas suggests the surprisingly high level of repeats within phytoplasma genomes to be of some significance. Several genes have been found to have multiple copies, for example

'*Ca.* Phytoplasma australiense' has two complete copies of the ABC transport system for peptide and nickel (13). Two strains of Onion yellows have been isolated that show both severe and mild symptoms in plants (OY-W and OY-M, respectively). OY-W was found to have tandem repeat of a 30 kb region of its genome, which contained genes essential for glycolysis (16).

Multicopy genes are often associated with another phenomenon of phytoplasma genomes: Potential Mobile Units (PMUs). Some PMUs do not contain all the necessary elements to be truly mobile, but some do including genes with DNA recombination function, a transposase and are flanked by inverted repeats (17). Within phytoplasma genomes PMUs tend to be clustered, and probably the best characterized PMU is PMU1 from AY-WB which was found to exist in two forms within AY-WB: a circular, extrachromasomal PMU1 (termed C-PMU1), and a linear, integrated form (L-PMU1) (18). Interestingly, the levels of C-PMU1 are higher in phytoplasmas infecting insect cells than those isolated from plants, and the increase of C-PMU1 in insect infecting cells suggests that this process helps adapt phytoplasmas to their environment (18). Adaptation to new environments provides an attractive explanation for why an organism with such a small genome contains multiple repeats of genes.

It is known that many plant-pathogenic bacteria use Type III and IV secretion systems to pump effectors and pathogenicity factors into plant cells. Whilst phytoplasmas lack these secretion systems, they do have Sec-dependent secretion systems, and based on genome sequencing, protein sequences that carry N-terminal signal peptides targeting them for secretion have been identified. It is postulated that the phytoplasmas secrete these small peptides into the phloem, and that they are then able to migrate into other plant cells to affect gene expression. A number of these genes have recently been expressed in transgenic plants and shown to act as effector proteins, interfering with transcription factors in the host plants to result in classical phytoplasma symptoms such as virescence, phyllody, witches' broom, and dwarfism (19–22). It has also been postulated that epigenetic methylation of plant DNA may be involved in specific down-regulation of certain floral development genes (10).

5. Transmission

Transmission of phytoplasmas between plants is by phloem-feeding insects of the order Hemiptera, primarily leafhoppers (Cicadelloidea), planthoppers (Fulgoroidea), and psyllids (Psylloidea) (23). However, there are still many phytoplasmas for which the vector(s) have not been confirmed, whilst others can be vectored by many species. Initial acquisition of phytoplasma occurs during feeding, and a period then occurs in which the insect vector is not able to transmit the phytoplasma—known as the latent period. In order for the vector to become infectious, phytoplasmas must pass through the midgut and basement membrane entering the insect's hemocoel. Phytoplasmas can move through the hemolymph and infect several parts of the insect, though it is penetration of the salivary glands that will allow transmission. Here the phytoplasma will replicate and reach a high titre that will allow for transmission during insect feeding (23–25). One of the other areas of the vector the phytoplasma can infect is the reproductive organs of the vectors, and it has been shown that transovarial transmission to progeny insects is possible. There are some instances where the presence of phytoplasmas in the insect has been reported to reduce the fitness of the vector (26), and also examples where they have been shown to increase the fecundity and longevity of the insect vector (27). The host range of phytoplasmas is influenced by their vector in that those such as the aster yellows type, which are transmitted by the polyphagous leafhopper *Macrosteles fascifrons* and other insects, can infect a broad host range (at least 191 plant species in 42 families), whilst those such as the coconut diseases that are transmitted by one or a few monophagous vector species tend to have a narrow host range. Protocols have been presented in this book for maintenance of insects, vector trapping, and the molecular identification of vector species (Chapters 5, 6, 7, and 8).

Experimentally, it is also possible to transmit phytoplasmas by grafting (see Chapter 2) or using plant parasitic dodders (*Cuscuta* sp.) (see Chapter 4), though attempts to transmit them by inoculating phytoplasma-containing sap into plants have not been successful. There have also been reports in which phytoplasma DNA has been detected in coconut embryos (28, 29) along with suggestions of seed transmission in alfalfa, tomato, oilseed rape, and lime. However, these indications of seed transmission of viable phytoplasmas have only been presented in conference proceedings to date and not in refereed papers, so it is unclear whether this is a means by which phytoplasmas are transmitted, and further studies are required to show whether it is just phytoplasma DNA that is present in seeds, or viable infectious organisms that can go on to produce symptoms in progeny plants.

6. Future Perspectives

Phytoplasmas cause a wide range of significant diseases, and whilst these have in the past been confined mainly to tropical and subtropical environments, there is an increasing risk of them moving to new crops and locations as the geographical range of their vectors increase. Furthermore, the plasticity of their genomes and presence of PMUs of DNA and virus-like sequences suggests that they may be able to generate new strains rapidly. Rigorous testing of plant material and regular surveys of insect vectors are the key to keeping these diseases under control. Significant advances have been made over recent years through the sequencing of phytoplasma genomes, and the molecular tools are now available to make rapid advances in our understanding of how they exert their pathogenic effects, especially as phytoplasmas have now been introduced into the model plant, *Arabidopsis thaliana*, and these genomic tools will also lead to improvements in diagnostic methods. In turn this understanding, combined with studies on the genetic and molecular basis of resistance and tolerance in plants should result in the development of improved strategies for managing phytoplasma diseases. It is hoped that the broad range of protocols presented in this book will enable both new and experienced researchers to continue investigating phytoplasmas in even greater detail so that we cannot only understand how these fascinating organisms survive and replicate in plant and insect hosts, but we can also develop appropriate and cost-effective means of controlling the major diseases of plants that they cause.

References

1. Doi Y et al (1967) Mycoplasma—or PLT group-like microorganisms found in the phloem elements of plants infected with mulberry dwarf, potato witches' broom, aster yellows, or paulownia witches' broom. Ann Phytopathol Soc Jpn 33:259–266
2. Lee I-M, Davis RE, Gundersen-Rindal DE (2000) Phytoplasmas: phytopathogenic mollicutes. Annu Rev Microbiol 56:1593–1597
3. Firrao G, Gibb K, Streten C (2005) Short taxonomic guide to the genus '*Candidatus* phytoplasma'. J Plant Pathol 87:249–263
4. Lee I-M et al (1998) Revised classification scheme of phytoplasmas based an RFLP analyses of 16S rRNA and ribosomal protein gene sequences. Int J Syst Bacteriol 48:1153–1169
5. Wei W et al (2007) Computer-simulated RFLP analysis of 16S rRNA genes: identification of ten new phytoplasma groups. Int J Syst Evol Microbiol 57:1855–1867
6. Wei W et al (2008) Automated RFLP pattern comparison and similarity coefficient calculation for rapid delineation of new and distinct phytoplasma 16Sr subgroup lineages. Int J Syst Evol Microbiol 58:2368–2377
7. Nei M, Li W-H (1979) Mathematical model for studying genetic variation in terms of restriction endonucleases. PNAS USA 76: 5269–5273
8. Christensen NM et al (2004) Distribution of phytoplasmas in infected plants as revealed by real-time PCR and bioimaging. Mol Plant Microbe Interact 17:1175–1184
9. Choi YH et al (2004) Metabolic discrimination of *Catharanthus roseus* leaves infected by phytoplasma using ^{1}H-NMR spectroscopy and multivariate data analysis. Plant Physiol 135: 2398–2410
10. Pracos P et al (2006) Tomato flower abnormalities induced by stolbur phytoplasma infection are associated with changes of expression of floral development genes. Mol Plant Microbe Interact 19:62–68
11. Oshima K et al (2004) Reductive evolution suggested from the complete genome sequence of a plant-pathogenic phytoplasma. Nat Genet 36:27–29
12. Bai X et al (2006) Living with genome instability: the adaptation of phytoplasmas to diverse environments of their insect and plant hosts. J Bacteriol 188:3682–3696
13. Tran-Nguyen LTT et al (2008) Comparative genome analysis of '*Candidatus* Phytoplasma australiense' (subgroup tuf Australia I; rp-A) and 'Ca. Phytoplasma asteris' strains OY-M and AY-WB. J Bacteriol 190:3979–3991
14. Kube M et al (2008) The linear chromosome of the plant-pathogenic mycoplasma '*Candidatus* Phytoplasma mali'. BMC Genomics 9:306
15. Christensen NM et al (2005) Phytoplasmas and their interactions with hosts. Trends Plant Sci 10:526–535

16. Oshima K, Nishida H (2007) Phylogenetic relationships among mycoplasmas based on the whole genomic information. J Mol Evol 65:249–258
17. Dickinson M (2010) Mobile units of DNA in phytoplasma genomes. Mol Microbiol 77: 1351–1353
18. Toruño TY et al (2010) Phytoplasma PMU1 exists as linear chromosomal and circular extrachromosomal elements and has enhanced expression in insect vectors compared with plant hosts. Mol Microbiol 77:1406–1415
19. Hoshi A et al (2008) A unique virulence factor for proliferation and dwarfism in plants identified from a phytopathogenic bacterium. PNAS USA 106:6416–6421
20. Bai X et al (2009) AY-WB phytoplasma secretes a protein that targets plant cell nuclei. Mol Plant Microbe Interact 22:18–30
21. Himeno M et al (2011) Unique morphological changes in plant pathogenic phytoplasma-infected petunia flowers are related to transcriptional regulation of floral homeotic genes in an organ-specific manner. Plant J 67:971–979
22. MacLean AM et al (2011) Phytoplasma effector SAP54 induces indeterminate leaf-like flower development in Arabidopsis plants. Plant Physiol 157:831–841
23. Weintraub PG, Beanland Le A (2006) Insect vectors of phytoplasmas. Annu Rev Entomol 51:91–111
24. Lefol C et al (1994) Propagation of flavescence doree MLO (Mycoplasma-Like Organism) in the leafhopper vector *Euscelidius variegatus.* J Invertebr Pathol 63:285–293
25. Hogenhout SA et al (2008) Phytoplasmas: bacteria that manipulate plants and insects. Mol Plant Pathol 9:403–423
26. Bressan A, Girolami V, Boudon-Padieu E (2005) Reduced fitness of the leafhopper vector *Scaphoideus titanus* exposed to Flavescence dorée phytoplasma. Entomol Exp Appl 115: 283–290
27. Beanland L et al (2000) Influence of aster yellows phytoplasma on the fitness of aster leafhopper (Homoptera: Cicadellidae). Ann Entomol Soc Am 93:271–276
28. Cordova I et al (2003) *In situ* PCR detection of phytoplasma DNA in embryos from coconut palms with lethal yellowing disease. Mol Plant Pathol 4:99–108
29. Nipah JO, Jones P, Dickinson MJ (2007) The Cape St Paul Wilt disease is transmitted to the embryos of infected West African tall coconut palms. Plant Pathol 56:777–784

Chapter 2

Techniques for the Maintenance and Propagation of Phytoplasmas in Glasshouse Collections of *Catharanthus roseus*

Jennifer Hodgetts, David Crossley, and Matt Dickinson

Abstract

Phytoplasma collections are a vital resource for researchers and diagnosticians studying phytoplasma diseases. They provide material as a point of reference and a research tool to increase our understanding of phytoplasmas and the diseases they cause. This chapter describes the techniques required to create and maintain collections of phytoplasma-infected *Catharanthus roseus* (Madagascar periwinkle).

Key words: *Catharanthus roseus*, Cuttings, Glasshouse maintenance, Grafting technique, Phytoplasma

1. Introduction

Catharanthus roseus (L.) G. Don (formerly known as *Vinca rosea*), commonly known as the Madagascar periwinkle has been found to be very susceptible to infections by phytoplasmas, although the reasons for this remain unknown. Because of this it has been used as an experimental host and for the maintenance of phytoplasma collections for research purposes (1–3). Periwinkles are a perennial tropical plant belonging to the family Apocynaceae which are distributed worldwide and widely cultivated (4). *C. roseus* is an evergreen plant with attractive dark glossy foliage and colored flowers meaning it is commonly grown in gardens and has long been used as an ornamental plant (5). However, it also has uses as a medicinal plant due to the discovery during the 1960s that leaf extracts had an antitumoral effect in rats. This is due to the production of numerous monoterpenoid indole alkaloids, two of which (vinblastine and vincristine) are now important commercial cancer therapy drugs (6, 7). Interestingly, studies have indicated a significant increase

Matt Dickinson and Jennifer Hodgetts (eds.), *Phytoplasma: Methods and Protocols*, Methods in Molecular Biology, vol. 938, DOI 10.1007/978-1-62703-089-2_2, © Springer Science+Business Media, LLC 2013

Fig. 1. Some of the typical symptoms caused by phytoplasma infection of *Catharanthus roseus*; (**a**) healthy *C. roseus*; (**b–f**) *C. roseus* infected with various phytoplasma strains showing (**b**) deformation of petals; (**c**) virescence of flowers; (**d–f**) various stages of phyllody.

of vinblastine in the roots of *C. roseus* infected with phytoplasmas compared to healthy plants (3).

Symptoms caused by phytoplasmas in the plant hosts (including *C. roseus*) can include, although are not limited to; virescence (the abnormal development of greening in floral tissue), phyllody (development of floral parts into leaf-like structures), sterility, elongation and etiolation of internodes, flower streaking and malformation, yellowing and upright posture of leaves, excessive branching of axillary shoots, proliferation, witches' broom (proliferation of shoots from a single point, typically in woody plants), general stunting, leaf curl or rolling, small and faintly colored flowers. The set of symptoms in infected plants may vary depending upon the phytoplasma in question, host plant species, the age of plant at infection, and the stage of infection. For example, within the Aster yellows group ('*Candidatus* Phytoplasma asteris') different subgroups produce distinct subsets of symptoms, and mild strains may induce no obvious symptoms. However, generally the symptoms are deleterious to the plant host (2, 8, 9) (Fig. 1).

The majority of phytoplasmas can be transmitted into *C. roseus* by means of grafting (this Chapter), dodder transmission (see Chapter 4) or insect transmission (see Chapter 5). However, there are notable exceptions, primarily phytoplasmas that are very host specific. In these cases, it is sometimes convenient to maintain the original host in glasshouse collections, for example some of the phytoplasmas which infect grasses such as Napier Grass Stunt (16SrXI).

However in other cases, the stature of the plant prevents this, for example Lethal yellowing diseases (16SrIV, XXII, and XXX) in *Cocos nucifera* (coconut palm), meaning that sources of these isolates are generally only available as natural infections in the environment. Phytoplasma titre (or concentration) is known to vary depending on the plant host, with woody hosts such as trees and grapevines tending to have particularly low levels. *C. roseus* tends to have much higher phytoplasma cell concentrations (10).

Common diseases which may afflict glasshouse grown *C. roseus* plants include (amongst others) red spider mite (*Tetranychus urticae*), powdery mildew, and sciarid fly. Perhaps the most damaging of these is red spider mite due to the slow, hard to detect build up of the disease. Often once the obviously visible symptoms of webbing have appeared any treatments are ineffective and the only option is to discard heavily infested plants. Milder infestations noticed earlier on can be treated with chemical applications or biological control with application of parasitic predators such as *Phytoseiulus persimilis* or *Neoseiulus californicus* (available commercially). However, it is not recommended to treat the plants prophylactically with broad ranging pesticides due to the risk of resistance developing. Sciarid fly infestations tend to result from overly wet compost and can be a more persistent problem if plants are housed on capillary matting in a glasshouse (to enable easier watering). There are no reports of commercially available pesticides or treatments having a derogatory effect on phytoplasmas (see Note 1), so when necessary these can be used to treat any disease outbreaks. However, we would not advocate routine application.

All phytoplasma-infected plant material should be kept in a glasshouse or plant growth rooms within a containment facility licensed by the local governing authority for quarantine pests. Local requirements may vary across countries, but tend to be tightly controlled to prevent the accidental release of a nonnative and/or quarantine disease to the local environment. Generally a license will be required, and the facilities inspected prior to this being granted, which can be a lengthy process—however, strict adherence to local requirements is vital.

To start a collection of phytoplasma-infected plants, plants can be collected from the natural environment (if possible and permitted) or alternatively acquired from existing collections. Many research groups maintain a few isolates which they primarily work on, and they are often willing to provide plants or cuttings to other researchers.

All of the methods detailed here are following standard horticultural procedures for propagation and maintenance of plants. The salient point with regard to phytoplasma-infected plants is the variable distribution within the host plant. This means that when propagating new plants (by whatever means) it is vital that the progeny plants are tested to ensure the phytoplasma has been transmitted before disposal of the older parent plant. The methods provided here are for grafting within the same species; however in some cases, it is possible to transmit phytoplasmas from the natural host into *C. roseus* by means of grafting across species. To do this, follow the protocol described here, although there is generally a higher success rate if the plant species are more similar in stem type. Often in cross-species grafts the scion will not continue to grow and transmission is achieved by the phytoplasma moving into the rootstock whilst in contact, with the scion later removed.

2. Materials

For maintenance of phytoplasma-infected plants glasshouses or controlled environment rooms, at least one of which meets local quarantine requirements for quarantine organisms, are required. Ideally stock plants of healthy *C. roseus* should be housed in a separate glasshouse or compartment to those infected with phytoplasmas. If this is not possible, then they should be grouped as physically separate as possible, for example on a different bench or at the end of one bench. The size of the glasshouse will determine how many plants can be maintained.

2.1. Growth of Healthy C. roseus from Seed

1. Plastic seed tray.
2. Clear plastic propagator lid (with vent) to fit seed tray.
3. Plastic plant pots (approximately 9 cm diameter) (see Note 2).
4. Plant pot saucers to fit pots (see Note 2).
5. Plant labels.
6. Permanent marker pen.
7. Dibber (see Note 3).
8. Proprietary seed and cutting compost.
9. Perlite.
10. Seeds of *C. roseus* (L.) G. Don (see Note 4).

2.2. Maintenance of Established C. roseus Plants

1. Plastic plant pots (various sizes) (see Note 2).
2. Proprietary multipurpose compost.
3. Perlite.
4. Secateurs or scalpel blades (e.g., Swann Morton No. 10A or 22).
5. 100% ethanol (if using secateurs).
6. Sharps disposal vessel (if using scalpel blades).
7. Plant fertilizer.

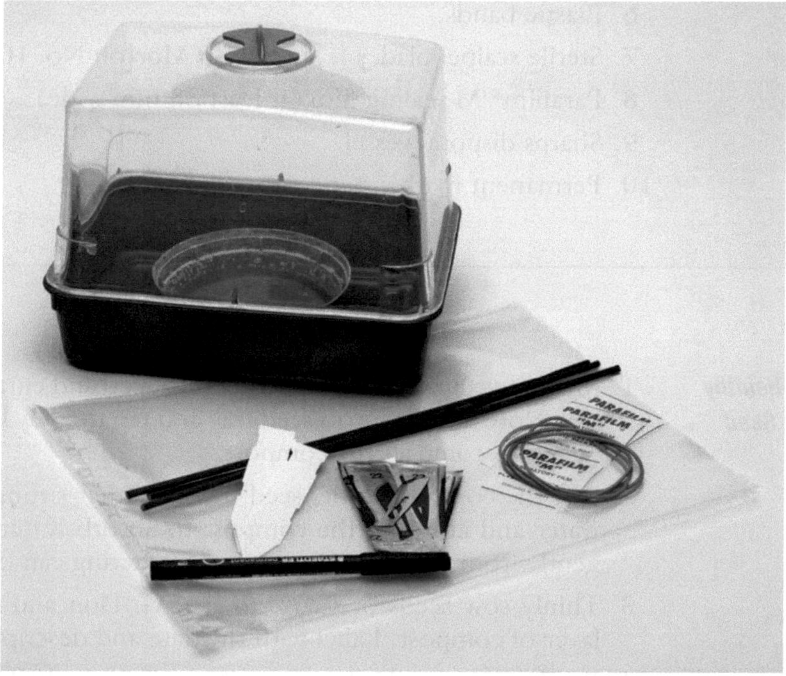

Fig. 2. The equipment required to take cuttings and conduct grafting, including seed tray and vented propagator lid, plant saucers, clear plastic bags, plant canes, plant labels, parafilm®, elastic bands, sterile scalpel blades, and permanent marker pen.

2.3. Propagation by Cuttings (Fig. 2)

1. Secateurs or sterile scalpel blades (e.g., Swann Morton No. 10A or 22).
2. 100% ethanol (if using secateurs).
3. Sharps disposal vessel (if using scalpel blades).
4. Proprietary seed and cutting compost.
5. Perlite.
6. Plastic plant pots (approximately 9 cm diameter) (see Note 2).
7. Plastic seed tray.
8. Clear plastic propagator lid (with vent) to fit seed tray.
9. Plant labels.
10. Permanent marker pen.
11. Hormone rooting powder or solution (optional).
12. Dibber (see Note 3).

2.4. Propagation by Grafting

1. Healthy *C. roseus* plants approximately 6–8 weeks old.
2. Phytoplasma-infected plant for propagation.
3. Plant pot saucers.
4. Small plant canes.
5. Clear plastic bags (approximately 20 × 30 cm).

6. Elastic bands.
7. Sterile scalpel blades (e.g., Swann Morton No. 10A or 22).
8. Parafilm® M sealing film (at least 38 mm wide).
9. Sharps disposal vessel.
10. Permanent marker pen.

3. Methods

3.1. Growth of Healthy C. roseus from Seed

1. Mix between 25 and 50% perlite into seed and cutting compost (see Note 5). Put into a seed tray and tap the tray lightly on the bench to compact the compost.
2. Wet the compost in the seed tray by either sitting in a tray of water and allowing the compost to absorb water or watering gently from above with a rose on a watering can (see Note 6).
3. Thinly sow seeds of *C. roseus* (L.) G. Don and cover with a layer of compost. Label with the date and description.
4. Put tray into a glasshouse at approximately 20°C and normal daylight until germinated.
5. Once the seedlings have developed the first set of true leaves (these develop after the seed leaves) they are ready to be "pricked out."
6. Prepare individual pots of the compost mix as in step 1. Do not fill pots to the very top but leave a gap of approximately 2 cm from the rim to allow easier watering of the plants. Use the dibber to make a hole in the center of the pot approximately 3 cm in diameter and 3 cm deep.
7. Remove each seedling into an individual pot. Holding the seedling only by the seed leaves (and not the stem or true leaves) use the dibber to loosen the compost around the roots, and then use the dibber to lift the roots and attached compost and support the weight.
8. Transfer the seedling into the hole made in the individual pot. Use the dibber to fill compost around the stem and gently firm in using your fingers. Maintain the compost level around the stem as it was in the seed tray.
9. Water the pots gently to settle the compost and top up the compost if required.
10. Place the plants in a glasshouse and grow on until well rooted.

3.2. Maintenance of Established C. roseus Plants

Both healthy and phytoplasma-infected plants (once established) should be maintained in this way. It is good practice to always maintain a stock of healthy parent plants which can then be used to

Fig. 3. Healthy *C. roseus* plants, *left*, an established parent plant and *right*, a young rooted cutting.

generate cuttings and/or seed as required. Young healthy plants raised from cuttings are used as rootstocks for maintaining and increasing stocks of phytoplasma-infected plants (Fig. 3). Generally it is wise to keep at least three established plants for each phytoplasma isolate because some phytoplasmas strains can cause plants to die very rapidly. If additional phytoplasma-infected plants are required, these can be created by either cuttings or grafting with equal success.

1. The glasshouse conditions are primarily determined for healthy growth of *C. roseus*. As most phytoplasmas are naturally found in temperate regions conditions suitable for *C. roseus* seem to also be appropriate for long-term maintenance of most phytoplasma strains; however, exceptionally some strains may need different conditions. The daytime temperature should be between 24 and 28°C, with a night time minimum of 16°C, which depending upon the country and time of year, may require artificial cooling and/or heating (see Note 7). The plants require 16 h of good intensity daylight, so supplementary lighting will be required for varying lengths of time depending upon the time of year and country. Humidity does not typically need controlling.
2. Plants should be watered with tap water close to ambient temperature. As a rule plants will likely need watering every 2 days, as *C. roseus* does not like to have water logged soil, however this will vary. Younger plants with less well established root

systems will require less water than established plants, and if overwatered can rot and die. To judge if a plant needs watering either touch the soil to feel how wet it is, or (with experience) the weight of a pot can be used. Depending upon the glasshouse setup, plants can either be watered from the top (directly in the pot) or bottom using either plant saucers or capillary matting (see Note 8).

3. Established plants should be fed weekly with a commercially available broad range plant fertilizer following the manufacturer's recommendations (see Note 9).
4. Plants should be spaced out across the benches as much as possible so that they are not touching, or touching minimally. This helps to limit the spread of disease.
5. Established plants should be pruned as required to maintain the plant at a workable size and improve the structure. *C. roseus* can be pruned quite severely and will regrow from brown woody stems. Generally a scheme of pruning two to three times a year is adequate, with one pruning session before the onset of winter. When pruning either use secateurs (which are disinfected by wiping with 100% ethanol between each plant) or scalpel blades. Make clean cuts without tearing the stems. Always cut just above a bud or leaf joint. When pruning a plant the desired shape is an "open goblet," with a structure of stems around the sides and the center open. This helps air flow though the plant which reduces disease. Any diseased or dying stems should be removed as required throughout the year without waiting for a pruning session (see Note 10).
6. Established plants can remain in larger pots (approximately 20 cm diameter or larger) for prolonged periods of time, and do not need frequent potting-on. Generally they can be repotted once a year into a pot a few centimeters bigger. Younger plants will need potting on more frequently until they become established, perhaps two to three times a year, but only gradually increase the size of the pot. When potting on plants use commercially available multipurpose compost mixed with approximately 10% perlite (see Notes 5 and 11).
7. If any common diseases afflict the plants, then treat with an appropriate pesticide/insecticide following manufacturer's recommendations. Regular monitoring of the plants to enable early detection of disease yields the best results.
8. Plants within a collection should be routinely monitored for the presence of the phytoplasma by any molecular method of you choosing (see Chapters 14, 15, 16, 17, 18, 19, 20, 21, 22, 23, 24, 25, and 26). This is especially important when creating new plants via grafting or cuttings to ensure the phytoplasma is successfully transmitted. Routine screening of the healthy plants should also be undertaken.

3.3. Propagation by Cuttings

Cuttings can be used to create new plants of both healthy and phytoplasma-infected *C. roseus*, with the process being completed in the same manner. However, if infected plants are being propagated, then this must be conducted within a quarantine facility.

1. Mix between 25 and 50% perlite into seed and cutting compost, fill individual pots, and tap lightly on the bench to compact the compost. Do not fill pots to the very top but leave a gap of approximately 2 cm from the rim to allow easier watering of the plants (see Note 5).
2. Select a stem from the phytoplasma-infected *C. roseus* or healthy *C. roseus* which requires propagating. If possible, select a young nonflowering shoot which shows some disease symptoms (if applicable) and is straight, 2–3 mm in diameter, and approximately 6–8 cm in length. Using a sterile scalpel blade gently cut the shoot to remove from the parent plant (see Note 12) (Fig. 4a left).
3. Prepare the shoot by carefully cutting away all the leaves (and flowers/buds if present) except the top three pairs, using a scalpel blade to leave clean cuts (Fig. 4a right).
4. (Optional) Dip the bottom 3 cm of the stem into hormone rooting compound (see Note 13).
5. Use a dibber to create a hole approximately 5 cm deep along the side of the plant pot.
6. Insert a prepared cutting into the hole until the top 3 cm are protruding above the compost and the leaves are close to (but slightly above) the compost level.

Fig. 4. Selected stems to act as cuttings or grafting scion; (**a**) suitable shoot removed from parent plant (*left*) and stripped of lower leaves (*right*); (**b**) pot of cuttings ready to grow.

7. Gently firm compost back around the shoot with your fingers.
8. Repeat steps 2–7 placing four or five cuttings around the edge of the pot. Ensure that the leaves are not touching each other (Fig. 4b).
9. Label the pot with the date and source of the cuttings (see Note 14).
10. Place the pot of cuttings in the seed tray and fit the propagator lid with the vent closed. Put into a glasshouse at approximately 20°C and normal day light.
11. Water gently as required (approximately twice per week) (see Note 15).
12. After 1 week open the vent on the propagator lid. If any cuttings start to rot or die, carefully remove these from the pot.
13. Once rooted (approximately 4 weeks) pot up individually (see Note 16). To do so, prepare individual pots as step 1.
14. Gently squeeze the sides of the pot of cuttings to loosen the compost. Place spread fingers over the top of the pot to support the stems, invert the pot and tap the bottom of the pot firmly. Tip onto its side, squeeze the side and the compost should slide out onto a bench.
15. Divide the compost so that each rooted cutting still has compost attached to its roots. Pot up individually following steps 7–10 in Subheading 3.1.
16. Once fully rooted in the new pot and the top growth is actively growing these can be used for grafting or as healthy material (see Note 17).

3.4. Propagation by Grafting

The method described below creates an "apical" graft union, which is the easiest manipulation and confers the best transmission rate due to two surfaces of the scion being exposed to the rootstock. The most common alternative is the "side wedge" or "top" graft where both the scion and rootstock stems are cut at a 45° angle and held together with film. This can be used to successfully transmit phytoplasmas; however, it is technically harder to achieve a good graft union site. Gather together all of the required equipment (Fig. 2) before commencing the grafting process (see Note 18).

3.4.1. Prepare the Rootstock

1. Firstly prepare the healthy plant which will be the rootstock (see Note 19).
2. Take a young (approximately 6–8 week old) well rooted healthy *C. roseus* plant (Fig. 5a) (see Note 20).
3. Using a sterile scalpel blade (see Note 21) carefully remove all of leaves from the plant at the stem. Use the blade to ensure a clean cut at each point rather than a tear which may introduce disease. Remove all stems apart from the main stem (see Note 22).

Fig. 5. Preparation of the healthy rootstock plant; (**a**) the rooted young healthy plant which will act as the rootstock; (**b**) the rootstock stripped of leaves; (**c**) the rootstock stem being cut in preparation for accepting the scion.

4. Remove the growing point of the main stem and then make a horizontal cut to reduce the height of the main stem to approximately 10 cm from the compost surface (Fig. 5b).
5. Cut vertically down the center of the stem for approximately 6 cm to open out the stem into a "V" shape (Fig. 5c) (see Note 23).

3.4.2. Prepare the Scion

1. Select a stem from the phytoplasma-infected *C. roseus* which requires grafting, this will be the scion.
2. If possible select a nonflowering shoot which shows some disease symptom and is straight, 2–3 mm in diameter, and approximately 6–8 cm in length. Using a sterile scalpel blade gently cut the shoot to remove from the parent plant (Fig. 4a left) (see Note 12).
3. Use the scalpel blade to remove the lower leaves from the stem (not leaving any material behind) leaving only the top 2–3 pairs of leaves (Fig. 4a right).
4. The lower 4–5 cm of the stem now needs cutting from two "sides" to expose the internal stem structure. Gently support the stem in one hand and then gradually cut into the stem to a depth of approximately 0.5–1 mm, tailoring the cut to a pointed shape at the nonleaf end of the stem. Repeat on the "opposite" side so that the shoot has two gradual cuts ending in a thin point with the phloem and xylem layers exposed along

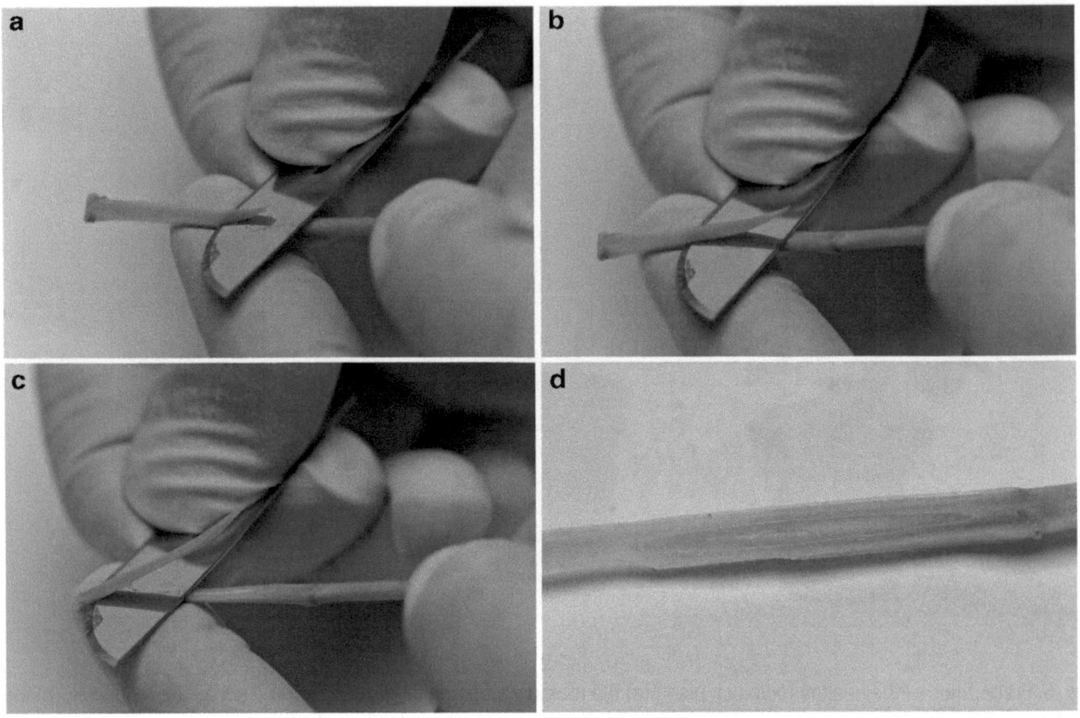

Fig. 6. The stages of preparing the scion for grafting showing; (**a**–**c**) cutting away of the scion outer stem layers; (**d**) the exposed internal structure of the scion stem.

the length of the cut. It may be necessary to make several passes to generate the cuts. The nonleaf end needs to be very thin to allow insertion into the rootstock (Fig. 6).

3.4.3. Create the Graft

1. Insert the cut scion into the cut main stem of the healthy plant; it may help to use the tip of the scalpel to open the cut in the rootstock stem. Ensure that the entire cut area of the scion is inside the cut rootstock stem. If necessary, increase the depth of the cut of the rootstock stem or trim the end of the scion (Fig. 7a).
2. Firmly hold and support the scion within the center of the stem and wrap the whole cut area with stretched parafilm®, ensuring to start wrapping the parafilm® on the rootstock stem beneath the cut area. Care is needed to support the stem whilst stretching, pulling, and wrapping the parafilm®. Wrap the parafilm® over the whole length of the scion (Fig. 7b, c) (see Note 24).
3. Label plant with designation (of rootstock and scion origin).
4. Place three canes equidistant around edge of pot (see Note 25).
5. Place a clear plastic bag over the canes and secure around the plant pot with an elastic band ensuring the bag is completely sealed (see Note 26) (Fig. 8).

Fig. 7. Grafting of the scion into the rootstock showing; (**a**) how to insert the scion into the rootstock; (**b**) wrapping of the scion rootstock join with parafilm®; (**c**) the completed graft union.

Fig. 8. A completed grafted plant ready to grow on.

6. Place the pot in a saucer to enable watering from the bottom and place in a glasshouse at approximately 20°C and normal daylight (see Note 27).
7. After 3–4 days cut 3 small holes near to the top of bag and leave to grow (see Note 28).
8. Once the grafted scion is actively growing well (approximately 3–4 weeks), the plant is ready for removal from the bag (see Note 29).
9. Cut several larger holes into the plastic bag. Leave overnight for the plant to acclimatize and then remove from bag and canes the following day (see Note 30).
10. Grow on and pot up the plant as required following Subheading 3.2.

4. Notes

1. Whilst there are no reports of pesticide treatments (e.g., foliar sprays or soil drenches) being related to the elimination of phytoplasmas from *C. roseus*, it may be prudent to empirically determine this for treatments commonly or likely to be used in the given country. To do this, generate phytoplasma-infected plants surplus to culture collection and experimental requirements and then treat these with the chemical as per manufacturer's instructions. Test the plant for presence of phytoplasmas before and at intervals after treatment to ensure that the phytoplasma remains present.
2. Terracotta pots and saucers may be used although it is easier to disinfect and sterilize plastic pots (by autoclaving) so for this reason plastic is recommended. The main advantage of terracotta is that it absorbs and maintains some moisture meaning the plant root ball does not dry out as easily. For this reason (and especially in hot climates), it may be desirable to use terracotta pots for large established parent plants in a collection.
3. A dibber is a pointed stick commonly made of wood or plastic which is used to make a hole in compost or soil to allow seeds or plants to be planted without damaging the roots. Dibbers are commonly available from garden centers; however, if one is not available, a sturdy plant label or pen may be used in its place.
4. Multiple sources of seeds for *C. roseus* are available from commercial seed suppliers. Once a healthy line has been established it is possible to collect and store seeds from these plants. Flowers should be allowed to mature on the plant and once full and ripe the pods should be carefully removed from the plant, opened and the seeds removed. Separate any debris and store

the dry seeds in a small paper envelope in dry, cool conditions (ideally a fridge over silica). Seeds collected and stored in this manner will not remain viable for as long as commercially available seeds and should be used within 6 months. The germination success rate may also be lower.

5. Perlite is added to compost to aid and improve drainage. Sharp grit or vermiculate may also be used for this purpose. These are typically added at a rate of 25–50% for young plants and 10% for established plants.

6. The compost is watered prior to the addition of seeds as if watered from above once sown the seeds can easily be washed away or redistributed to the edges of the tray.

7. During the winter months, (if desired) bubble-wrap type insulation may be attached to the interior sides of the greenhouse glass to reduce heat loss and minimize heating costs, however ensure this does not touch the plants.

8. Water should be close to ambient temperature to avoid shocking the plants. It may be necessary to store water within the glasshouse (e.g., in a water butt or watering can) in some instances, for example in a cold winter if the water is piped from outside directly for watering. When watering, try to avoid splashing water onto the plants leaves as this may cause scorch damage. If possible, water plants in the morning to enable surplus water to drain away or evaporate during the course of the day. Excessive water will increase the humidity which, if very high, can promote disease within a glasshouse. If plants are housed on capillary matting (which can enable easier watering), then try to avoid moving the pots excessively. This breaks the root connections with the matting and limits water uptake by the plant until they become reestablished. Do not leave plants sitting in excessive water if plant saucers are used.

9. Do not be tempted to overdose the plants with fertilizers as this can be harmful. Plants potted up into fresh compost do not required feeding for 6 weeks due to the presence of fertilizers in the fresh compost. Some strains of phytoplasma in particular can cause yellowing or bleaching of the leaves which appears like nutrient deficiency; however, if the plants are being fertilized regularly, then this is not the case and does not require additional treatment.

10. Pruning before winter decreases the leaf load and removes any dead or diseased tissue. Plants can be prone to more disease in winter so this is an important preventative measure. Always prune to a bud or leaf so that sections of stem are not left to die back. Sometimes stems of an established plant can die back whilst the rest of the plant is healthy. If this occurs, remove the entire stem cleanly from as close to the plants base as possible. Plants usually recover and grown on well.

11. *C. roseus* does not like root disturbance so be very gentle and careful when repotting plants and never root prune specimens.
12. If a nonflowering shoot is not present on the plant, then select a shoot with the fewest flowers and buds and gently remove the flowers at the stem and cut the growing tip with the buds from the shoot. Ideally try and match the diameter of the scion stem with that of the rootstock; however, if this is not possible, the scion stem can be smaller than the rootstock, although not the other way round. Stems showing some phytoplasma symptoms generally give good transmission of the phytoplasma, however not all phytoplasma strains cause symptoms in *C. roseus*. In this case, select a healthy young shoot. Older shoots tend to propagate less well. Be careful and delicate when handling the stems to not crush them.
13. Hormone rooting compound can be used if desired. *C. roseus* does root well without this but it may increase the speed of rooting.
14. Each phytoplasma and plant within a collection should have a unique designation, and when propagating it is a good idea to record which plant is used as the source as you will likely have multiple plants of a given phytoplasma isolate.
15. Be very careful to not overwater cuttings as this will cause the stems to rot and die. Do not allow the pots to stand in water.
16. To tell if the cuttings are rooted, look at the bottom of the pot for fine white roots to begin emerging from the drainage holes. This will indicate the cuttings are well rooted and ready to be potted on.
17. If a young rooted cutting is intended to be used as a rootstock for grafting, then it is desirable for the plant to have one main stem. In this instance, leave the growing point of the plant intact, as this will enable one strong dominant stem to develop. However, if the plant is for a general collection or a parent plant, the growing point should be carefully removed once the plant is a reasonable size (roughly 10–12 cm in height) to promote the development of lateral stems, which will give the plant a better rounded shape and structure.
18. The primary advantage of using the "side wedge" or "top" graft technique is that in rootstock plants with two main stems it is easy to graft in two difference scions. This may be to increase the "dose" of phytoplasma or to create a plant where two phytoplasmas are intentionally mixed together. If this is the desired outcome, then when generating rootstock plants remove the growing tip of the main stem when the plant is small (approximately 5 cm height) which will promote multiple lateral shoots to develop.

    Once started the grafting process should be completed without any large gaps in time once the parent/rootstock plant and scion have been cut. This is to limit any drying out of the cut surfaces, which if they became dry would mean the graft may not take as well.

19. Great care should be taken when grafting, especially when cutting down the rootstock plant and slicing the scion, to avoid injury.
20. The rootstock should be an established cutting from the healthy parent plants maintained, or grown from seed. These should have been routinely tested to confirm the absence of phytoplasmas; however, if desired, a sample of leaves from the rootstock plant can be collected as the plant is being prepared for grafting and used for confirmation that the plant is indeed free from phytoplasmas. The rootstock plant should be at most 12–14 weeks of age. Once the main stem has gone woody the grafts are much harder to create and take less well so the plant should not be used at this point.
21. Use one blade for each plant (healthy rootstock plant and each phytoplasma-infected plant to be grafted). This is to prevent the spread of any other disease (e.g., viruses) which may be in the plants. Either scalpel blades alone or a blade on a handle may be used for the grafting process with the choice a personal preference. Most people find that only a blade allows the greatest dexterity for making the maneuvres.
22. The extreme removal of all leaves and stems (other than the main stem) from the rootstock plant enables easier transmission and spread of the phytoplasma out from the scion and into the whole rootstock plant. If the rootstock plant remains larger at the time of grafting, then there is a greater area for the phytoplasma to spread into which can delay the spread of the phytoplasma throughout the entire plant.
23. Grafting as low down the rootstock as possible ensures that the grafted scion is structurally stable when the plant in fully grown.
24. The parafilm® prevents the graft union from drying out which would cause the graft to fail. This needs to be firmly applied with the film stretched so that it sticks to itself as it wraps around the stem.
25. The purpose of the canes is to prevent the plastic bag from touching the plant as this can cause scorch and damage the plant.
26. The size of the bag required will vary depending on the size of the plant, pot, and canes. Ensure that the elastic band completely seals the plastic bag against the pot. This is necessary to maintain high humidity for the first days after grafting; if this is not maintained, then the graft often fails to take.
27. This now needs to be in a quarantine glasshouse, which can be the same one that the main parent plants are kept in.

28. Occasionally the elastic band may perish before the plant is ready for removal from the bag. If this is the case, replace with a new band. The holes allow fresh air to enter the bag but still maintain an environment of high humidity. If the scion is excessively wilted, then leave the bag intact for up to 1 week. On occasion the scion may fail to take and will die, usually within the first 1–2 weeks. In this case, discard the plant as the phytoplasma is unlikely to have been transmitted. Once experienced, most grafters can achieve a 100% success rate of grafts.
29. To determine that the scion is actively growing, note that the scions shoot should be growing away from the growing tip increasing in stem length and producing new leaves. The rootstock may also have started reshooting from the base of the main stem at this point.
30. The grafted plant requires acclimatization to the environmental conditions before complete removal from the bag to prevent shock causing damage (or death) to the plant.

Acknowledgements

We would like to acknowledge Professor Phil Jones and the horticultural staff at the University of Nottingham and the Food and Environment Research Agency for training and guidance provided on the creation and maintenance of phytoplasma collections. We would also like to thank all individuals and organizations that have provided us with phytoplasma-infected plant samples.

References

1. Seemuller E et al (1998) Current status of molecular classification of the phytoplasmas. J Plant Pathol 80:3–26
2. Choi YH et al (2004) Metabolic discrimination of *Catharanthus roseus* leaves infected by phytoplasma using ^{1}H-NMR spectroscopy and multivariate data analysis. Plant Physiol 135: 2398–2410
3. Favali MA et al (2004) *Catharanthus roseus* L. plants and explants infected with phytoplasmas: alkaloid production and structural observations. Protoplasma 223:45–51
4. Pati PK, Kaur J, Singh P (2011) A liquid culture system for shoot proliferation and analysis of pharmaceutically active constituents of *Catharanthus roseus* (L.) G. Don. Plant Cell Tissue Organ Cult 105:299–307
5. Aslam J et al (2010) *Catharanthus roseus* (L.) G. Don. an important drug: it's applications and production. Pharm Glob 1:1–16
6. Abdul JA et al (2007) Antioxidant potentials and ajmalicine accumulation in *Catharanthus roseus* after treatment with gibberellic acid. Colloids Surf 60:195–200
7. Gupta S et al (2007) Construction of genetic linkage map of the medicinal and ornamental plant *Catharanthus roseus*. J Genet 86: 259–268
8. Lee I-M et al (2004) '*Candidatus* Phytoplasma asteris', a novel phytoplasma taxon associated with aster yellows and related diseases. Int J Syst Evol Microbiol 54:1037–1048
9. Kakizawa S (2008) Genome sequencing and analysis of cell surface membrane proteins of phytoplasmas. J Gen Plant Pathol 74:457–548
10. Berges R, Rott M, Seemüller E (2000) Range of phytoplasma concentrations in various plant hosts as determined by competitive polymerase chain reaction. Phytopathology 90:1145–1152

Chapter 3

Micropropagation and Maintenance of Phytoplasmas in Tissue Culture

Assunta Bertaccini, Samanta Paltrinieri, Marta Martini, Mara Tedeschi, and Nicoletta Contaldo

Abstract

Maintenance of phytoplasma strains in tissue culture is achievable for all strains transmitted to periwinkle (*Catharanthus roseus*), and also for other naturally infected plant host species. Shoots of 1–3 cm length are grown in a solid medium containing Murashige and Skoog (MS) micro- and macroelements and 0.12 mg/L benzylaminopurine. The continued presence of phytoplasmas in infected shoots of periwinkle that have been maintained in micropropagation for up to 20 years can be shown by diagnostic methods such as nested PCR tests using the 16S rDNA gene (*see* Chapters 14, 15, 16, 17, 18, 19, 20, 21, 22, 23, 24, 25, and 26 for phytoplasma diagnostic methods).

Key words: 16S rDNA, Collection, Micropropagation, Strain identification

1. Introduction

Phytoplasmas are prokaryotes lacking cell walls, and they have been implicated in several plant diseases causing substantial losses to agricultural crops and severe damage to landscape plants. Phytoplasma identification is based mainly on the 16S ribosomal RNA gene, and on other genes such as the translation elongation factor *tuf*, protein translocase subunit *secY*, ribosomal protein operon *rp*, 16S–23S intergenic sequence, protein translocase subunit *secA*, and *groEL* genes (1–8). However, the maintenance of different strains in culture for use as reference strains for studies related to host–pathogen interactions, the development of new diagnostic methods, and the elimination of phytoplasmas from crops is a basic requirement for research in this field.

Phytoplasma-infected shoots of different species can be maintained over the long-term using micropropagation. The protocol

Matt Dickinson and Jennifer Hodgetts (eds.), *Phytoplasma: Methods and Protocols*, Methods in Molecular Biology, vol. 938, DOI 10.1007/978-1-62703-089-2_3, © Springer Science+Business Media, LLC 2013

described here applies to *Catharanthus roseus* (periwinkle): specific media may need to be prepared for other plant species (9–12). Some phytoplasmas are quite virulent, especially when maintained in micropropagation and the longevity of shoots can vary substantially. However, endophytic organisms such as other bacteria or yeasts can contaminate the explants, so antibiotics or other chemicals specifically eliminating such organisms should be used. It is important to avoid the use of tetracycline in any step to avoid unwanted elimination of the phytoplasmas from the shoots (13). However, unintentional elimination of phytoplasmas may also occasionally occur with the use of some antibiotics that are not intended for use to eliminate phytoplasmas from tissue-cultured material, as reported for both phytoplasmas and viruses (14–17). The uneven distribution of phytoplasmas is found in the shoots used in micropropagation, as is well demonstrated by studies on fruit tree shoots infected by phytoplasmas. Thus, micropropagated material free from phytoplasmas can be obtained from infected mother tissues, but this can be assessed only by careful analysis with very sensitive (molecular) detection methods, (see Chapters 21, 22, and 23). Periwinkle shoots are grown in a solid medium based on Murashige and Skoog medium (18), slightly modified (19), and maintained in a growth chamber with controlled light and temperature under previously described conditions (20, 21). Micropropagation of infected shoots of periwinkle is a reliable method to maintain viable phytoplasma strains long-term: verification of long-term phytoplasma maintenance in tissue culture has been demonstrated by PCR/RFLP methods (22); however, shoots in tissue culture should be routinely checked for the presence of phytoplasmas.

2. Materials

1. Periwinkle shoots (*C. roseus*) infected with phytoplasmas transmitted by dodder, insects, or grafting (see Note 1).
2. Glass tubes (2 × 20 cm with metal caps) or Magenta® containers.
3. Weighing balance (scale to 4 decimal points).
4. Weighing balance (scale to 2 decimal points).
5. Macroelements mix: 170 mg NaH_2PO_4, 1,650 mg HH_4NO_3, 440 mg $CaCl_2{\cdot}2H_2O$, 1,900 mg KNO_3, 370 mg $MgSO_4{\cdot}7H_2O$, 170 mg KH_2PO_4, 37.3 mg NaFeEDTA, 6.2 mg H_3BO_3, 22.3 mg $MnSO_4{\cdot}4H_2O$, 8.6 mg $ZnSO_4{\cdot}4H_2O$, 80 mg adenine sulfate, 100 mg *myo*-inositol, for 1 L of solid medium.
6. Microelements solution: prepare solutions of 5 mg/100 mL of each of KI, $CoCl_2{\cdot}6H_2O$, $CuSO_4{\cdot}5H_2O$, $Na_2MO_4{\cdot}2H_2O$,

glycine, nicotinic acid, pyridoxine, and thiamine hydrochloride. From these stock solutions, mix together 16.6 mL KI, 0.5 mL $CoCl_2 \cdot 6H_2O$, 5 mL $CuSO_4 \cdot 5H_2O$, 5 mL $Na_2MO_4 \cdot 2H_2O$, 60 mL glycine, 10 mL nicotinic acid, 10 mL pyridoxine, 8 mL thiamine hydrochloride for 1 L of solid medium.

7. Autoclave.
8. Deionized water.
9. Sterile (autoclaved) water.
10. Sterile growth chamber (or growth cabinet).
11. Pipettes (1 mL, 10 mL).
12. Water bath.
13. Heater or microwave oven.
14. Laminar flow hood.
15. Sterile razor blades.
16. Sterile forceps.
17. Sterile plastic or glass Petri dishes.
18. Commercial bleach.
19. Sterile 100 mL and 200 mL flasks with screw caps.
20. pH meter.
21. Sucrose.
22. Agar.
23. Benzylaminopurine (BAP): 0.05 mg/mL.
24. 1.5 L sterile flask.
25. Sodium hydroxide.

3. Methods

All micropropagation operations must be carried out in a sterile environment and in particular using a sterile vertical-laminar-flow hood. All instrument employed must be heat-sterilized (e.g., with a Bunsen burner) (Fig. 1). Sterile plastic- and glassware should be used in all the steps (see Note 2).

3.1. Micropropagation Medium

1. Weigh the components of the macroelements mix, and the microelements solutions. Combine in a 1.5 L sterile flask, and add 750 mL of deionized water.
2. Adjust the pH to 5.5–5.7 by adding sodium hydroxide and mixing carefully while measuring the pH.
3. Add 30 g of sucrose and 4 g of agar, mix well, and adjust the volume to 1 L.

Fig. 1. All micropropagation operations must be carried out in a sterile vertical-laminar-flow hood. Displayed is the required equipment: glass tubes containing micropropagation medium, forceps, razor blades, sterile plastic dishes, alcol for better sterilization of instruments.

4. Add 2.4 mL of BAP stock solution.
5. Heat the solution in a water bath to dissolve the components.
6. Fill the tubes or the Magenta® containers with appropriate volumes of medium (~5 mL per tube, or at least 10 mL per Magenta® pot), and close them with caps (see Note 3).
7. Autoclave at 121°C for 20 min.
8. Allow containers to gradually cool and leave medium to solidify in a standing position (see Note 4).

3.2. Shoot Sterilization

1. Prepare 3–5 *C. roseus* shoots 3–5 cm long per phytoplasma strain/plant by carefully taking off all the leaves and covering the shoots with 80 mL of bleach:sterile distilled water (1:3) in a 100 mL sterile flask with a screw cap.
2. Gently mix for 1–3 min.
3. Open the flask in the sterile hood, and transfer the shoots to a 100 mL sterile flask containing 80 mL of sterile distilled water using sterile forceps.
4. Close the cap and gently mix for 1–3 min.
5. Repeat steps 3 and 4.
6. Gently dry off the shoots then place a shoot in a sterile Petri dish. Remove any remaining leaflets and cut 1–2 mm off the top and the bottom of the stems.

Fig. 2. Periwinkle shoots infected by different phytoplasmas during maintenance by micropropagation.

7. Cut shoots into lengths of about 2 cm and put in sterile tubes containing sterile solidified medium.
8. Mark the tubes with the acronym of the phytoplasma strain or in a way that strain and source can always be identified.
9. Place tubes in a sterile growth chamber at 24 ± 1°C, 3,000 lux for 16 h/day.
10. Every 8–12 weeks, transplant shoots onto fresh medium (Fig. 2) (see Note 5).

3.3. Testing Phytoplasma Presence in Micropropagated Shoots

The maintenance of shoot infection with the correct phytoplasma should be tested once a year (e.g., by PCR/RFLP analyses) on 1 g (or less if not available) of micropropagated tissue following extraction using appropriate methodology and diagnostic procedures as described in Chapters 14, 15, 16, 17, 18, 19, 20, 21, 22, 23, 24, 25, and 26 of this book.

4. Notes

1. Plants should be PCR-positive for the presence of an identified phytoplasma and ideally symptomatic (23–25).
2. The methodology described is for micropropagation of phytoplasma-infected periwinkle shoots. Healthy shoots can also be micropropagated but the survival on the medium described in this protocol will not necessarily be the best achievable. For other phytoplasma-infected species, it may be necessary to prepare different, species-specific, micropropagation media.

3. Use glass tubes about 2×20 cm with plastic or metal caps for periwinkle shoots. For other species in which extensive branching may be produced, use polycarbonate Magenta® containers. All containers employed must be sterile; to reuse containers, wash with 30% bleach, and rinse with sterile water then sterilize in an autoclave.
4. Once prepared maintain the media at room temperature but never above 30°C, in a clean environment (or sterile is better if available). Use after 1–2 days to verify that sterilization was successful. Do not use if any signs of contamination are present.
5. To transplant shoots use sterile forceps to gently remove the shoot from the tube and place into a fresh tube with new medium. Gently insert the stem into the medium short distance. If the shoot or media shows any signs of contamination, it is necessary to discard the tube. If this is not possible (e.g., this may be the only shoot of a given isolate), then you may try to clean-up the shoot. To do so, remove the tip and repeating the shoot sterilization process (Subheading 3.2) before putting the shoot in a fresh tube. This may remove contamination.

References

1. Botti S, Bertaccini A (2003) Variability and functional role of chromosomal sequences in 16SrI-B subgroup phytoplasmas including aster yellows and related strains. J Appl Microbiol 94:103–110
2. Hodgetts J et al (2008) Phytoplasma phylogenetics based on analysis of *secA* and 23S rRNA gene sequences for improved resolution of candidate species of '*Candidatus* Phytoplasma'. Int J Syst Evol Microbiol 58:1826–1837
3. Langer M, Maixner M (2004) Molecular characterisation of grapevine yellows associated phytoplasmas of the stolbur-group based on RFLP-analysis of non-ribosomal DNA. Vitis 43:191–199
4. Lee I-M, Zhao Y, Bottner KD (2006) *SecY* gene sequence analysis for finer differentiation of diverse strains in the aster yellows phytoplasma group. Mol Cell Probes 20:87–91
5. Lee I-M et al (2010) Phylogenetic analysis and delineation of phytoplasmas based on *secY* gene sequences. Int J Syst Evol Microbiol 60: 2887–2897
6. Martini M et al (2002) Genetic variability among flavescence dorée phytoplasmas from different origins in Italy and France. Mol Cell Probes 16:197–208
7. Martini M et al (2007) Ribosomal protein gene-based phylogeny for finer differentiation and classification of phytoplasmas. Int J Syst Evol Microbiol 57:2037–2051
8. Mitrović J et al (2011) The *groEL* gene as an additional marker for finer differentiation of '*Candidatus* Phytoplasma asteris'-related strains. Ann Appl Biol 159:41–48
9. Jarausch W, Lansac M, Dosba F (1994) Micropropagation for maintenance of mycoplasma-like organisms in infected *Prunus marianna* GF 8-1. Acta Hortic 359:169–176
10. Jarausch W, Lansac M, Dosba F (1996) Long-term maintenance of nonculturable apple-proliferation phytoplasmas in their micropropagated natural host plant. Plant Pathol 45:778–786
11. Laimer da Mâchado M et al (2001) Improved detection of viruses and phytoplasmas in fruit tree tissue cultures. Acta Hortic 550:463–470
12. Laimer M et al (2002) FAIR CT 97-3889: Validierung von diagnostischen methoden an *in vitro* Pflanzer zur Zertifizierung von Obstgehőlzen. Erwerbs Obstbau 44:76–81
13. Ishiie T et al (1967) Suppressive effects of antibiotics of tetracycline group on development of mulberry dwarf disease. Ann Phytopathol Soc Jpn 33:267–275
14. Bertaccini A et al (2004) Micropropagation and establishment of mite-borne virus-free garlic (*Allium sativum*). Acta Hortic 631: 201–206

15. Bertaccini A et al (2004) Improved molecular methods for detection of *European stone fruit yellows* (ESFY) phytoplasmas from *in vitro* shoots of fruit trees. Acta Hortic 657:495–500
16. Plavsic B et al (1986) Kinetin treatment of stolbur diseased tomato plants (*Lycopersicon esculentum* L.) and the possibility of its application in chemotherapy. Acta Bot Croat 45:27–32
17. Veronesi F et al (2000) PCR indexing of phytoplasma-infected micropropagated periwinkle treated with PAP-II, a ribosome inactivating protein from *Phytolacca americana* leaves. Acta Hortic 530:113–120
18. Murashige T, Skoog F (1962) A revised medium for rapid growth and bio assays with tobacco tissue cultures. Physiol Plant 15: 473–497
19. Bertaccini A, Marani F (1986) BYMV-free clones of eight gladiolus cultivars obtained by meristem-tip culture. Acta Hortic 177:299–308
20. Bertaccini A, Davis RE, Lee I-M (1992) In vitro micropropagation for maintenance of mycoplasma-like organisms in infected plant tissues. HortSci 27:1041–1043
21. Bertaccini A et al (2000) Micropropagation of a collection of phytoplasma strains in periwinkle and other host plants. Proceedings of the 13th International Congress of IOM. Fukuoka, Japan, July 14–19, abstract 101
22. Bertaccini A et al (1992) Sensitive detection of mycoplasmalike organisms in field-collected and *in vitro* propagated plants of *Brassica*, *Hydrangea* and *Chrysanthemum* by polymerase chain reaction. Ann Appl Biol 121:593–599
23. Duduk B, Bertaccini A (2011) Phytoplasma classification: taxonomy based on 16S ribosomal gene, is it enough? Phytopathogenic Mollicutes 1:3–13
24. The IRPCM Phytoplasma/Spiroplasma Working Team—Phytoplasma taxonomy group (2004) '*Candidatus* Phytoplasma', a taxon for the wall-less, non-helical prokaryotes that colonize plant phloem and insects. Int J Syst Evol Microbiol 54:1243–1255
25. Lee I-M et al (1998) Revised classification scheme of phytoplasmas based on RFLP analyses of 16S rRNA and ribosomal protein gene sequences. Int Syst Bacteriol 48:1153–1169

Chapter 4

Dodder Transmission of Phytoplasmas

Jaroslava Přibylová and Josef Špak

Abstract

Transmission of phytoplasmas from naturally infected plant host species using the parasitic plant *Cuscuta* spp. (dodder) to *Catharanthus roseus* (Madagascar periwinkle) is an effective way to maintain a wide range of phytoplasmas for further research. Here, we describe transmission via dodder from an infected medicinal plant *Rehmannia glutinosa* var. *purpurea* and from a symptomatic redcurrant plant (*Ribes* spp.) to *C. roseus* indicator plants using a "stable bridges" method. In both cases, typical symptoms of phytoplasma disease on periwinkle plants were obtained: virescent flowers with an increased number of axillary shoots and smaller leaves after transmission from *R. glutinosa*, and greening petals (virescence) after transmission from *Ribes* spp. Phytoplasmas could be detected in donor and recipient plants by electron microscopy and by polymerase chain reaction (PCR) assays using universal phytoplasma primer pairs. Restriction fragment length polymorphism analyses of PCR fragments can also be used to confirm the identity of the phytoplasmas from donor and recipient plants.

Key words: Cuscuta, *Catharanthus roseus*, Periwinkle, *Ribes* spp.

1. Introduction

Catharanthus roseus (L.) G. Don (Madagascar periwinkle), family *Apocynaceae*, is used in plant pathology as an experimental host for phytoplasmas (1). It acts as a convenient host for an extremely wide range of phytoplasmas. A key step in the use of *C. roseus* for maintenance of phytoplasma cultures is the successful transmission of the phytoplasma from the naturally infected plant host species to periwinkle, which can be achieved by various methods including the use of the parasitic plant dodder (*Cuscuta* spp.). Use of a dodder bridge enables pathogen transmission between two botanically unrelated plants, whether woody or herbaceous in habit. The phytoplasma can be transmitted from an infected donor plant to a recipient periwinkle, and then back to a healthy host plant to induce the original disease symptoms.

Matt Dickinson and Jennifer Hodgetts (eds.), *Phytoplasma: Methods and Protocols*, Methods in Molecular Biology, vol. 938, DOI 10.1007/978-1-62703-089-2_4,

One of the advantages of periwinkle as an experimental host for phytoplasmas is the development of distinctive and characteristic symptoms in infected plants. Periwinkle provides the opportunity to compare symptoms induced by different phytoplasmas in genetically identical hosts. Strains of known characteristics are required as standards and as positive and negative experimental controls. Collections of phytoplasma strains are maintained in many laboratories as a prerequisite for research and must be maintained under special quarantine and insect control regimes (as per local requirements). Various phytoplasma strains in a collection may require different temperature regimes for optimum maintenance (2). Research on phytoplasmas is greatly facilitated if the organisms are available in an herbaceous host such as *C. roseus*. The availability of phytoplasmas in the experimental host periwinkle offers several advantages, in particular for work at the molecular level, and provides the possibility of all-year-round testing. For example, periwinkle is a much better source for phytoplasmal nucleic acids, proteins, and other immunogens than trees because, in periwinkle, the phytoplasma is usually present at a higher titer and phytoplasma components are easier to obtain. For many phytoplasmas that are not available in periwinkle or herbaceous plants, neither antibodies nor cloned DNA fragments are available for detection, differentiation, and characterization (1). Growth conditions, along with the time of infection and colonization of phytoplasma may influence symptom development (3), and it is also known that uneven distribution of phytoplasmas within a host plant can cause difficulties with detection procedures (4).

One effective method of transmission of phytoplasmas is by using *Cuscuta* spp. (dodder). All species of the genus *Cuscuta* (*Convolvulaceae*) are holoparasites. The genus is found throughout the temperate to tropical regions of the world. Plants are characterized by rootless green, yellow, or white stems with scale-like leaflets and very low levels of chlorophyll; some species such as *Cuscuta reflexa* can photosynthesize slightly (5). The plants have close physical linkages with their hosts and are completely dependent on the host for a supply of assimilates and water (6). A dodder seedling must form connections with a host within several days after germination. At points of contact with the host, the coiled dodder stem produces a haustorium, a specialized organ that penetrates host tissues. Searching hyphae grow between and through the host cells and forms connections with host vascular tissues (7). Dodders form direct connections with host xylem and, although the precise nature of phloem connections is still unresolved, evidence also supports direct phloem connections. For example, the phloem-specific dye carboxyfluorescein readily moves from the host into dodder (8). In addition to direct vascular connections, dodder appears to have cytoplasmic continuity with its host through plasmodesmata. Vaughn (7) has reported that searching hyphal cells of the dodder haustorium have plasmodesmata that span walls of adjacent dodder hyphae and host cortex cells.

A well-known aspect of host–dodder connections is the transmission of viruses: a single dodder plant simultaneously parasitizing two hosts may transmit plant viruses from one host to the other (9–12). Early investigations using dodder were usually for diagnostic purposes or to study mycoplasma-like organisms inside the dodder cells (13). Experimental transmission of a phytoplasma from an infected plant to a healthy plant of the same or different species can be accomplished in the laboratory or glasshouse. Phytoplasma cells that are present in the host's phloem enter the dodder plant through the haustoria, multiply within and pass through the dodder strand, and eventually arrive at haustoria that have penetrated the healthy plant. The phytoplasma cells pass from the dodder into the recipient plant's phloem, completing the transmission (2). Two basic methods of transmission can be used: stable dodder bridges and cut strands of infected dodder (14). Transmission efficiency by dodder of different phytoplasma strains is variable, particularly when using detached cuttings of infected dodder. Phytoplasmas are more actively transmitted when the dodder connection grows from the source of inoculum towards the recipient plant (rather than in the opposite direction): the movement of the phytoplasmas in dodder stems seems to be mostly in the direction of the growing points (15). Transmission depends on both type of phytoplasma and *Cuscuta* vector: for example, rubus stunt and cotton phyllody were transmitted at high efficiency by *Cuscuta europea* and *Cuscuta campestris*, respectively, whereas the other phytoplasmas were transmitted only at a low rate (16).

2. Materials

2.1. Dodder Transmission

Rehmannia glutinosa var. *purpurea* and redcurrant plants are used as examples in this chapter (see Note 1) and therefore this determines some of the required materials (items 1–4), the infected and healthy (control) plant species. These should be varied as per your application.

1. Two *R. glutinosa* plants showing disease symptoms were taken from the field and maintained under glasshouse conditions.
2. *R. glutinosa* plants raised from seeds in a glasshouse, as healthy controls (17).
3. Symptomatic rooted cuttings taken from a redcurrant bush with severe twisting of shoots, branch malformation, and producing a poor yield of berries found in a private garden in Southern Bohemia, Czech Republic. The original symptoms appeared on new growing shoots and persisted throughout the life of the plant. Cuttings were maintained in an insect-proof glasshouse.

4. Redcurrant plants grown from seed, as healthy controls.
5. *Medicago sativa* seeds.
6. *C. roseus* plants raised from seed, maintained in the glasshouse.
7. *C. campestris* raised from seed, parasitizing on *R. glutinosa* and the redcurrant plant, cultivated in a glasshouse under controlled conditions (see Note 2).
8. Access to appropriate controlled environment glasshouse or room.
9. Seed trays.
10. Seed compost.

3. Methods

3.1. Dodder Transmission

1. Mix *C. campestris* and *M. sativa* seeds and sow these near the surface of the compost in a seed tray, in a controlled environment glasshouse (25°C temperature, 70–90% relative humidity, supplementary intensive light).
2. After 10–14 days, transfer seedlings of *M. sativa* parasitized with dodder shoots to pots with infected plants (*R. glutinosa* or redcurrant) and maintain in a glasshouse. Allow dodder stems to attach to the infected plants.
3. Connect up newly developed dodder strands from each infected plant with 2–3 healthy periwinkle plants (see Note 3). The stable dodder bridges method for transmission of possible infectious agents is shown in Fig. 1.
4. Keep the dodder bridges intact until symptoms develop on the periwinkles, or until the dodder bridges break (see Note 4).
5. Use healthy periwinkle plants, and *R. glutinosa* and redcurrant plants raised from seed, as controls.
6. After transmission has been confirmed (see Chapters 14, 15, 16, 17, 18, 19, 20, 21, 22, 23, 24, 25, and 26 for appropriate diagnostic methods), remove the dodder strands from the periwinkle plants and keep the periwinkles in an insect-proof glasshouse under controlled conditions (see Note 5) for further study and symptom appearance (see Note 6).

4. Notes

1. Not all symptoms resembling phytoplasma diseases are really due to phytoplasmas; in particular, physiological malfunction could be the cause.

Fig. 1. Dodder (*Cuscuta campestris*) transmission of phytoplasma from an infected redcurrant plant (*Ribes* sp.) to periwinkle plants (*Catharanthus roseus*).

2. Use of a suitable dodder species for transmission is crucial for successful transmission and you may need to try several different species to achieve successful transmission. If dodder transmission fails, insect vectors or grafting may be attempted to achieve transmission.
3. Place one end of a strand of dodder around the infected plant's stem and place the other end around a healthy plant's stem. The dodder must parasitize both the donor (infected) plant and the recipient (healthy) plant.
4. Transmission time can differ in herbaceous and woody plants. Distribution of phytoplasmas is irregular in woody plants, so the time taken for symptoms to appear on periwinkle plants can be variable.
5. Maintain various phytoplasma strains in a controlled-environment insect-proof glasshouse under suitable conditions; different temperatures may be required.

6. It may be useful to transmit the transmitted phytoplasma from the infected periwinkle plants back onto healthy plants of the original species to maintain the isolate.

Acknowledgements

This work was supported by a grant from the Grant Agency of the Czech Republic (no. 522/98/0792) and by projects QH91224 of the National Agency for Agricultural Research and AV0Z50510513 of the Academy of Sciences of the Czech Republic.

References

1. Marcone C, Ragozzino A, Seemüller E (1997) Dodder transmission of alder yellows phytoplasma to the experimental host *Catharanthus roseus* (periwinkle). Eur J Forest Pathol 27: 347–350
2. Davis RE, Lee I-M (1991) Mycoplasmalike organisms as plant disease agents—a review. ATCC Q Newsl 11:7–11
3. Kamińska M, Korbin M (1999) Graft and dodder transmission of phytoplasma affecting lily to experimental hosts. Acta Physiol Plant 21:21–26
4. Davies DL et al (1992) Parry's disease of pears is similar to pear decline and is associated with mycoplasma-like organisms transmitted by *Cacopsylla pyricola*. Plant Pathol 41:195–203
5. Machado MA, Zetsche K (1990) A structural, functional and molecular analysis of plastids of the holoparasites *Cuscuta reflexa* and *Cuscuta europaea*. Planta 181:91–96
6. Jeschke WD et al (1994) Modelling the flow and partitioning of carbon and nitrogen in the holoparasite *Cuscuta reflexa* Roxb. and its host *Lupinus albus* L. Methods for estimating net flows. J Exp Bot 45:791–800
7. Vaughn KC (2003) Dodder hyphae invade the host: a structural and immunocytochemical characterization. Protoplasma 220:189–200
8. Birschwilks M et al (2006) Transfer of phloem-mobile substances from the host plants to the holoparasite *Cuscuta* sp. J Exp Bot 57:911–921
9. Bennet CW (1940) Acquisition and transmission of viruses by dodder (*Cuscuta subinclusa*). Phytopathology 30:2
10. Bennet CW (1944) Studies of dodder transmission of plant viruses. Phytopathology 34:905–932
11. Johnson F (1941) Transmission of plant viruses by dodder. Phytopathology 31:649–656
12. Hosford RM (1967) Transmission of plant viruses by dodder. Bot Rev 33:387–406
13. Siller W et al (1987) Occurrence of mycoplasma-like organisms in parenchyma cells of *Cuscuta odorata* (Ruiz et Paw.). J Phytopathol 119:147–159
14. Bennett CW (1967) Plant viruses: transmission by dodder. In: Maramorosch K, Koprowski H (eds) Methods in virology, vol 1. Academic, New York, pp 393–401
15. Carraro L et al (1991) Transmission characteristics of the clover phyllody agent by dodder. J Phytopathol 133:15–22
16. Marcone C et al (1999) Dodder transmission of pear decline, European stone fruit yellows, rubus stunt, Picris echioides yellows and cotton phyllody phytoplasmas to periwinkle. J Phytopathol 147:187–192
17. Přibylová J et al (2001) Association of aster yellows subgroup 16SrI-B phytoplasmas with a disease of *Rehmannia glutinosa* var. *purpurea*. Plant Pathol 50:776–781

Chapter 5

Insect Maintenance and Transmission

Heather Kingdom

Abstract

Phytoplasmas are plant pathogens of huge economic importance due to responsibility for crop yield losses worldwide. Institutions around the world are trying to understand and control this yield loss at a time when food security is high on government agendas. In order to fully understand the mechanisms of phytoplasma infection and spread, more insect vector and phytoplasma colonies will need to be established for research worldwide. Rearing and study of these colonies is essential in the research and development of phytoplasma control measures. This chapter highlights general materials and methods for raising insect vector colonies and maintenance of phytoplasmas. Specific methods of rearing the maize leafhopper and maize bushy stunt phytoplasma and the aster leafhopper and aster yellows phytoplasma strain witches' broom are also included.

Key words: *Dalbulus maidis*, Insect cages, Insect colonies, *Macrosteles quadrilineatus*, Rearing leafhoppers, Rearing planthoppers, Rearing psyllids

1. Introduction

Raising insect colonies is essential for the study of phytoplasma transmission. Aster yellows phytoplasma strain witches' broom (AY-WB) readily infects the model plant *Arabidopsis thaliana* producing leafy flowers, stunted growth, and proliferation of young leafy tissue. This symptom is typical of the witches' broom phenotype produced in other AY-WB-infected plants (1). This system has potential to provide a model in which to study phytoplasma spread. Maintaining vector colonies allows transmission efficacy to be tested. Vector age of *Macrosteles quadrilineatus* and temperature affect the transmission efficacy of aster yellows phytoplasma strains (2). Furthermore, the transmission of chrysanthemum yellows phytoplasma by *Eusclidius variegatus* has been shown to depend on the phytoplasma distribution within the insect (3, 4).

Insect vector colonies also facilitate the study of the vector interactions with their plant hosts and allow the study of phytoplasma

Matt Dickinson and Jennifer Hodgetts (eds.), *Phytoplasma: Methods and Protocols*, Methods in Molecular Biology, vol. 938, DOI 10.1007/978-1-62703-089-2_5, © Springer Science+Business Media, LLC 2013

manipulations of these interactions. For example, the bolt and severe strains of aster yellows phytoplasma have different effects on the fitness of the vector leafhopper *M. quadrilineatus* (5). Meanwhile, two strains of flavescence dorée phytoplasma are known to affect the survival and fecundity of the leafhopper vector *Euscelidius varegatus* (6).

These discoveries were all made using laboratory-reared colonies of vectors. Some phytoplasma vectors, such as the psyllid *Cacopsylla melanoneura*, are more difficult to rear through their entire life cycle (7). Despite this, by rearing only part of the life cycle, much research can be carried out. *C. melanoneura* produces fewer viable eggs on '*Candidatus* Phytoplasma mali'—infected apple shoots in laboratory conditions (7), whereas *C. melanoneura* collected from '*Ca.* Phytoplasma mali'-infected orchards efficiently vectors the phytoplasma to healthy apple trees (8, 9). Some leafhoppers, planthoppers, and psyllids are difficult to rear entirely in laboratory conditions due to their poor breeding in captivity or range of host plants. Micropropagation techniques can provide a useful solution in rearing insects such as these for only part of their life cycle (10). By rearing colonies of vectors through part or all of their life cycle, phytoplasmas can be maintained in laboratory conditions and research into their control can progress.

Leafhopper, planthopper, and psyllid colonies can be raised from field-collected adults or from egg-laden plant material sent from other facilities (11). In either case, a few individuals are usually enough to start a colony. Colonies of this kind can yield many thousands of individual insects (12, 13). Colonies initiated from field collections must be timed with the insect appearance, whereas material sent from other facilities can be set up at any time if conditions are set for continuous generations (14). Many leafhoppers, planthoppers, and psyllids are univoltine or bivoltine, producing only one or two generations per year, but if kept under continuous summer conditions many generations can be produced per year (15).

All insects and phytoplasmas should be kept in controlled growth rooms within a containment facility building licensed by the local authority. In the UK, for example, the Food and Environment Research Agency (Fera) permits the keeping of insects and phytoplasmas under License to Import, Move and Keep Prohibited Invertebrates and a Plant Health License, respectively (16). Importing plant and insect material is strictly controlled in many countries. Gaining permission to do so can take time and will probably require inspections of facilities. These tight restrictions are in place to ensure that the insects and pathogens do not escape. This is especially relevant when studying nonnative insects and phytoplasmas. Specific requirements of the insect system of study and the local licensing authority should be adhered to when establishing insect and phytoplasma colonies.

2. Materials

1. Growth chamber. The full requirements of this facility will depend on the requirements outlined in licensing authority guidelines for keeping quarantine insects and pathogens.

   Main requirements include sealed access, filtered ventilation, temperature control, timer-controlled fluorescent lighting, water drainage, and humidity monitoring (Fig. 1).

Fig. 1. Growth chambers provide a controlled environment to safely rear quarantine insects and phytoplasmas. Essential elements of such a chamber are sealed containment access, secured filtered ventilation, temperature regulation, time-controlled fluorescent lighting, water supply and drainage, and humidity monitoring. Our growth chambers have all these features and are housed within a high containment building. Areas surrounding the growth chambers are equipped with UV high-powered insect killers and lit with red light, which is invisible to contained insects.

2. General use glasshouse, suitable for growing plants for maintenance of insect colonies. A more contained facility may be required to house quarantine phytoplasma-infected plants.
3. Dark room with portable table lights (see Note 1).
4. Aspirator (see Note 2).
5. Insect colony cage (see Note 3).
6. Insect acquisition/transmission cage (see Note 4).

3. Methods

3.1. Raising Plants

This must be started before collection of insects from the field or obtaining eggs on plant material from other facilities. Mature healthy plants will be needed to maintain any insects, eggs gained at the start of colony generation. Plant generation time will depend on plant species needed for the particular insect and phytoplasma of interest. Outlined below are methods for generating four different plant species.

3.1.1. *Arabidopsis*

1. Sow *A. thaliana* (e.g., ecotype Columbia-0) seeds thinly onto packed and water-soaked low-nutrient compost in small pots (see Note 5).
2. Place pots at 4°C for 4 days to vernalize (see Note 6).
3. Move pots to a controlled-environment growth room with a 10 h day under fluorescent light at a constant temperature of 20°C.
4. After 10 days transplant the seedlings (approximately 10 mm tall) to individual pots in trays to grow for a further 3 weeks (see Note 7).
5. Allow plants to grow for a further 3–6 weeks, depending on age of plants required. Water three to four times a week (see Note 8).

3.1.2. Aster and Lettuce

1. Sow aster or lettuce seed onto lightly packed compost in 24-well seedling trays and cover with a 2 mm layer of sieved compost.
2. Allow plants to grow in glasshouse conditions for 6 weeks (4 weeks for lettuce) before transplanting plugs into larger pots.
3. Grow plants on in the same conditions for a further 3 weeks when they can be used for colonies. Plants should be approximately 15 cm tall, bushy, and free of pests, and disease such as thrips and mildew.

3.1.3. Maize

1. Sow 3 maize seeds into 14 cm-diameter pots, pushing them into packed compost with your index finger. Distribute compost over holes and leave to grow in glasshouse conditions for 4 weeks.
2. Stake plants up to stabilize, and grow on for a further 1–3 weeks to the desired age/height for colonies. Approximately 20–40 cm tall is a good size for colonies.

3.2. Raising Insect Colonies from Material from Other Facilities

1. Remove all packaging from material. Pot up each plant in suitable compost, water it and allow the plant to acclimatize in a colony cage in a growth chamber (see Note 9).
2. After 2–3 days, add two fresh healthy plants generated earlier into the cage, as a fresh food source for hatching nymphs.
3. Add two more fresh plants after a further 3 days and allow initial start-up plant to die back in the cage (see Note 10).
4. Once the initial plant has died back completely, remove it from the cage.
5. Add further plants when required to maintain at least two plants in good condition at all times.
6. Allow hatched nymphs to go through all five instars and eclose before moving any insects on to new cages (see Note 11).
7. Count the number of eclosed adults and determine the sex ratio (see Note 12).
8. If the ratio is approximately even or female-biased and >200 insects are present, set up a new cage in the dark room and remove half the insects to the new cage with two new plants to start a new colony (see Note 13).
9. Allow the first generation of insects to oviposit on fresh plants in the two new colonies. Add more plants to cages as required to maintain at least two plants in good condition at all times.
10. Allow eggs to hatch, nymphs to go through all five instars and eclose, and then repeat from step 7.
11. At the split of this second generation of adults, start to allow the first colony created to die back (see Note 14).

3.3. Generating Phytoplasma-Infected Insect Colonies

After you have insect colonies established, you are ready to start a colony of phytoplasma-infected insects.

1. Collect some infected plants from the field or arrange to be sent phytoplasma-infected plants from another facility. Ensure all plants are complete with a hydrated root ball.
2. Pot up phytoplasma-infected plants immediately after arrival and individually place in cages within growth chambers and allow to acclimatize for 2 days (see Note 15).

3. Add approximately 30 adult insects to each cage containing an infected plant. Adults should be 2:1 or 1:1 female:male ratios (see Note 16).
4. Leave insects on infected material for as long as required for acquisition (see Note 17).
5. Keep phytoplasma-infected plants with the insects in the cage until the plant dies back completely. When plants look very sick, add another phytoplasma-infected plant or one fresh uninfected plant into the cage.
6. When the insect acquisition time has elapsed, relocate the insects to a new cage with a healthy fresh plant (see Note 18).
7. Allow the newly infected insects to transmit phytoplasmas to the new healthy plants and monitor plants for symptom development.
8. When symptoms have developed, relocate insects to a large colony cage with fresh healthy plants.
9. Move newly infected plants to a holding cage until required for more insect acquisition (see Note 19).
10. Allow the insects in the colony cage to oviposit, and nymphs develop and eclose. At this stage plants will be infected and the plants and new emerged adults can be split to form an additional colony. Adults should be 2:1 or 1:1 female:male ratios.
11. Continue splitting the colonies after a new generation and cleaning old cages as you require and space allows.

3.3.1. Maintaining Aster Yellows Phytoplasma Strain Witches' Broom (AY-WB)

This phytoplasma can be maintained in china aster, lettuce, and *Arabidopsis* using the vector leafhopper *M. quadrilineatus* (see Note 20).

1. Collect AY-WB-infected plants from the field or another facility, pot them up, water and allow to acclimatize for 2 days in individual acquisition cages.
2. Add 50 adult *M. quadrilineatus* to each cage containing a mature lettuce or aster (>20 cm tall). Adults should be 2:1 or 1:1 female:male ratios. For mature *Arabidopsis* (>6 cm rosette diameter), add 20 adult *M. quadrilineatus* to each cage.
3. Relocate insects after 7 days for acquisition to healthy aster or lettuce plants in transmission cages. Monitor these plants for symptoms of yellowing developing over the next 10–14 days (see Note 21).
4. Add fresh insects to acquisition cages and repeat steps 2 and 3 (see Note 22). Dispose of AY-WB-infected plants after second acquisition round.
5. When symptoms have developed on transmission plants, relocate insects to a large colony cage with fresh healthy plants.

6. Move newly AY-WB-infected plants to a holding cage until needed for more insect acquisition (see Note 19).
7. Allow the insects in the colony cage to oviposit, and nymphs develop and eclose (see Note 23).
8. Continue splitting the colonies after new generations and cleaning old cages as you require and space allows.

3.3.2. Maintaining Maize Bushy Stunt Phytoplasma in Maize Using Dalbulus maidis

Maize bushy stunt phytoplasma (MBSP)-infected maize should be obtained, complete with root ball, from the field or another facility. Plants should be 7–10 weeks old so they are not too tall to work with.

1. Pot plants up, water and allow to acclimatize for 2 days in individual acquisition cages (see Note 24).
2. Add 50 adult *Dalbulus maidis* to each cage containing a one maize plant. Adults should be 2:1 or 1:1 female:male ratios.
3. Relocate insects after 14 days' acquisition to healthy maize in transmission cages. Monitor these plants for symptoms of reddening developing after 14–21 days (see Note 25).
4. Dispose of original MBSP-infected plants only when *D. maidis* have successfully acquired and transmitted MBSP to new maize plants.
5. When symptoms have developed on transmission plants, relocate insects to a large colony cage with fresh healthy plants.
6. Move newly MBSP-infected maize to a holding cage until needed for more insect acquisition (see Note 19).
7. Allow the insects in the colony cage to oviposit, and nymphs develop and eclose (see Note 26).
8. Continue splitting the colonies after new generations and cleaning old cages as you require and space allows.

4. Notes

1. This is ideal for handling leafhoppers, planthoppers, and psyllids inside cages, as it provides a condition in which the insect flight response towards light can be used to handle, separate, and determine sex of insects. To handle insects outside cages, lights will need to be behind a screen. Direct access to light bulbs is hazardous as the insects cook very quickly on hot light bulbs.

   Optional extras:

   - Collection container(s) can be added to the aspirator replacing the collection tip so that insects can be collected.

Rose Entomology (PO box 1474, Benson, AZ 85602, USA) supplies a good range. This is useful in the field when collecting insects from plants and later adding to a cage to set up a colony.

- A pump can be used to suck up insects by adding tube to the input valve or blow insects out of aspirator by placing tube over the output valve of a small pump. A good example is the small air diaphragm pump B100 SE from Charles Austen Pumps Ltd (Royston Rd, Byfleet, Surrey, KT14 7NY, UK).

2. An aspirator provides a useful way of handling and moving small fragile insects carefully. Mouth and pump aspirators can be purchased or home-made and are very similar in design. A simple one can be made with two 1 mL pipette tips, 70 cm of clear 6 mm-diameter plastic tube, some gauze, and a filter (Fig. 2). The filter is optional but minimizes contact with insect brochosomes and other cage dust. To make an aspirator: cut the tips off the 2 pipette tips at a slant and file down any sharp burrs with a file; cut the tube into a 10 and 60 cm length; place

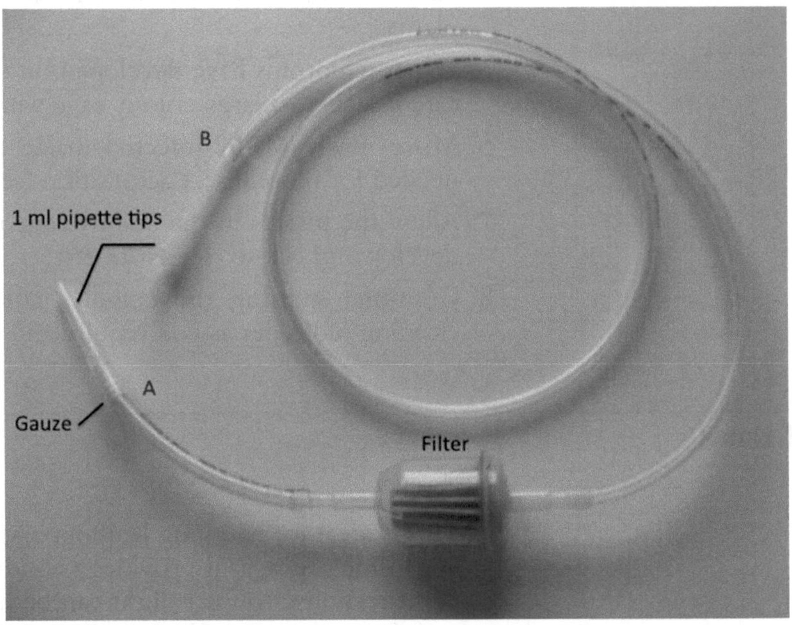

Fig. 2. Simple aspirators for handling insects are easy to make. The moth aspirator can be adapted to be a pump aspirator by removing the tip (point A) and placing it onto the input valve of a pump. This aspirator can also be used to collect insects in the field. Simply disconnect the pump and replace the tip at one end, and replace the tip at point B with an insect collection container. Collection containers can be bought with multiple tube containers. To change the tubes as you collect, gently tap the bottom of the tube on a table and as the insects fall to the bottom of the tube remove the tube and seal. Push in a fresh tube and continue collecting.

2.5 cm square of gauze over one end of 10 cm length of tube and push into the wide end of one pipette tip; push the front filter nozzle into the other end of the 10 cm tube and the back filter nozzle into one end of the 60 cm tube; then push the slant tip end of the remaining pipette tip into the other end of the 60 cm tube.

3. Colony cages can be of any size and design to suit requirements. Good quality cages are made with aluminum frames and dacron organdy mesh (17). Perspex with ventilation gauze areas or plastic framed net cages can also be used, available from BugDorm Store (MegaView science Co. No. 656-2 Fuya Rd, Taichung 40762, Taiwan). Cages in our insectary were made by Polaris Instruments Ltd. (discontinued) and designed by Dr Ian Bedford at the John Innes Centre. Cages consist of a dark plastic water tray, aluminum frame with clear plastic sides and roof. Cages are equipped with gauze ventilation panels on either side, an electric fan on the rear, and two magnetic doors on the front (Fig. 1). Some plastic mesh cages get very dirty and decay if left in standing water for long periods. To prolong the life of any cage, regular cleaning is required.
4. These cages are for single plants and short-term use. Many plastics are not suitable to use due to manufacturing treatments harmful to insects. Plastic drinks bottles are best avoided; butyrate tubes can be used (Polymer Plastics Company LC, 550 Mallroy Way, Carson City, NV 89701, USA). Butyrate tubes can be cut to height to make very sturdy cages. Lids can be made from plastic lids with a round window cut for ventilation and sealed with gauze, or a large piece of gauze can be held over the top of the cage with an elastic band. Jetran tubing and fitted plastic lids are also useful (The Bag N Box Man Ltd. Unit 1, West St, Shutford, Banbury, Oxfordshire, OX15 6PH, UK). Jetran tubing is much more flexible than butyrate, can be cut to any length and is light and hence cheap to post. Cut a hole in the lid and glue a circle of gauze to patch and allow ventilation. Once made, wash cages twice with warm water and mild detergent and allow to air-dry. This eliminates static energy within the cage that causes insects to stick to the side of cages and die.
5. *Arabidopsis* seeds are very small and often kept in microfuge tubes. If this is the case, there can be a static charge in the tube that causes seeds to come out in large groups. To avoid this, gently tap the tube with a fingernail to allow a few seeds to scatter.
6. This cold treatment encourages uniform seedling germination and growth.
7. Prepare pots in tray with compost and water. Make 1 cm-deep wells in wet compost with the handle of a fine paintbrush for

seedlings to fit into. To transplant, gently separate a seeding from others with forceps and place into prepared wells in compost in tray. Gently bury root with forceps and a paintbrush, taking care not to damage seedling.

8. *Arabidopsis* does best in moderately damp conditions. A 10 h day at 20°C allows uniform rosette growth and delays flower stem bolt. Any plants with symptoms of pests such as aphids or disorders such as mildew should be removed and destroyed immediately.
9. When material arrives from other centers, unpack material as soon as possible to prolong the life of the plant and the insect eggs.
10. Nymphs will find it much easier to travel to fresh plants if they are touching old plants in the cage. This will ensure higher nymph survival.
11. Eclosure is the final molt nymphs go through to emerge as adults. Timing of eclosure depends on insect species and environmental conditions.
12. Move the cage into a darkroom/dark box with a light, then set the table lamp to shine close to one side of the plastic cage. After 5–10 min the insects will settle on the light patch of the cage. Here they can be counted and the sex ratio determined. Sex of leafhoppers, planthoppers, and psyllids is quickly determined by the presence of a dark ovipositor on the underside of the abdomen. Ovipositors are the easiest way to distinguish female and male hoppers and psyllids without the use of a hand lens. However, young female ovipositors are paler than those of older females so use caution in identification. Males also have a more pointed abdomen end with more hair-like setae. If in doubt use a hand lens. Also worth note during handling is that much insect behavior alters with age: adults can become more flight-responsive with age and males may travel to lights more frequently than females.
13. Splitting colonies generates more insects whilst minimizing risk of losing all insects in colony failure or contamination. Sometimes a heavy bias towards males can occur in populations. In this case, do not split colonies, simply keep the colony whole until a new generation emerges. The sex ratio should even out and the colony can then be split.
14. These insects will be older, and the cage will need cleaning of honeydew and mould so that it can be used for the next generation. Killing-off and cleaning old colonies ensures that insects are kept clean, fit and free of disease. Make regular checks on the health of colonies. If insects develop fungal infections, they often become lethargic and can be covered in a fine blue-green powder—kill off such colonies to avoid spread

of the fungus. If plants become infested with spiders or spider mites or aphids or mildew, either try to control the infestation by removal and replacement of plants or kill off the colony. The ideal cycle of colonies kept will depend on the cage type available, space available for cages, plant longevity, and length of insect life cycle time.

15. Ideally have at least four plants each within individual cages. In the case of using grafted plants, as soon as phytoplasma infection is established throughout the plants, place plants in individual cages.
16. Exact number of insects depends on species of insect, and size and state of plant. The aim is to allow the plants to survive long enough so that the insects have time to feed and acquire the phytoplasma from the phloem. Larger plants can tolerate many more insects.
17. Insect acquisition time of phytoplasmas varies between species and phytoplasmas. In many cases 2 weeks is enough; however, plants may not live that long and additional plants may be needed.
18. At this point, if the infected plant is still viable, more insects can be added to acquire phytoplasmas; if not, the cage should be closed down and cleaned.
19. Removal of insects from these plants will help them to live longer to be used later to create/sustain more colonies.
20. Healthy *M. quadrilineatus* or many other related leafhoppers may produce greater numbers on oats so you need fewer colonies. To raise oats: sow 15–20 oat seeds onto packed compost in 14 cm-diameter pots; cover the seeds with 1 cm depth of compost and leave to grow in glasshouse conditions for 4 weeks; stake plants up to stabilize them and grow on for a further 1–3 weeks to desired age/height for colonies. Approximately 20–40 cm tall is good for colonies.
21. Use a tall china aster variety for best results, such as cv. Bridesmaid (Sutton Seeds, Woodview Road, Paignton, Devon, UK). Lettuce cvs Paris Island and Lobjoits Green (Ball Colegrave, West Adderbury, Banbury, Oxfordshire, UK) give good AY-WB yield and symptoms.
22. Lettuce and aster plants should still be alive for a second acquisition. *Arabidopsis* is more fragile and may not survive for second round of acquisition.
23. *M. quadrilineatus* takes 4.5 weeks from oviposition to eclosure in 16 h days at 24°C and 22°C night. At this stage plants will be infected and the plants and new emerged adults can be split to form an additional colony. Adults should be 2:1 or 1:1 female:male ratios.

24. Maize inoculation cages need to be tall, well-ventilated, and well-sealed. The plants can push lids off as they grow.
25. Use maize (*Zea mays* var. *rugosa*) cv. Early Sunglow (Park Seed Co. Greenwood, South Carolina, USA) for good phytoplasma symptoms.
26. *D. maidis* takes 4 weeks from oviposition to eclosure in 16 h days at 26 and 22°C night. At this stage plants will be infected and the maize plants and newly emerged adults can be split to form an additional colony.

Acknowledgments

I acknowledge Angela Strock (Ohio Agricultural Research and Development Center, Wooster, Ohio State University, OH 44691) and El-Desouky Ammar (Subtropical Insects Research Unit, USDA-ARS, Fort Pierce, FL. 34945) for tuition in maintenance and handling of leafhoppers, and Insectary staff and Dr Saskia Hogenhout (John Innes Centre, Norwich Research Park, Norwich, NR4 7UH, U.K.) for tuition and guidance. The project was funded by a BBSRC studentship.

References

1. Sugio A et al (2011) Phytoplasma protein effector SAP11 enhances insect vector reproduction by directly manipulating plant development and defense hormone biosynthesis. PNAS USA. doi:10.1073/pnas.1105664108
2. Murral DJ et al (1996) Effects of temperature and vector age on transmission of two Ohio stains of aster yellows phytoplasma by the aster leafhopper (Homoptera: Cicadellidae). J Econ Entomol 89:1223–1232
3. Galetto L et al (2009) Variation in vector competency depends on chrysanthemum yellows phytoplasma distribution within *Euscelidius variegatus*. Entomol Exp Appl 131:200–207
4. Marzachí C, Bosco D (2005) Relative quantification of chrysanthemum yellows (16Sr I) phytoplasma in its plant and insect host using real-time polymerase chain reaction. Mol Biotechnol 30:117–127
5. Beanland L et al (2000) Influence of aster yellows phytplasma on the fitness of aster leafhopper (Homoptera: Cicadellidae). Ann Entomol Soc Am 93:271–276
6. Bressan A et al (2005) Effects of two strains of Flavescence dorée phytoplasma on the survival and fecundity of experimental leafhopper vector *Euscelidius variegatus* Kirschbaum. J Invertebr Pathol 89:144–149
7. Malagnini V et al (2010) A study of the effects of '*Candidatus* Phytoplasma mali' on psyllid *Cacopsylla melanoneura* (Hemiptera: Psyllidae). J Invertebr Pathol 103:65–67
8. Tedeschi R, Bosco D, Alma A (2002) Population dynamics of *Cacopsylla melanoneura* (Homoptera: Psyllidae), a vector of apple proliferation phytoplasma in northwestern Italy. J Econ Entomol 95:544–551
9. Tedeschi R, Alma A (2004) Transmission of apple proliferation phytoplasma by *Cacopsylla melanoneura* (Homoptera: Psyllidae). J Econ Entomol 97:8–13
10. Hughes GL et al (2008) In vitro rearing of *Perkinsiella saccharicida* and the use of leaf segments to assay *Fiji disease virus* transmission. Phytopathology 98:810–814
11. Maramorosch K (1999) Leafhopper and planthopper rearing. In: Maramorosch K, Mahmood F (eds) Maintenance of human, animal, and plant pathogen vectors. Science Publishers, Enfield, pp 199–211
12. Ammar E-D, Hall DG (2011) A new method for short-term rearing of citrus psyllids

(Hemiptera: Psyllidae) and for collecting their honeydew excretions. Fla Entomol 94:340–342

13. Skelley LH, Hoy MA (2004) A synchronous rearing method for the Asian citrus psyllid and its parasitoids in quarantine. Biol Control 29:14–23
14. Zhang J et al (2004) Molecular and symptom analyses of phytoplasma strains from lettuce reveal a diverse population. Phytopathology 94:842–849
15. Ebbert MA, Nault LR (2001) Survival of *Dalbulus* leafhopper vectors improves after exposure to maize stunting pathogens. Entomol Exp Appl 100:311–324
16. UK Department of Food and Rural Affairs (2009) Explanatory leaflet on the issue of licences for the import, movement and keeping of prohibited plants, plant material, plant pests, soil and growing medium. http://www.fera.defra.gov.uk/plants/forms/documents/plantHealth/DEF-PHI1-FERA-1.pdf. Accessed 11 Nov 2011
17. Nault LR, Ammar ED (1989) Leafhopper and planthopper transmission of plant viruses. Annu Rev Entomol 34:503–529

Chapter 6

Capturing Insect Vectors of Phytoplasmas

Phyllis Weintraub and Jürgen Gross

Abstract

Insect vectors of phytoplasmas are limited to leafhoppers, planthoppers, and psyllids. While populations can be monitored by a number of passive techniques in the field, the capture of live insects is necessary for manipulation and study. A number of physical methods for capturing these insects already exist, but more innovative traps equipped with infochemical lures for species-specific monitoring and mass trapping are being developed.

Key words: Beat sheet, Emergence cages, Infochemicals, Kairomones, Mass trapping, Monitoring, Pheromones, Push-pull method, Sticky traps, Vacuuming, Volatile organic compounds

1. Introduction

There are three mechanisms for phytoplasmas to infect plants: insect feeding, vegetative propagation/grafting, or connection between infected and noninfected plants by parasitic plants such as *Cuscuta* spp. Insect feeding is the most important means of transmission at a local level. All known vectors are species of Hemiptera, which has important consequences in the epidemiology of the pathogen: both immature and adults feed on the same plants; they feed specifically in certain plant tissues such as the phloem where phytoplasmas reside; and feeding is nondestructive and does not elicit plant defense responses. The vectors are limited to the Cicadellidae, Fulgoromorpha, and Psyllidae (1). In the competent vector, the phytoplasma must be able to penetrate the midgut cells and migrate to and replicate in specific cells in the salivary gland to be transmitted. The period of time from acquisition to transmission competence is known as the latent period and can be days to months in length. There are documented cases of insects acquiring phytoplasmas but being unable

Matt Dickinson and Jennifer Hodgetts (eds.), *Phytoplasma: Methods and Protocols*, Methods in Molecular Biology, vol. 938, DOI 10.1007/978-1-62703-089-2_6, © Springer Science+Business Media, LLC 2013

to transmit them (2). For this reason, to determine the ability of any species to transmit phytoplasmas, transmission trials with live insects must be performed.

As symptoms of phytoplasma infection in a plant take some time to develop, the association of insects with a plant at the time symptoms are observed may be irrelevant because the vector may no longer be associated with the plant. Therefore, long-term monitoring must be conducted to determine when the potential vectors arrive. Some vectors, such as *Scaphoideus titanus*, are limited to feeding on a single genus of plants (3) whereas others, such as *Cacopsylla pruni*, are found on two entirely different hosts in different localities (4), thus requiring different forms of monitoring. Psyllids are usually captured live by using beating/beat sheets (5) although sweep nets are also used. Leafhoppers are usually captured live by the use of light traps (6), sweep nets (7), or vacuum sampling (8). Malaise traps are used for determining the directional movement of vectors (9). Emergence cages are used to determine plant hosts for nymphal Fulgoromorpha (10). All vectors can be monitored by yellow or clear sticky traps.

Current findings on the chemical ecology of Psyllidae and Fulgoromorpha may help to design new and innovative methods for trapping these insects (Fig. 1). It was shown recently that these insects use chemical cues for orientation and host identification. These so-called infochemicals (or semiochemicals) can influence target organisms in general (e.g., green leaf volatiles), may be specific for a specific group of insects (e.g., many kairomones), or may even be species-specific (e.g., pheromones). For instance, the psyllid species *Cacopsylla picta* and *C. melanoneura* are able to distinguish their specific reproduction and overwintering hosts by means of chemical signals produced by their host plants (11). Analyzing the chemically mediated interactions between *C. picta*, the main vector of the phytoplasma causing apple proliferation disease, ('*Candidatus* Phytoplasma mali'), and its host plants (reproduction and overwintering hosts), this phytoplasma was found to lure its highly adapted vector *C. picta* to infected apple trees by changing their odor (12, 13). The phytoplasma stimulates apple trees to produce higher amounts of a sesquiterpene (β-caryophyllene; Fig. 2b1) that preferentially attracts newly hatched adults of *C. picta* (emigrants) just before they start their emigration to their overwintering host plants (12, 14). By intensified feeding on infected plants, the probability of phytoplasma acquisition increases. In contrast, nonvector species like the hawthorn psyllid, *C. melanoneura*, or the plum psyllid, *C. pruni*, did not react to this sesquiterpene. After overwintering, *C. picta* remigrates to apple plants (remigrants), but prefers to oviposit mainly on uninfected plants (15).

This behavior results in a high probability of transmission of the phytoplasmas to uninfected plants. The identity of the infochemical(s) that regulate this egg-laying behavior still remains unknown. Other psyllid species, such as *C. pruni* (J. Gross,

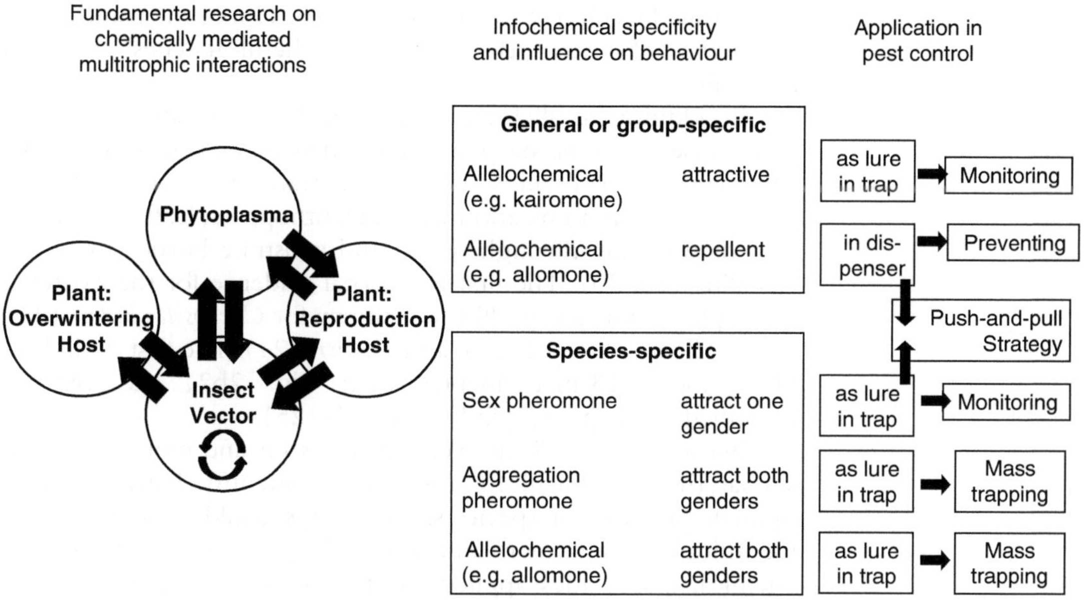

Fig. 1. The research on chemically mediated multitrophic interactions between phytoplasmas, their host plants, and insect vectors reveals the possibility to identify infochemicals influencing the behavior of insect vectors. Such behavior-modifying compounds may be used in biotechnical pest control. Methods of application are the use of attractive components as a lure in traps for monitoring or, in cases where both genders are attracted, for mass trapping. Repellent compounds could be used as a spray or in dispensers for making the crop unattractive or unsuitable for the vector. Attractive and repellent compounds can be combined in so-called push-and-pull strategies (see Note 16).

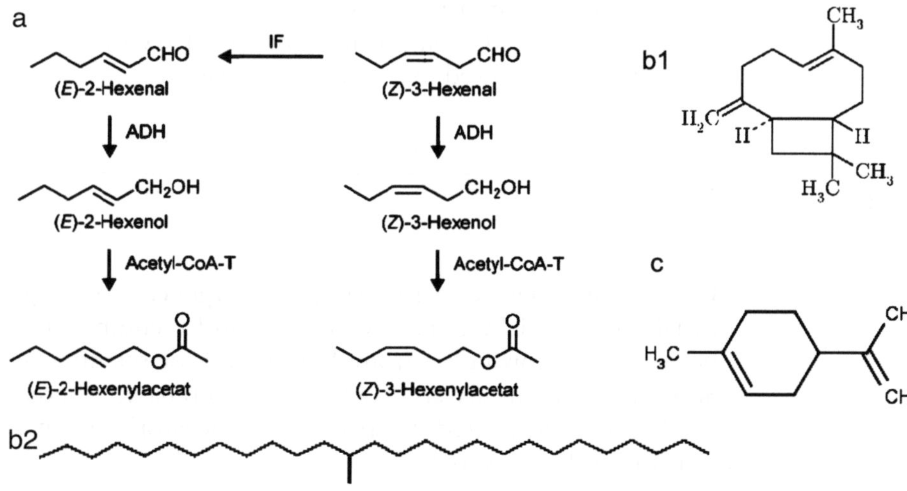

Fig. 2. Examples of volatile organic compounds which modify the behavior of phytoplasma transmitting insects. (**a**) General attractive compounds (green leaf volatiles); (**b**) species-specific attractive compounds, 1:β-caryophyllene (allomone, attractive for *Cacopsylla picta*) (12), 2:13-methylheptacosane (sex pheromone, attractive for *Cacopsylla pyricola* males) (25); (**c**) limonene (repellent for *Trioza apicalis*) (33).

unpublished results), *Diaphorina citri* (16, 17), or the planthopper *Hyalesthes obsoletus* (10), also use chemical cues for host recognition.

In addition to allelochemicals, which are attractive for some vector insect species, sex pheromones may play a role in the courtship behavior of psyllids. For a long time, it was thought that intraspecific communication within planthoppers, leafhoppers, and psyllids was only through sound and substrate-borne vibrational signaling (18–20). The first behavioral evidence for the existence of sex pheromones in psyllids was shown for *Cacopsylla bidens* (21), *Cacopsylla pyricola* (22, 23), and *D. citri* (24). The first psyllid sex pheromone (13-methylheptacosane; Fig. 2b2) was recently identified in the pear psyllid *C. pyricola* (25).

Because *Cacopsylla* are very small insects and morphologically similar, they are difficult for nonexperts to identify. Thus, the development and use of species-specific traps could simplify correct identification by farmers and further help reduce the amount of chemical insecticides applied by determining the ideal date for spraying. Based on these facts, it is possible to create group- or species-specific traps, lured with attractive components for capturing psyllids for monitoring purposes (14, 26) (Fig. 2). The cost of chemical baits, their attraction radius (effectiveness), and chemical longevity in the field are important considerations when developing new trapping systems (27). When allelochemicals produced by infected plants are attractive to both genders of vectoring psyllids, as is the case in apple proliferation-infected trees (12, 14), it could also be possible to develop mass trapping systems for a more sustainable control of these insects in the near future (Fig. 1) (see Note 1). This chapter presents a variety of passive and active traps for the capture and monitoring of potential vectors or phytoplasmas which can be used in a variety of situations.

2. Materials

Equipment (such as beating sheets, emergence cages, light traps, malaise traps, sticky traps, sweep nets, and vacuum samplers) can be purchased through entomological and/or scientific suppliers. Some allelochemicals are available for purchase, whereas others need to be synthesized by specialized laboratories or in conjunction with other researchers.

2.1. Mechanical Traps

1. Beat sheet. Prepare a square sheet (0.5–1 m^2) of sturdy material, such as muslin, and reinforce the corners to create triangular pockets that are attached on either side of each corner at a length of ~10 cm. Purchase two lengths of dowelling (~1 cm diameter)

to create a criss-crossed frame fitting the corners of the sheeting. Any similar device, such as a light-colored umbrella, light-weight light-colored tray, or light-colored ground cloth, can be substituted.

2. Vacuum trap. Modify a gas-driven portable leaf blower by switching the air intake and exhaust ports. Attach a 1–1.5 m long, 15–30 cm diameter flexible or rigid hose to the air intake port. Put 3–4 pop rivets evenly spaced around the distal opening. Purchase or make collection bags of fine nylon mesh to fit the diameter of the hose. Secure collection bags in the hose with rubber bands over the pop rivets.
3. Emergence cages. Square or rectangular cages can be constructed of any size with a metal, plastic, or wooden frame, keeping it lightweight to be easily transported. Five sides should be covered with insect exclusion screening (50-mesh) to prevent vectors from escaping but allow some air movement to prevent overheating. A sticky trap may be inserted to capture emerging insects.
4. Sweep nets. For use with low vegetation, a 1 m handle is suitable; for trees use a net with a handle extendable up to 4 m.

2.2. Trapping Using Visual or Chemical Signals

1. Traps equipped with visual signals. Traps can be made of yellow or clear plastic with enough rigidity to not fold in on itself. Trap size may vary up to ~50 cm square, but should be consistent throughout monitoring period. Water-soluble or oil-based glues can be purchased or cooking oil can be used.
2. Traps baited with attractive chemicals. Known attractive chemicals can be purchased if commercially available (some volatile organic compounds (VOCs) emitted by plants) or may have to be synthesized in adequate amounts by a chemical laboratory (pheromones). The release of chemical lures can be assured by passive substrates (filter paper, membrane dispensers, rubber septum dispensers, tape dispensers, sol–gel formulations, etc.), or active systems like micro-pumps or piezoelectric micro-sprayers (28, 29). They can be used for the release of any mixture or single substances of attractive infochemicals or odors at constant ratios and rates; there are many different types available, which should be adjusted to the requirements necessary for trapping the target species. Release of very volatile compounds can be delayed by adding paraffin oil to the active compounds. Many different trap types are commercially available and can be purchased, or must be developed, depending on the species-specific behavior of the target vector species. They have to be equipped additionally with the appropriate system for releasing the attractants and the specific mixture of compounds attractive for the target species.

3. Methods

Monitoring of insect populations must be carried out on a regular basis: weekly, bi-weekly, monthly, and preferably before the first insects are observed in the field until after their population has declined. There must be replicate sampling in each field, usually a minimum of ten traps or plants to obtain accurate data (see Note 2).

3.1. Beating

For use on woody bushes and trees to collect live vectors, primarily psyllids (see Note 3).

1. Place beating sheet beneath the branch to be sampled.
2. Use a short piece of rubber or plastic hose to hit the branch without causing injury to the tree or bush.
3. Collect samples by aspiration.

3.2. Vacuum Traps

For use with all types of plants and vectors.

1. Vacuum a section of the plant.
2. Secure the collection bag and place in a cooler for transportation to the laboratory for specimen identification and manipulation.

3.3. Emergence Cages

For use in determining suitable host plants, primarily for Fulgoromorpha.

1. Place trap beneath target plant and secure it so that emerging vectors cannot escape.
2. Sticky traps may be placed in cage to collect specimens.

3.4. Sweep Nets

For use with all plants, for live specimens.

1. For monitoring purposes, use a fixed number of swings in at least four locations in a field (7).
2. Aspirate desired specimens.

3.5. Light Traps

For night-time monitoring of vector species, for live or dead specimens.

1. Specimens can be collected in a bucket or bottle beneath the light source.
2. Collection vessel may be filled with 70% ethanol or polyethylene glycol to preserve specimens.

3.6. Malaise Traps

For monitoring directional movement of vectors, for live or dead specimens.

1. Traps may contain collection bottles for two-, four-, or nondirectional movement of vectors.
2. Collection vessel may be filled with 70% ethanol or polyethylene glycol to preserve specimens.

3.7. Sticky Traps

For general monitoring over long periods of time, dead specimens only.

1. Place trap in desired location for as long as required (see Note 4).
2. Specimens may be removed with hexanes, butane, or other solubilizing chemicals.

3.8. Traps Baited with Attractive Chemicals

3.8.1. Identification of Potential Infochemical Lures

1. Put the organism (plant or insect) in an adequate glass container, or wrap branches of larger plants in polyethylene terephthalate (PET) bags (Ø 20 cm).
2. Pump commercially available purified air or charcoal-cleaned air (granulated charcoal, 4–8 mm) through the container or bag for an adequate time period. Control the air flow by a flowmeter.
3. Collect the VOCs from the plant or insect headspace, choosing a method which is appropriate for your experimental setting (see Note 5).
   - Dynamic headspace sampling technique: collect VOCs in collection filters (e.g., 5 mg charcoal, Gränicher + Quartero, Daumazan, France) and elute the extract containing the VOCs by rinsing the filter with a solvent containing an internal standard (IS) for later quantification, or collect VOCs on thermodesorption tubes (glass or metal) filled with an appropriate adsorbent material (see Note 6).
   - Static headspace sampling technique: solid phase microextraction (SPME) (see Note 7).
4. To analyze the VOCs collected from plant or insect headspace, samples are either injected as solvent extracts (see above) into a gas chromatograph (GC), or are thermally desorbed from their adsorbent (thermodesorption tube or SPME fiber) by heating the matrix to 250–300°C (see Note 8).
5. Choose a GC-column for separation with properties belonging to the chemical structures of the collected VOCs. For volatile analysis, use a fused silica capillary column with an appropriate stationary phase, such as nonpolar dimethyl polysiloxanes (e.g., DB-5, DB-1), or more polar polyethylene glycol polymers (e.g., DB-Wax, Carbowax®). Use Helium 6.0 as the mobile phase.
6. Use an appropriate temperature programme for optimum separation of the collected compounds.

7. Use a mass spectrometer (MS) coupled with the GC for the identification of the separated compounds by comparing the characteristic ion fragmentation pattern (mass spectrum) with data from mass spectra libraries (see Note 9).
8. Confirm the identification by injection of synthetic samples on the same column, or preferably on two columns with different polarities, by comparing retention times and mass spectra (see Note 10).

3.8.2. Evaluating Potential Behavior-Modifying Compounds

1. Use gas chromatography coupled with electro-antennographic detection (GC-EAD) to identify compounds detected by the insect antenna (see Note 11).
2. Behavioral tests can be carried out by using dynamic or static olfactometers, wind tunnel, or a Kramer sphere, depending on the specific behavior of target organisms (see Note 12).
3. Pump purified air through two containers containing the test VOC and attach each outlet with a test arm of the olfactometer. Control the air flow with a flow meter (see Note 13).
4. Count every individual that passes a final mark on one of the test arms within a specific period of time (normally 5–15 min) (see Note 14).
5. Analyze the data statistically (e.g., by dependent paired *t*-test or *G*-test, if more than one psyllid was tested simultaneously).
6. Choose an attractive compound or mixture as a lure for infochemical traps.

3.8.3. Development of Infochemical Traps

1. Choose a passive substrate or active system (e.g., dispenser) for providing the chemical compound/mixture in the trap over the flight period of the target organism.
2. Check the release rate of the dispensers by weighing them over the test period under natural conditions (see Note 15).
3. Adjust the release rates of chemical compounds according to appropriate requirements of the target species.

3.8.4. Modifying Specialized Traps

1. Purchase a commercially available trap and modify it with the system developed above for providing the infochemical.
2. Add sticky surface, insecticide, or fluid (70% ethanol or polyethylene glycol to preserve specimens).
3. Inspect the traps at regular intervals for caught specimens.
4. Monitor the charging level of the dispenser during the flight period of the target species and refill when necessary.

4. Notes

1. Besides monitoring of insect vectors by traps baited with attractive chemicals, mass trapping for control purposes may also be possible when the chemical lure is attractive for both genders of the target species. This can be the case when aggregation pheromones, plant kairomones, or allomones are used as the chemical lure (Fig. 1). The efficacy of mass trapping methods can be explored through field experiments and by computer simulations. The costs of chemical baits, their attraction radius, and chemical longevity in the field must be taken into consideration when developing a mass trapping system. For further reading, the Web site of Byers (27) provides software, Java applets, and other tools regarding this method.
2. Certain types of monitoring, sweep net sampling, vacuum sampling must be carried out after dew has evaporated, as insects caught by these methods will adhere to the bags and are difficult to dislodge. Conversely, under conditions of moderate to strong wind, insects move deep into plants and are difficult to dislodge. Monitoring by beating, sweep net and vacuuming should be avoided during these times.
3. For receiving quantitatively comparable results, the monitoring of population dynamics of vector insects can be carried out according to (30). For taking one sample, ten comparable branches of each host plant are strongly tapped with a pole (about 1 m long) three times per branch/plant. The dislodged insects can be collected in a beating sheet held beneath, equipped with a removable polyethylene collection bag. After sampling, the collection bag can be placed in a cooler for transportation to the laboratory where collected insects can be counted and stored in vials for later identification or experiments.
4. The length of time in the field is determined by the amount of dust that accumulates on the trap.
5. Further information on practical approaches to plant volatile analysis for researchers inexperienced with analytical techniques can be found described in detail in (31).
6. For an overview on commonly used adsorbents, see ref. (31). The closed loop stripping analysis (CLSA) can be used when only very small amounts of volatiles will be emitted by the organism.
7. SPME is a fast and simple method for collecting volatiles at very low detection limits, but due to its specific characteristics it is suitable for sampling VOCs for qualitative rather than quantitative analysis.

8. When using a two-stage thermal desorber, the thermally released volatiles will be concentrated in a cold trap prior to their injection into the gas chromatograph, while the SPME fiber can be directly desorbed within the injector.
9. Suitable libraries include NIST/EPA/NIH Mass Spectral Library, National Institute of Standards and Technology, Wiley.
10. In case of resulting complex volatile patterns, bioinformatics like multivariate analysis should be implemented in the statistical interpretation of differences between plant VOC or insect pheromone mixtures (32).
11. This will reduce the number of VOCs to those which may be responsible for vector attraction to the host plant, and so also reduce the amount of the effort needed for identification of behavior-modifying compounds.
12. For some psyllid and planthopper species, dynamic Y-shaped olfactometers have been proven to work (10, 12). They consist mostly of a Y-shaped glass tube mounted on an angular board.
13. According to the target species-specific behavior in the olfactometer, test either one, five (10), or ten specimens (12, 21) of the same gender together. Do not test different genders simultaneously.
14. In case the infection status of the vector needs to be determined later, place each specimen after the behavioral trial into a separate vial filled with ethanol (70–96%). Extract the DNA and identify the phytoplasma as required.
15. Ideally, the release rate of the attractants should be relatively constant and remain effective during the whole flight period of the target species. However, during periods of higher ambient temperatures, the release rate may increase depending on the volatility of the components. Thus, it is important to monitor the charging level of the dispenser during the flight period of the target organism often and refill when necessary. In order to reduce the release of volatile compounds provided by passive substrate, dilute them in paraffin oil. To avoid temperature effects in general, use an active system such as a micro-pump or a piezoelectric micro-sprayer with constant release rates.
16. The so-called push-pull strategies consist of a behavioral manipulation of an insect vector by the integration of two different stimuli. A repellent makes the protected resource unattractive or unsuitable to the vector (push stimulus), while luring it toward an attractive source or attractant (pull stimulus), e.g., located in a sticky trap. By doing fundamental research on multitrophic interactions between plants, phytoplasma and insect vectors, the isolation and identification of attractive and repellent infochemicals may also open the path to this very innovative control strategy.

References

1. Weintraub PG, Beanland L (2006) Insect vectors of phytoplasmas. Annu Rev Entomol 51:91–111
2. Vega FE et al (1993) Detection of a plant pathogen in a nonvector insect species by the polymerase chain reaction. Phytopathology 83:621–624
3. Maixner M (1993) *Scaphoideus titanus*, a possible vector of grapevine yellows in New York. Plant Dis 77:408–413
4. Jarausch B, Jarausch W (2010) Psyllid vectors and their control. In: Weintraub PG, Jones P (eds) Phytoplasmas: genomes, plant hosts and vectors. CAB International Oxfordshire, Wallingford, UK, pp 250–271
5. Carraro L et al (2004) *Macropsis mendax* as a vector of elm yellows phytoplasma of *Ulmus* species. Plant Pathol 53:90–95
6. Pilkington LJ et al (2004) Reducing the immigration of suspected leafhopper vectors and severity of Australian lucerne yellows disease. Aust J Exp Agric 44:983–992
7. Rice Mahr SE, Wyman JA, Chapman RK (1993) Variability in aster yellows infectivity of local populations of the aster leafhopper (Homoptera: Cicadellidae) in Wisconsin. J Econ Entomol 86:1522–1526
8. Weintraub PG, Horowitz AR (1996) Spatial and diel activity of the pea leaf miner (Diptera: Agromyzidae) in potatoes, *Solanum tuberosum*. Environ Entomol 25:722–726
9. Irwin ME et al (2000) Diversity and movement patterns of leaf beetles (Coleoptera: Chrysomelidae) and leafhoppers (Homoptera: Cicadellidae) in a heterogeneous tropical landscape: implications for redressing the integrated pest management paradigm. In: Ekbom B, Irwin ME, Robert Y (eds) Interchanges of insects between agricultural and surrounding landscapes. Kluwer, Dordrecht, The Netherlands, pp 141–168
10. Sharon R et al (2005) *Vitex agnus-castus* is a preferred host plant for *Hyalesthes obsoletus*. J Chem Ecol 31:1051–1063
11. Gross J, Mekonen N (2005) Plant odours influence the host finding behaviour of apple psyllids (*Cacopsylla picta*; *C. melanoneura*). IOBC/WPRS Bull 28(7):351–355
12. Mayer CJ, Vilcinskas A, Gross J (2008) Pathogen-induced release of plant allomone manipulates vector insect behavior. J Chem Ecol 34:1518–1522
13. Mayer CJ et al (2009) *Cacopsylla melanoneura* has no relevance as vector of apple proliferation in Germany. Phytopathology 99:729–738
14. Mayer CJ, Vilcinskas A, Gross J (2008) Phytopathogen lures its insect vector by altering host plant odor. J Chem Ecol 34:1045–1049
15. Mayer CJ, Vilcinskas A, Gross J (2011) Chemically mediated multitrophic interactions in a plant–insect vector–phytoplasma system compared with a partially nonvector species. Agric For Entomol 13:25–35
16. Patt JM, Sétamou M (2010) Responses of the Asian citrus psyllid to volatiles emitted by the flushing shoots of its rutaceous host plants. Environ Entomol 39:618–624
17. Wenninger EJ, Stelinski LL, Hall DG (2009) Role of olfactory cues, visual cues, and mating status in orientation of *Diaphorina citri* Kuwayama (Hemiptera: Psyllidae) to four different host plants. Environ Entomol 38: 225–234
18. Percy DM, Taylor GS, Kennedy M (2006) Psyllid communication: acoustic diversity, mate recognition and phylogenetic signal. Invertebr Syst 20:431–445
19. Percy DM, Boyd EA, Hoddle MS (2008) Observations of acoustic signaling in three sharpshooters: *Homalodisca vitripennis, Homalodisca liturata*, and *Graphocephala atropunctata* (Hemiptera: Cicadellidae). Ann Entomol Soc Am 101:253–259
20. Tishechkin DY (2005) Vibrational communication in Psylloidea (Hemiptera). In: Drosopoulos S, Claridge MF (eds) Insects sounds and communication: physiology, behaviour, ecology, and evolution. CRC Taylor & Francis, Boca Raton, FL, pp 358–363
21. Soroker V et al (2004) The role of chemical cues in host and mate location in the pear psylla *Cacopsylla bidens* (Homoptera: Psyllidae). J Insect Behav 17:613–626
22. Horton DR, Guédot C, Landolt PJ (2007) Diapause status of females affects attraction of male pear psylla, *Cacopsylla pyricola*, to volatiles from female-infested pear shoots. Entomol Exp Appl 123:185–192
23. Horton DR, Landolt PJ (2007) Attraction of male pear psylla, *Cacopsylla pyricola*, to female-infested pear shoots. Entomol Exp Appl 123:177–183
24. Wenninger EJ, Stelinski LL, Hall DG (2008) Behavioral evidence for a female-produced sex attractant in *Diaphorina citri*. Entomol Exp Appl 128:450–459
25. Guédot CN et al (2009) Identification of a sex attractant pheromone for male winterform pear psylla, *Cacopsylla pyricola*. J Chem Ecol 35: 1437–1447

26. Gross J, Mayer CJ (2010) Nuove prospettive per il monitoraggio e la difusa biomolecolare. Frutta e vite 34:39–41
27. Byers JA (2011) Mass trapping. http://www.chemical-ecology.net/masstrap.htm. Accessed 14 Nov 2011
28. Zada A, Falach L, Byers JA (2009) Development of sol–gel formulations for slow release of pheromones. Chemoecology 19:37–45
29. El-Sayed AM (2011) The Pherobase: database of insect pheromones and semiochemicals. http://www.pherobase.com. Accessed 14 Nov 2011
30. Müther J, Vogt H (2003) Sampling methods in orchard trials: a comparison between beating and inventory sampling. IOBC/WPRS Bull 26(5):67–72
31. Tholl D et al (2006) Practical approaches to plant volatile analysis. Plant J 45:540–560
32. Van Dam NM, Poppy GM (2008) Why plant volatile analysis need bioinformatics—detecting signal from noise in increasingly complex profiles. Plant Biol 10:29–37
33. Nehlin G, Valterova I, Borg-Karlson A-K (1994) Use of conifer volatiles to reduce injury caused by carrot psyllid, *Trioza apicalis*, Förster (Homoptera, Psylloidea). J Chem Ecol 20:771–783

Chapter 7

Insect Vector Transmission Assays

Domenico Bosco and Rosemarie Tedeschi

Abstract

Phytoplasmas are transmitted in a persistent propagative manner by phloem-feeding vectors belonging to the order Hemiptera, suborder Homoptera. Following acquisition from the infected source plant, there is a latent period before the vector can transmit, so transmission assays consist of three basic steps: acquisition, latency, and inoculation. More than 90 vector species (plant-, leafhoppers, and psyllids) have been discovered so far but many others are still undiscovered, and their role in spreading economically important crop diseases is neglected. Therefore, screening for vectors is an essential step in developing rational control strategies targeted against the actual vectors for phytoplasma-associated diseases. The mere detection of a phytoplasma in an insect does not imply that the insect is a vector; a transmission assay is required to provide conclusive evidence. Transmission experiments can be carried out using insects from phytoplasma-free laboratory colonies or field-collections. Moreover, transmission assays can be performed by feeding vectors on an artificial diet through Parafilm®, after which phytoplasmas can be detected in the sucrose feeding medium by PCR. Transmission trials involve the use of different techniques according to the biology of the different vector species; planthoppers, leafhoppers, and psyllids.

Key words: Artificial feeding medium, Feeding, Hemiptera, PCR, Transmission, Vector species

1. Introduction

Phytoplasmas are phloem-limited pathogens associated with many diseases in both cultivated and wild plants. In nature, phytoplasmas are transmitted by insect vectors in a persistent, propagative manner. The insects acquire the phytoplasmas by feeding on an infected plant and then transmit them to a healthy plant after a latent period, during which the phytoplasmas move through and multiply in the vector body. The insects often remain infective for the entire lifespan. Since phytoplasmas are phloem-limited, their vectors, belonging to the order Hemiptera, are phloem feeders. Most phytoplasma vectors are confined into three main taxonomic groups: leafhoppers (Auchenorrhyncha: Cicadellidae), planthoppers (Auchenorrhyncha: Fulgoromorpha), and psyllids (Sternorrhyncha: Psyllidae).

Matt Dickinson and Jennifer Hodgetts (eds.), *Phytoplasma: Methods and Protocols*, Methods in Molecular Biology, vol. 938, DOI 10.1007/978-1-62703-089-2_7, © Springer Science+Business Media, LLC 2013

Phytoplasmas can be experimentally transmitted by grafting, by dodder (1), and in at least one case by root bridges (2). Possible seed transmission of phytoplasmas has also been suggested, but although phytoplasma DNA can be detected in embryos, there is as yet no evidence that the pathogen is seed transmitted through to the seedling to cause disease in progeny plants (3). Insect transmission is by far the most important manner of phytoplasma spread under field and natural conditions. Only selected species can act as vectors and the transmission of phytoplasmas by insects certainly involves, at several levels, elements of host-pathogen specificity (4). Vector insects can be polyphagous, oligophagous, or strictly monophagous according to their ability to feed and reproduce on many, few, or one host plant, respectively. Similarly, phytoplasmas may be generalists, infecting several different plant species, or specialists, infecting one or a few related plant species. As a consequence, a generalist phytoplasma can be transmitted by several vector species. Plant-specialist phytoplasmas can be transmitted by a narrow range of vector species or by a specific vector (4).

When screening for phytoplasma vectors, PCR assays of field-collected insects may provide indications on the possible role of a given species in transmitting phytoplasmas. However, since phytoplasmas may be acquired but not injected with feeding (5), the mere detection of a phytoplasma in an insect does not confirm that the insect is a vector; a transmission assay is the conclusive evidence required. Transmission experiments are mandatory to (1) identify new vectors (2) describe the transmission characteristics, such as transmission efficiency, duration of acquisition, incubation, and inoculation periods, and (3) understand the epidemiology of a phytoplasma disease (is the vector feeding and breeding on the infected crop or is it a visiting insect that erratically transfers the phytoplasma from a weed or a natural reservoir into a crop species?). Even though nearly 100 phytoplasma vector species have been listed by Weintraub and Beanland (6), many vectors of phytoplasma diseases are still undiscovered and their role in spreading economically important diseases has been neglected.

This chapter aims to provide the technical information necessary to perform transmission trials using leaf- and planthoppers as well as psyllids. Following acquisition from the infected source plant, there is a latent period before the vector can transmit, so transmission assays consist of three fundamental steps: acquisition, latency, and inoculation. During the acquisition phase, the insect vector ingests phytoplasma cells through feeding on an infected plant. Acquisition can be "natural", when insects collected in the field, thus naturally infected, are used in transmission trials, or "controlled" when healthy putative vectors are caged on infected plants. In this latter case, the period of time given to the insects to acquire the phytoplasma is called the Acquisition Access Period (AAP). Usually, the AAP lasts from a few to several days to ensure an high acquisition efficiency. Even if a few hours are enough for some species to acquire a phytoplasma ((7, 8), author's unpublished results), when the optimal AAP is unknown, long acquisition times should provide maximum efficiency. The latency period (LP), also called "incubation period", is the time interval between acquisition and the beginning of infectivity. During this phase, the phytoplasmas invade the insect body via the hemolymph, multiply, and reach the salivary glands. The LP varies from 12 days to well over a month depending on the insect species, phytoplasma strain/species, and abiotic factors such as temperature (5). During this phase, the insects should be reared on a suitable plant to ensure the highest survival rate. Once the LP is completed, the insects inject the phytoplasmas directly into the sieve tubes of a healthy plant. Therefore, putative vectors should be isolated, individually or in batches, onto healthy plants to assess phytoplasma transmission. The period of time given to the putative vector to transmit the phytoplasma is called the Inoculation Access Period (IAP). A few hours can be sufficient to transmit the pathogen, but longer times can provide higher transmission efficiencies.

Based on the procedure we can distinguish (1) controlled transmission experiments (in which all the steps are carried out under controlled conditions in climatic chambers or greenhouses) and (2) transmission experiments with field-collected insects (in which only the inoculation step is carried out under controlled conditions). First, the basic techniques for vector rearing are described, and then methods for transmission experiments using laboratory-reared or field-collected insects are detailed. Finally, techniques used to assess phytoplasma transovarial transmission in the vector are presented. The different techniques are discussed for leaf-, planthoppers, and psyllids in the light of their different biology.

2. Materials

1. Phytoplasma-free insect vector colony.
2. Phytoplasma-infected source plants. Source plants can be obtained by graft inoculation from diseased to healthy material, by a previous insect transmission, or by micropropagation of phytoplasma-diseased shoot cultures (9) (see Chapters 2, 3, and 4).
3. Sweep net or beating tray.
4. Potted seedlings of test plants.
5. Climatic chambers, greenhouse or screenhouse.
6. Rearing cages (plexiglass and/or net cages) (see Fig. 1a, b).

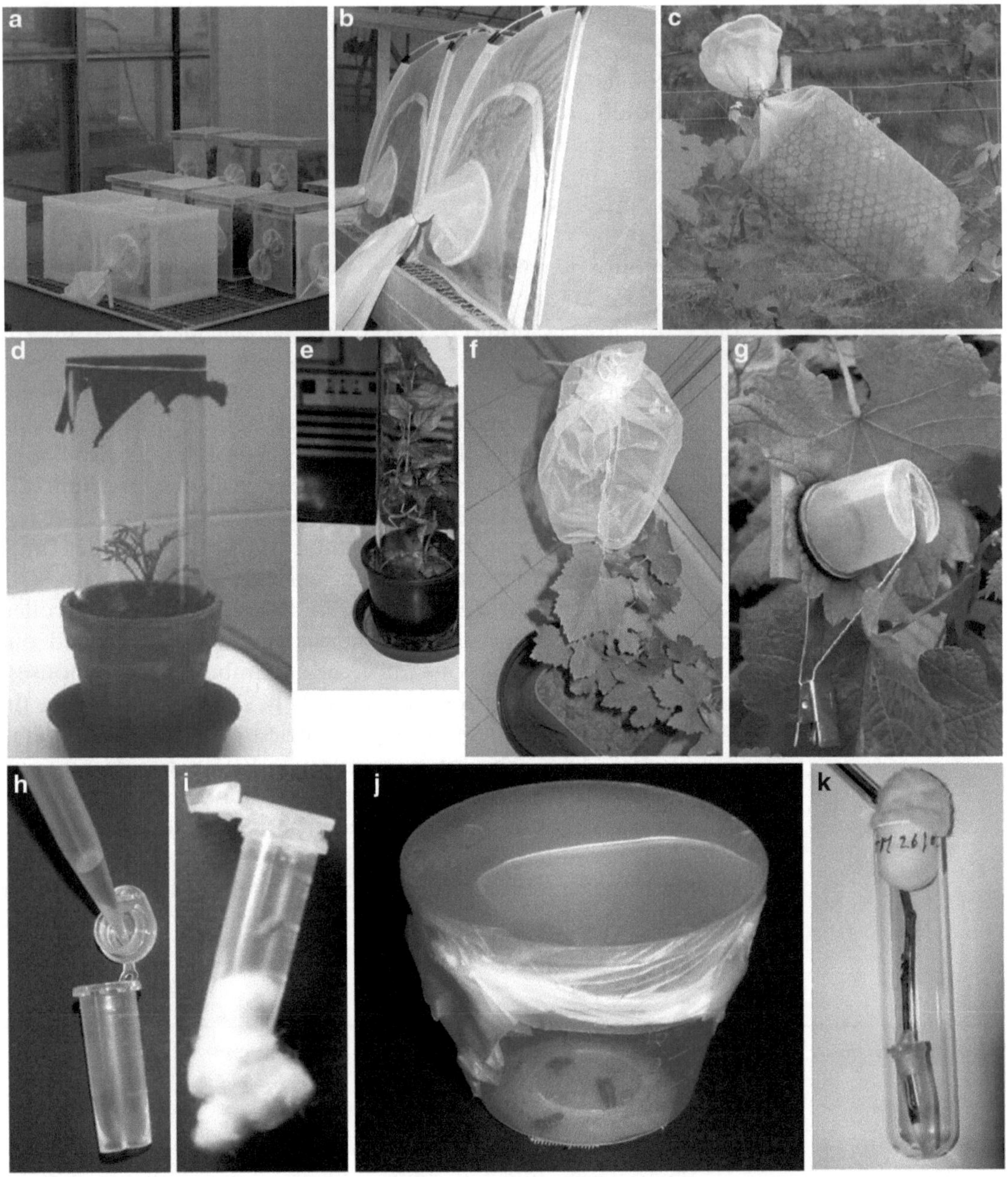

Fig. 1. (**a**, **b**) Plastic and net cages for vector rearing (**c**): net cage isolating a single branch for phytoplasma acquisition in the field (**d**, **e**): glass cylindrical cages for phytoplasma inoculation in the laboratory (**f**): net cage isolating a single branch of a test plant for phytoplasma inoculation in the laboratory (**g**): clip cage (**h**, **i**): microcentrifuge tube cage for phytoplasma inoculation to feeding medium with a single insect (**j**): cup-like cage for phytoplasma inoculation to feeding medium with insect groups (**k**): glass tube with an apple twig for egg laying in transovarial experiments.

7. Insect aspirator.
8. Insecticides.
9. Insect feeding media, e.g.,
   (a) 5% Sucrose in TE (10 mM Tris–HCl pH 8.0, 1 mM EDTA) (10).
   (b) 10% Sucrose, 0.2% fructose, 0.375% K_2HPO_4, 0.028% $MgCl_2$, pH 7.5.
10. Microcentrifuge tubes or cup feeding chambers.
11. Parafilm® membrane.

3. Methods

3.1. Controlled Transmission Experiments

The availability of a laboratory colony of the vector depends mainly on the biology of the insect. Most leafhopper vectors (family Cicadellidae) are relatively easy to grow under controlled conditions since they breed continuously (multivoltine) or once a year (monovoltine) on one (monophagous), few (oligophagous), or many (polyphagous) host plants. Planthoppers are much more difficult to rear since nymphs are root feeders (11). In this case, the rearing cage must include the soil with the roots of the host plants and attention must be paid to the disturbance or predation by other arthropods in the soil (e.g., ants) and to the watering of potted plants to avoid drowning of nymphs. A rearing technique has been described for the planthopper *Hyalesthes obsoletus* (12). Some psyllids (e.g., *Cacopsylla* spp.) are migrating insects that lay eggs and develop on a host plant and then migrate to shelter plants for aestivation and overwintering. In this latter case, even though, in theory, the complex cycle can be reproduced under controlled conditions, in practice, permanent rearings of these species are not feasible. In some cases, e.g., leafhoppers laying eggs on woody hosts, it is possible to collect branches with eggs in the field, leave them to hatch under controlled conditions, and establish an annual rearing. For *Scaphoideus titanus*, the vector of Flavescence dorée, grapevine branches with eggs are collected in winter, and stored in a fridge until needed (eggs remain viable after several months' storage); newly hatched nymphs fed on grapevine or broad bean plants can be used for transmission experiments. In our experience, broad bean plants are more suitable than grape, even though both plant species can be used (together or alternatively). As an option, *S. titanus* nymphs can be easily reared in small batches in cylindrical plastic boxes (10 cm height, 5 cm diameter), the top covered with a net, and containing a 1-cm layer of technical agar solution (8%) on which a disk of grapevine leaf is laid and replaced twice a week (13). This system may avoid the use of plants when a limited number of individuals are needed.

A laboratory colony of an insect vector can be established on a suitable host plant on which a multivoltine vector may breed several times per year. Basically, we can describe two kinds of continuous rearings: mixed-age rearing and age-structured rearing. In a mixed-age rearing, a number of insects of all stages feed and breed on several plants. New plants are provided as soon as the old ones are aging and dying and, when the insect population is too high, excess insects must be removed to ensure that plants can carry the population. When needed, nymphs are taken from the cage for transmission experiments. An age-structured rearing consists of an oviposition chamber where a number of females (together with males) lay eggs on host plants for a short period (from a few days to 1 week). After the oviposition, host plants are moved to new cages where the eggs complete their embryonic development and give rise to the nymphs. Each cage contains plants exposed to ovipositing females at the same time and therefore all the nymphs are of approximately the same age. With this rearing, a lot of coetaneous nymphs can be conveniently used in transmission experiments. When dealing with monovoltine species with an obligate diapause (generally in the egg stage), continuous rearing is not feasible. For these latter cases, the natural life cycle can be reproduced under controlled conditions or the insects can be obtained as described above for *S. titanus*.

1. Before starting the transmission experiments, a molecular analysis (PCR) should be done to ascertain the presence of phytoplasmas in the source plants (see Chapters 14, 15, 16, and 17).
2. Cage a number of nymphs (preferably third to fifth instar nymphs) from the phytoplasma-free colony onto source plants to feed for the AAP (see Note 1).
3. After a suitable AAP, transfer insects (nymphs and adults) surviving the AAP onto suitable host plants to complete the latent period (LP).
4. At the end of the LP, cage insects on healthy test plants for an IAP. Insects can be either caged in large groups on a number of test plants inside a plexiglass and nylon cage or a net cage, or caged singly or in small groups (e.g., 3–5 insects) on individual plants inside small cages (see Notes 2 and 3).
5. In order to define the actual length of the LP under given environmental conditions, serially transfer the insects after the AAP, singly or in groups, to test plants for successive one or a few day IAPs. The shorter the successive IAPs, the more precise will be the estimation of the LP duration.
6. At the end of the IAP, spray with insecticides to free the plants from insects, and maintain in a greenhouse or climatic chamber to allow phytoplasma symptoms to develop (see Note 4).

7. After a variable time lapse from the inoculation, analyze the test plants by PCR to check for the presence of phytoplasmas (see Note 5).

3.2. Transmission Experiments with Field-Collected Insects

1. Collect a number of adults in or around phytoplasma-infected fields using a sweep net or a D-Vac equipment (plant and leafhoppers) or a beating tray (psyllids).
2. Cage field-collected insects on test plants as described above. A long IAP is advisable (e.g., weeks or until insect death) since it will not be known whether infected insects from the field have already completed the LP.
3. Further steps in the transmission experiment are the same as described above.
4. Transmission experiments with field-collected insects may provide evidence that the insect is a vector of the phytoplasma but do not provide evidence of the nature of the source plant and of the duration of the LP.

3.3. Transmission to Artificial Feeding Medium

Transmission experiments are generally performed by caging infectious insects onto susceptible test plants. However, since sucking insects can also feed through membranes, it is sometimes useful to test vectoring ability on artificial media. This can be particularly useful when (1) available test plants are poor hosts for the potential vector; (2) a high number of insects, e.g., field-collected insects, have to be tested for the infection rate under field conditions; and (3) screening for unknown vectors and therefore host plants that are not known. In this latter case, a lot of insects can be easily tested avoiding the need to produce and maintain a large number of plants for a long period in the greenhouse for disease development. Moreover, after vector feeding, the medium can be immediately analyzed by PCR for phytoplasma detection, so long incubation times in the plant are avoided. A number of phytoplasma transmission tests have been performed on artificial media (4, 10, 14–18) using plant- and leafhoppers. Even though only successful transmission to the plant is the final evidence of vectoring ability, transmissions to artificial feeding media may help in determining transmission capability. Due to the "artificial system", quantitative data from transmission experiments should be interpreted with caution since transmission capability/efficiency may be either overestimated (host plant may be partly tolerant and only some inoculation events may result in transmission while feeding media may contain phytoplasma cells injected with saliva that provide positive results in PCR assays) or underestimated (since the artificial feeding medium does not support phytoplasma multiplication, phytoplasmas are present in the diet at a low number, are rapidly degraded and feeding media may provide negative results in PCR assays).

1. Prepare the feeding chamber by either (1) filling the lid of a microcentrifuge tube with 200 μL of feeding medium (see Fig. 1h) and closing the lid with parafilm. Then cut the bottom off the tube and isolate an insect inside the microcentrifuge tube, closing it with a small cotton wool ball (see Fig. 1i); or (2) stretching a first parafilm sheet on the top of a plastic cup, adding feeding medium on the parafilm and closing with a second parafilm sheet (we use 45 mm diameter cups filled with 800–1,000 μL of diet). A number of leafhoppers can then be isolated inside the cup (we generally cage 5 leafhoppers per cup) through a hole that is then closed with a small cotton wool ball (see Fig. 1j).
2. Tubes and cups should be maintained with the cap facing a light source to attract the insects to the feeding medium (see Note 6).
3. At the end of the inoculation period, collect the feeding medium with a pipette and perform a DNA extraction procedure that avoids the presence of sugars in the template for PCR.
4. Pellet phytoplasma cells out of the feeding solution by centrifugation at 12,000 × *g* for 15 min.
5. Extract the genomic DNA by adding 10 μL of 0.5 M NaOH, followed by the addition of 20 μL of 1 M Tris–HCl (pH 8.0) containing 1% sodium dodecyl sulfate and 20 mM EDTA.
6. Incubate the mixture at 65°C for 15 min, precipitate with 2 volumes of ethanol, redissolve in 30 μL of TE, and store at -80°C (10) (see Note 7).
7. Use 2 μL of the extract as a template in direct PCR, followed by a nested PCR (see Chapters 14, 15, 16, and 17) (see Notes 8 and 9).

3.4. Evaluation of Transmission Experiment Results

Results of transmission experiments are based on the evaluation of the symptoms developed by the test plants exposed to the inoculation by the vectors and/or by PCR assay of the test plants to detect the phytoplasmas. When dealing with transmission experiments through Parafilm® into the artificial diet, PCR assays (generally nested PCR to increase sensitivity) are obviously mandatory to achieve transmission results. When transmission trials are carried out using groups of insects, the actual proportion of infectious insects can be estimated using the maximum likelihood estimator of p, $p\hat{} = 1 - Q^{1/k}$ where Q is the observed fraction of noninfected plants and k is the number of insects per plant, assuming that the vectors acted independently (19). Obviously, the estimation will be more precise in the smaller the groups. Group-transmission experiments allow convenient testing of a large number of field-collected vectors without losing information on their infectivity rate.

3.5. Transovarial Transmission Experiments

Transovarial transmission has been proven for the leafhopper *Matsumuratettix hiroglyphicus*, vector of the sugarcane white leaf phytoplasma (20), and for *Cacopsylla pruni* as a vector of '*Ca.* Phytoplasma prunorum' (21). For other phytoplasma-vector associations, only the presence of phytoplasma cells or DNA in the progeny of leafhoppers fed on phytoplasma-infected plants has been proven (22, 23). Transovarial transmission trials can be carried out with either field-collected or phytoplasma-free insect vectors from laboratory colonies.

1. For vectors from laboratory colonies, nymphs should be first caged onto source plants for the AAP and then maintained on the same or different plants until emergence.
2. Isolate one virgin female together with one or more male/s or individual mated females on potted seedlings of a phytoplasma-immune plant species for mating and/or oviposition (see Note 10).
3. After a sufficient number of eggs have been laid, remove the adults and test the females for phytoplasma presence by PCR. Only the eggs laid by infected females are used for further steps.
4. At different times after oviposition, sample some further eggs, nymphs, and newly emerged adults and test by PCR for phytoplasmas.
5. Cage the remaining adults onto healthy potted seedlings for transmission trials. At the end of the IAP, all the adults should be tested for the presence of phytoplasma.
6. After a variable time lapse from the IAP, observe test plants for symptoms and analyze by PCR to check for the presence of phytoplasmas (see Note 11).

4. Notes

1. The use of nymphs instead of adults is advisable since transmission of phytoplasmas requires a relatively long latent period to be completed in the insects before they can transmit. The use of adults will result in a high mortality before the latent period can be completed. On the other hand, first and second instar nymphs are very small and delicate, so the manipulation of older nymphs is advisable. For Flavescence dorée phytoplasma (FD) it has been reported that newly hatched nymphs fail to acquire the phytoplasmas from grape because their short stylets do not reach the phloem (24). The AAP may last from one to several days according to the suitability of the source plant for the vector insect (if the source plant is a good host plant for the vector, longer AAPs result in higher acquisition efficiencies; if the

source plant is a poor host for the insect, longer AAPs result in high mortality and shorter AAPs are needed). AAPs shorter than 1 day are not advisable since they can result in very low acquisition efficiency. AAPs may take place in a greenhouse or in a climatic chamber and nymphs can be caged on the whole plant/s when using small potted plants as a source; alternatively, the AAP may take place in the field inside net cages isolating a single branch of the source plant (see Fig. 1c).

2. These can be glass or plexiglass cylinders, topped with insect-proof nets (see Fig. 1d, e) or net cages isolating a single branch of a larger test plant (see Fig. 1f). The duration of the IAP may vary and, as a general rule, longer IAPs result in higher transmission rates. As stated for the AAP, if the test plant is a poor host for the insect, longer IAPs result in high mortality and shorter IAPs are needed to avoid high mortality. Nevertheless, if one wishes to maximize transmission efficiency, the vector insects can be maintained on the same test plant/s until death. Successive transfers of infectious insects on different test plants may result in a high number of infected plants, provided that the LP is completed. A few test plants should be exposed to healthy insects of the same species (from the lab-reared colony) as a negative control. These plants, representing the healthy control, will also allow a reliable and comparative evaluation of the symptoms expression by test plants exposed to infectious insects.

3. To study phytoplasma movement in the plant, vectors can be caged on a restricted site of the foliage (e.g., apical or basal leaves). To achieve this, test plants can be covered with aluminum foil except for one or a few apical or basal leaves (25). Alternatively, vectors can be caged on a single leaf using clip-cages (26) (see Fig. 1g).

4. If test plants are expected to be used as source plants in new transmission experiments (e.g., for the routine maintenance of a phytoplasma strain in the lab), nonpersistent insecticides (e.g., dichlorvos) must be applied or insecticide application should be avoided. Insects removed from test plants can be tested by PCR for phytoplasma presence, immediately or after storage in pure ethanol or at −20°C. In this way, a correlation between acquisition and transmission efficiency can be obtained (number of phytoplasma-positive vs. number of infective insects).

5. When dealing with routine transmission of a phytoplasma on test plants that develop clear symptoms of infection, detection of phytoplasmas is not needed. Test plants may become infected and develop symptoms at variable times post inoculation. Herbaceous hosts generally develop symptoms between 10 days and 2 months post inoculation and therefore should be checked by PCR at these time points, while for perennials,

plants should be kept for at least 1-year postinoculation and then analyzed by PCR. PCR assays are sensitive enough to reveal infection before the symptoms show; in our experience, an aster yellows phytoplasma was detectable in daisy plants as soon as 4 days postinoculation while symptoms were manifested only from 12 days onwards (25).

6. Besides feeding, leafhoppers also lay eggs through the Parafilm® into the feeding medium.
7. These quantities of reagents are detailed for 200 μL of artificial diet and should be adjusted accordingly when using greater volumes of diets (e.g., in the cups).
8. Several different liquid diets of different complexity have been proposed for leafhoppers but, for most work with vector transmission of phytoplasmas a simple feeding solution, buffered sucrose, can be conveniently used (27). It is also possible to feed insect vectors on a solid diet, consisting of a solidified aqueous solution of 5% sucrose and 4% low-melt agarose wrapped with Parafilm® (28), but then a protocol for the isolation of phytoplasmal DNA from this medium must be validated.
9. For some sap-sucking insects, Teflon membranes with pores of about 1 μm have been used instead of Parafilm® with good results (29). So far, for phytoplasma vectors the use of these Teflon hydrophobic membranes has not been reported, and preliminary results obtained in our laboratory seem to indicate that leafhoppers do not feed through these membranes. This may be due to the size of the pores (1 μm) being too small for insertion of the stylets. Membranes with larger pores result in liquid diet leaking through the pores and are therefore not suitable.
10. When dealing with a monophagous insect that cannot be reared on an alternative immune host, infectious adults can be caged on twigs (see Fig. 1k) of the susceptible plant species but, after the oviposition, twigs hosting the eggs are moved into cages with healthy potted seedlings to allow development of the nymphs, thus avoiding acquisition through the plant (21).
11. To assess the presence of phytoplasmas, eggs, nymphs and newly emerged adults can be tested singly or in batches. For eggs, that may represent a poor DNA target, batches are more feasible or, alternatively, single eggs can be tested by PCR for both phytoplasma and an internal insect control gene, to avoid false negative results (30). The presence of phytoplasmas in all the stages originating from the same infected female provides evidence of phytoplasma DNA inheritance to the progeny, while successful transmissions by the progeny adults is the final evidence of vertical infectivity transmission.

References

1. Marcone C et al (1999) Dodder transmission of pear decline, European stone fruit yellows, Rubus stunt, *Picris echioides* yellows and cotton phyllody phytoplasmas to periwinkle. J Phytopathol 147:187–192
2. Ciccotti AM et al (2008) Natural and experimental transmission of '*Candidatus* Phytoplasma mali' by root bridges. Acta Hortic 781:459–464
3. Nipah JO, Jones P, Dickinson MJ (2007) Detection of lethal yellowing phytoplasma in embryos from coconut palms infected with Cape St Paul wilt disease in Ghana. Plant Pathol 56:777–784
4. Bosco D, D'Amelio R (2010) Transmission specificity and competition of multiple phytoplasmas in the insect vector. In: Weintraub PG, Jones P (eds) Phytoplasmas: genomes, plant hosts and vectors. CAB International, Wallingford, UK, pp 293–308
5. Marzachì C, Milne RG, Bosco D (2004) Phytoplasma-plant-vector relationships. In: Pandalai SG (ed) Recent research developments in plant pathology, vol 3. Research Signpost Trivandrum, Kerala, India, pp 211–241
6. Weintraub PG, Beanland LA (2006) Insect vectors of phytoplasmas. Annu Rev Entomol 51:91–111
7. Golino DA, Oldfield GN, Gumpf DJ (1987) Transmission characteristics of the beet leafhopper transmitted virescence agent. Phytopathology 77:954–957
8. Carraro L et al (1996) Studies on the transmission of a phytoplasma from *Chrysanthemum leucanthemum* L. by the leafhopper *Euscelidius variegatus* Kbm. IOM Lett 4:127–128
9. Jarausch W, Lansac M, Dosba F (1996) Long-term maintenance of nonculturable apple-proliferation phytoplasmas in their micropropagated natural host plant. Plant Pathol 45:778–786
10. Melamed S et al (2003) Identification and characterization of phytoplasmal genes, employing a novel method of isolating phytoplasmal genomic DNA. J Bacteriol 185: 6513–6521
11. Holzinger WE, Emelianov AF, Kammerlander I (2002) The family Cixiidae Spinola 1839 (Hemiptera: Fulgoromorpha)—a review. In: Holzinger WE (ed) Zikaden leafhoppers planthoppers and cicadas (Insecta: Hemiptera: Auchenorrhyncha), vol 4. Oberösterreichisches Landsmuseum, Linz, Austria, pp 113–138
12. Sforza R et al (1999) Field observations, laboratory rearing and descriptions of immatures of the planthopper *Hyalesthes obsoletus* (Hemiptera: Cixiidae). Eur J Entomol 96:409–418
13. Mazzoni V et al (2009) Disruption of the reproductive behaviour of *Scaphoideus titanus* by playback of vibrational signals. Entomol Exp Appl 133:174–185
14. Zhang J et al (1998) A rapid method for detection and differentiation of aster-yellows phytoplasma-infected and inoculative leafhoppers. Phytopathology 88(suppl):S84
15. Tanne E et al (2001) Detection of phytoplasma by polymerase chain reaction of insect feeding medium and its use in determining vectoring ability. Phytopathology 91:741–746
16. Ge Q, Maixner M (2003) Comparative experimental transmission of grapevine yellows phytoplasmas to plants and artificial feeding medium. 14th Meeting of the International Council for the Study of Virus and Virus-Like Diseases of the Grapevine (ICVG), Locorotondo, Italy, 12–17 Sept 2003, pp 109–110
17. Bressan A et al (2006) Insect injection and artificial feeding bioassays to test the vector specificity of Flavescence dorée phytoplasma. Phytopathology 96:790–796
18. Pinzauti F, Trivellone V, Bagnoli B (2008) Ability of *Reptalus quinquecostatus* (Hemiptera: Cixiidae) to inoculate Stolbur phytoplasma to artificial feeding medium. Ann Appl Biol 153:299–305
19. Swallow WH (1985) Group testing for estimating infection rates and probabilities of disease transmission. Phytopathology 75:882–889
20. Hanboonsong Y et al (2002) Transovarial transmission of sugarcane white leaf phytoplasma in the insect vector *Matsumuratettix hiroglyphicus* (Matsumura). Insect Mol Biol 11:97–103
21. Tedeschi R et al (2006) Possible phytoplasma transovarial transmission in the psyllids *Cacopsylla melanoneura* and *Cacopsylla pruni*. Plant Pathol 55:18–24
22. Alma A et al (1997) Identification of phytoplasmas in eggs, nymphs and adults of *Scaphoideus titanus* Ball reared on healthy plants. Insect Mol Biol 6:115–121
23. Kawakita H et al (2000) Identification of mulberry dwarf phytoplasmas in the genital organs and eggs of leafhopper *Hishimonoides sellatiformis*. Phytopathology 90:909–914
24. Carle P, Moutous G (1965) Observation sur le mode de nutrition sur vigne de quatre espèces de cicadelles. Ann Epiphyt 16:333–354

25. Saracco P et al (2006) Quantification over time of chrysanthemum yellows phytoplasma (16Sr-I) in leaves and roots of the host plant *Chrysanthemum carinatum* (Schousboe) following inoculation with its insect vector. Physiol Mol Plant Pathol 67:212–219
26. Ammar ED (1976) A clip-on cage for holding leafhoppers and planthoppers on living plants. Anz Schadling Pflanzen Umweltsc 49:136–137
27. Markham PG (1982) Insect vectors. In: Daniels MJ, Markham PG (eds) Plant and insect mycoplasma techniques. Halsted Press, Wiley, New York, pp 307–326
28. Ranger CM et al (2004) Glandular trichome extracts from *Medicago sativa* deter settling by the potato leafhopper *Empoasca fabae*. J Chem Ecol 30:927–943
29. Davidson EW et al (2000) Improved artificial feeding system for rearing the whitefly *Bemisia argentifolii* (Homoptera: Aleyrodidae). Fla Entomol 83:459–468
30. Bosco D, Mason G, Accotto GP (2004) TYLCSV DNA, but not infectivity, can be transovarially inherited by the progeny of the whitefly vector *Bemisia tabaci* (Gennadius). Virology 323:276–283

Chapter 8

Molecular Identification of Phytoplasma Vector Species

Sabrina Bertin and Domenico Bosco

Abstract

The correct identification of the insect species involved in phytoplasma transmission is an essential condition for managing phytoplasma diseases and employing control strategies. The taxonomy of leafhoppers, planthoppers, and psyllids traditionally relies on morphological characters. Unfortunately, the identification of the distinctive traits requires skills and experience possessed by only a few specialist entomologists. In this chapter we provide protocols for the molecular identification of phytoplasma-vector species, mainly based on the polymerase chain reaction (PCR) amplification of mitochondrial and ribosomal DNA. Protocols for the application of molecular identification keys to dried specimens stored in insect collections are also provided. The same total DNA preparations can serve as a PCR template for either insect species or phytoplasma identification. The molecular identification methods can be applied not only to males, but also to nymphs and females for which a morphological taxonomic tool is generally unavailable. We suggest that taxonomic databases of planthoppers, leafhoppers, and psyllids should include species-specific DNA sequences as soon as they become available.

Key words: Hemiptera, Mitochondrial DNA, Molecular identification, PCR, Ribosomal DNA, Vector species

1. Introduction

Phytoplasmas are phloem-inhabiting pathogens associated with a large number of diseases in both cultivated and wild plants. In nature phytoplasmas are transmitted by insect vectors in a persistent, propagative manner. The insects acquire the phytoplasma by feeding on an infected plant and then transmit it to a healthy plant after a latent period, during which the phytoplasma moves through and multiplies in the vector body. The insects often remain infective for their entire life span (1).

The phytoplasma vectors belong to the order Hemiptera and are confined to three main taxonomic groups: leafhoppers (Auchenorrhyncha: Cicadellidae), planthoppers (Auchenorrhyncha:

Matt Dickinson and Jennifer Hodgetts (eds.), *Phytoplasma: Methods and Protocols*, Methods in Molecular Biology, vol. 938, DOI 10.1007/978-1-62703-089-2_8, © Springer Science+Business Media, LLC 2013

Fulgoromorpha), and psyllids (Sternorrhyncha: Psyllidae). These insects possess metabolism and feeding behavior that enhance their efficiency in phytoplasma transmission. They are hemimetabolous; thus nymphs and adults have a similar feeding habit and both can transmit phytoplasmas. Furthermore they are mostly phloem feeders and their sucking activity is not destructive, preventing the damage of sieve tissues and the elicitation of defense responses (1).

The family Cicadellidae includes the largest number of vector species, which are responsible for the transmission of a wide variety of phytoplasma groups. More than 75% of all confirmed phytoplasma vectors belong to the subfamily Deltocephalinae. The other leafhopper vectors belong to the subfamilies Agallinae, Aphrodinae, Cicadellinae, Coelidiinae, Iassinae, Idiocerinae, Macropsinae, Scarinae, and Typhlocybinae. Within the superfamily Fulgoromorpha, three families of vector species are numbered: Cixiidae, Delphacidae, and Derbidae. These families include species transmitting phytoplasmas from the coconut lethal yellows group ('*Candidatus* Phytoplasma palmae', 16SrIV genetic group) and stolbur group ('*Candidatus* Phytoplasma solani', 16SrXII-A genetic group). Within the Sternorrhyncha, only two genera in the Psyllidae family have been confirmed as vectors: *Cacopsylla* spp., which transmit apple proliferation phytoplasmas (16SrX genetic group) to pome and stone fruit trees, and *Bactericera trigonica* Hodkinson, which transmits a stolbur phytoplasma to carrots (1).

The correct identification of the insect species involved in phytoplasma transmission is an essential condition for managing phytoplasma diseases and employing control strategies. The taxonomy of the phytoplasma vectors traditionally relies on morphological characters. Generally, the external body morphology allows the identification at family and genus level, whereas the genitalia morphology is used to determine the species. Unfortunately, the identification of the distinctive traits is often awkward and requires experience. Such experience is afforded only to a few specialist entomologists, who focus on specific insect taxa. Even for experts, in some cases morphological differences do not allow the distinction of closely related taxa: for example, the relationship of some subfamilies and tribes of the family Cicadellidae remains unclear. Furthermore, the taxonomy at species level is mostly based on genital morphology of adult males, thus hampering the identification of juvenile stages and females.

In this perspective, DNA-based approaches offer valuable support to the traditional taxonomic methods and nowadays are widely employed for insect species identification (2). Most of these approaches are based on the polymerase chain reaction (PCR) and provide valuable diagnostic markers, taking advantage of both coding and noncoding regions. Different DNA regions can be informative for taxonomic studies by showing diverse evolutionary rates and highlighting high or low levels of variability. Popular markers such as random amplified polymorphic DNA (RAPD), amplified fragment length polymorphism (AFLP), and microsatellites (SSRs) are highly polymorphic and perform well in population genetics. Mitochondrial and ribosomal DNA as well as some nuclear genes are more conserved and have repeatedly provided insights into the genetic basis of evolution of recently diverged species and genera (3). Therefore, these markers are good candidates as molecular taxonomic keys for species discrimination. Notably, the mitochondrial gene coding for the cytochrome-*c* oxidase I (COI) has been chosen recently as a reference point for a universal system of animal species identification (DNA barcoding) (4, 5). The reliability of molecular taxonomic keys depends on their conservation within the species. Although evolution of mitochondrial and ribosomal DNA yields variability mostly between species, some polymorphism can be found even within species, especially when distant populations are compared. In order to exclude any possible intraspecific variation, these molecular markers should be validated on the widest range of vector populations having different features such as geographical origin and host plants. In this context, dried insects from museum collections may contribute to widen the population samples.

The present chapter aims to provide the information necessary to develop and apply molecular taxonomic keys to phytoplasma-vector species (see Note 1). The basic techniques for insect collection (see Note 2) and storage (see Note 3) are described elsewhere in this book. DNA extraction methods are described. The molecular markers are focussed on those DNA regions which are known to be the most taxonomically informative at the species level; the analysis of intraspecific variability and biotype/haplotype identification is not addressed in this chapter. Sets of primers that worked on insect DNA are reviewed. Finally, the chapter reports the first case studies in which phytoplasma vectors have been easily identified by means of molecular markers.

2. Materials

2.1. DNA Extraction

2.1.1. CTAB Method

1. Grinding buffer: 2% w/v cetyltrimethyl ammonium bromide (CTAB), 1.4 M NaCl, 20 mM EDTA pH 8.0, 100 mM Tris–HCl pH 8.0, 0.2% β-mercaptoethanol.
2. Carborundum or quartz sand, sterile.
3. Micropestle to fit 1.5 mL microfuge tube.
4. Microfuge tubes (1.5 mL).

5. Vortexer.
6. Incubator at 60°C.
7. Microfuge.
8. Chloroform/isoamyl alcohol: 24 parts chloroform, 1 part isoamyl alcohol.
9. Isopropanol (cold).
10. Ethanol (70%).
11. Vacuum drier.
12. Sterile water.
13. TE buffer (sterile): 10 mM Tris–HCl pH 8.0, 1 mM EDTA pH 8.0.

2.1.2. Dried Specimen Method

1. Digestion buffer: 3 mM $CaCl_2$, 2% sodium dodecyl sulfate (SDS), 40 mM dithiothreitol (DTT), 250 mg/mL proteinase K, 100 mM Tris–HCl pH 8.0, 100 mM NaCl.
2. Microfuge tubes (2 mL).
3. Incubator at 55°C and 65°C.
4. Ethanol (100%).
5. Ethanol (85%, cold).
6. Phenol/chloroform: equal parts phenol and chloroform.
7. Chloroform.
8. Microfuge.
9. Isopropanol.
10. Glycogen.
11. Sodium acetate solution: 3 M, pH 5.2.
12. Vortexer.
13. Sterile water.
14. TE buffer (sterile): 10 mM Tris–HCl pH 8.0, 1 mM EDTA pH 8.0.

2.2. Molecular Markers

DNA-based approaches are widely employed at different evolutionary levels in both phylogeny and population genetic studies (see Note 4). Mitochondrial DNA (mtDNA) (see Note 5), ribosomal DNA (rDNA) (see Note 6), and a few nuclear genes (see Note 7) are commonly accepted for species identification.

1. Primer pairs designed on the mtDNA regions (Table 1).
2. Primer pairs designed on the rDNA regions (Table 2).
3. Two alternative primer pairs for *EF-1α*:

   EF-1α 44: 5′ GCTGAGCGSGAGCGTGGTATCAC 3′.

Table 1
A range of primer pairs designed on the mitochondrial DNA regions used for species identification

Amplified section	Primer pair	Sequence (5′–3′)	Approximate amplicon length (bp)	References
COI	C1-J-1709 C1-N-2776	AATTGGWGGWTTYGGAAAYTG GGTAATCAGAGTATCGWCGNGG	1,070	(23)
COI	C1-J-1718 C1-J-2191	GGAGGATTTGGAAATTGATTAGTTCC CCCGGTAAAATTAAAATATAAACTTC	470	(22)
COI	C1-J-2183 L2-N-3014	CAACATTTATTTTGATTTTTTGG TCCAATGCACTAATCTGCCATATTA	830	(22)
COI	C1-J-2195 L2-N-3014	TTGATTTTTTGGTCATCCAGAAGT TCCAATGCACTAATCTGCCATATTA	820	(22)
COI	HCO LCO	GGTCAACAAATCATAAAGATATTGG TAAACTTCAGGGTGACCAAAAAATCA	650[a]	(25)
COII	TL2-J-3033 TK-N-3796	TCTAATATGGCAGATTAGTGCA ACTATTAGATGGTTTAAGAG	760	(23)
COII	TL2-J-3037 C2-N-3661	ATGGCAGATTAGTGCAATGG CCACAAATTTCTGAACATTGACCA	620	(22)
COII	C2-J-3399 TK-N-3796	ACAATTGGTCAYCAATGATAYTG ACTATTAGATGGTTTAAGAG	400	(23)
*cyt*b	CB-J-10612 CB-N-10920	CCATCCAACATCTCAGCATGATGAAA CCCTCAGAATGATATTTGTCCTCA	310	(22)
*cyt*b	CB-J-10933 CB-N-11367	TATGTACTACCATGAGGACAAATATC ATTACACCTCCTAATTTATTAGGAAT	430	(22)
ND5	N5-J-7077 N5-N-7793	TTAAATCCTTWGARTAAAAYCC TTAGGTTGRGATGGNYTAGG	720	(23)
CR	MEL_fw MEL_rev	TTTTATCCACTCTTAAAGCTTG TGATAGAGCTTTTTGAATTCTC	381 and 436	(11)
CR	AFF_fw AFF_rev	TTTAACCACCTCAAACTCAA CGTAAAATTCTTGGCGA	499	(11)

The target genes code for the subunits I and II of the cytochrome-*c* oxidase (COI and COII), cytochrome b (*cyt*b), the subunit 5 of the NADH dehydrogenase system (ND5), and the control region (CR)
[a]The COI fragment amplified by HCO-LCO primers has been formally designated as reference point for the "barcoding of life" database

EF-1α 52: 5′ CACRGGTACTGTTCCAATACC 3′ (approximate amplicon length: 600 bp).

EF-1α 46.5: 5′ GCHTTTGTGCCYATYTCTGG 3′.

EF-1α 53.5: 5′ GGCTTGGAAGGAACCARRATGATRAT 3′ (approximate amplicon length: 550 bp).

Table 2
A range of primer pairs designed on the ribosomal DNA regions used for species identification

Amplified section	Primer pair	Sequence (5′–3′)	Annealing site	References
ITS1	ITSa ITSb	TCCGTAGGTGAACCTGCGG GCTGCGTTCTTCATCGATGC	18S 5.8S	(28)
ITS2	5.8Sfw 28Srv	TGTGAACTGCAGGACACATG ATGCTTAAATTTAGGGGGTA	5.8S 28S	(27)
ITS1 + 5.8S + ITS2	18Sfw 28Srv	CCTTTGTACACACCGCCCGT ATGCTTAAATTTAGGGGGTA	18S 28S	(27)
28S D2-D3	28SIfw 28SIrv	AGTCGKGTTGCTTGAKAGTGCAG TTCGGGTCCCAACGTGTACG	28S	(30)
28S D3-D6	28SIIfw 28SIIrv	GGGACCCGTCTTGAAACAC ACCCTCCTACTCGTCAAGG	28S	(30)
28S D6-D7	28SIIIfw 28SIIIrv	CCGCTAAGGAGTGTGTAA GAAGTTACGGATCTARTTTG	28S	(30)
28S D8	28SIVfw 28SIVrv	CCTCGGACCTTGAAAATCC TGTCTCCTTACAGTGCCAGA	28S	(30)
28S D9-D10	28SVfw 28SVrv	GTAGCCAAATGCCTCGTCA CACAATGATAGGAAGAGCC	28S	(30)

The target regions are internal transcribed spacers 1 (ITS1), internal transcribed spacers 2 (ITS2), and D2-D10 variable region of the 28S gene

3. Methods

3.1. DNA Extraction

The methods for extracting the phytoplasma DNA from vectors are suitable also for identifying the insect species. Therefore, a single procedure can be suggested for both diagnostic and taxonomic purposes (see Note 8).

3.1.1. CTAB Method for Fresh Samples

This method (6) was modified from (7).

1. Macerate a single adult in 500 μL of grinding buffer with a micropestle in a 1.5 mL microfuge tube using sterile carborundum or quartz sand.
2. Vortex the suspension, then incubate at 60°C for 30 min.
3. Centrifuge the suspension at 15,000 × *g* for 5 min at room temperature. Transfer the supernatant to a new 1.5 mL microfuge tube.

4. Add 1 vol of chloroform/isoamyl alcohol, invert repeatedly and then centrifuge at 15,000×g for 5 min at room temperature.
5. Add one volume of cold isopropanol.
6. Centrifuge for 20 min at 15,000×g and 4°C. Discard the isopropanol.
7. Wash the DNA with 300 μL of 70% ethanol.
8. Centrifuge for 5 min at 15,000×g and 4°C. Discard the ethanol.
9. Dry the DNA pellet in vacuo and then resuspend it in 20–100 μL of sterile water or TE buffer.

3.1.2. Alternative Method for Dried Specimens

This method is suggested for dry specimens from, e.g., entomological collections (8) (see Note 9)

1. Place the whole specimen in a 2 mL Eppendorf tube and fully immerse it in digestion buffer (500–700 μL, dependent on specimen size).
2. Incubate the sample overnight (16–20 h) at 55°C with gentle agitation.
3. Remove the specimen from the digestion buffer, place in 100% ethanol for 2–4 h to stop further digestion, allow it to air dry, and replace it in the collection.
4. Purify nucleic acids from the digestion buffer using an equal volume of phenol/chloroform. Mix the solutions well, and centrifuge at room temperature at maximum speed for enough time to separate the organic and aqueous phases.
5. Transfer the aqueous phase to a new tube and add an equal volume of chloroform. Mix the solutions well, and centrifuge at room temperature at maximum speed for enough time for phase separation.
6. Transfer the aqueous phase to a new tube and precipitate the DNA by adding 0.6 vol of isopropanol, 20 mg of glycogen, and 0.1 vol of 3 M sodium acetate. Gently vortex the mixture immediately.
7. Centrifuge at room temperature at 15,000×g for 25 min.
8. Discard the supernatant. Wash the pellet twice with 1.5 mL of ice-cold 85% ethanol.
9. Discard the ethanol. Allow the DNA to air-dry at 65°C.
10. Resuspend the pellet in 100 μL of sterile water or TE buffer.

3.2. Species Identification: General Methods

We suggest two different protocols for species identification from field-collected and dried specimens, respectively (Fig. 1) (see Note 10).

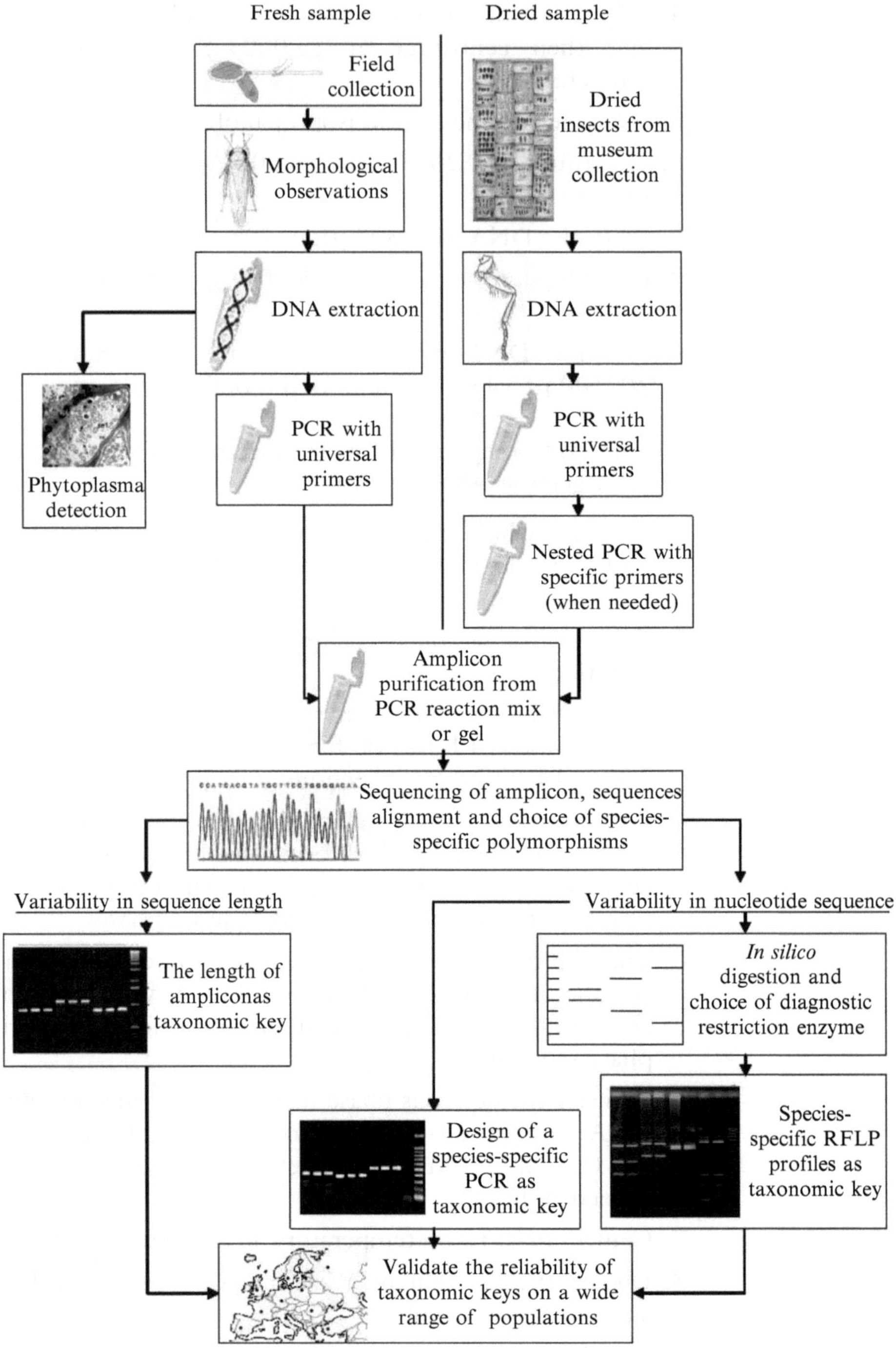

Fig. 1. Diagram summarizing the steps for species identification of fresh (field-collected) and dried museum specimens.

3.2.1. Field-Collected Specimens

1. Sample collection (see Note 11).
2. Preliminary morphological identification to identify the insect at the family/genus level (when applying an available molecular key) or at the species level (when developing a new molecular key).
3. DNA extraction from single specimens (see Subheading 3.1.1).
4. PCR amplification of the potentially informative DNA regions using conserved primers, already described in the literature.
5. Purification of the amplified DNA (see Note 12).
6. Sequencing of the purified fragment (see Note 13).
7. Sequence alignment (see Note 14).
8. The polymorphisms among sequences are recorded and only those that are conserved within the species and differ among species are retained. The intraspecific polymorphisms are discarded (see Note 15). Polymorphisms between species may be:
   - Variability in sequence length. If constant and repeated size variability is recorded, such polymorphisms can be directly exploited for species differentiation. Length polymorphisms are typical of ribosomal ITS.
   - Variability in nucleotide sequence. If only sequence polymorphism is recorded (i.e., the sequence length does not vary), species-specific primers can be designed based on the differential sites (see Note 16).
9. Perform species-specific PCR and diagnostic RFLP assays (see Note 17).
10. When developing a new molecular taxonomic key, test its reliability on samples from different sites. Species-specific PCR or PCR-RFLP assays must be repeated on the widest range of populations, so as to verify that polymorphisms are conserved within the species. All keys showing some intraspecific variations should be discarded.

3.2.2. Dried Museum Specimens

1. DNA extraction from single specimens (see Subheading 3.1.2).
2. PCR amplification of the potentially informative DNA regions (see Note 18).
3. Continue with steps 5–10 of Subheading 3.2.1.

3.3. Species Identification: Examples

Molecular taxonomic keys have been developed for a few insect species which are suspected or known to be phytoplasma vectors. To date, three detailed works have been published. Two of these deal with the most common European species belonging to the genera *Hyalesthes* and *Reptalus*, family Cixiidae (9, 10). These planthoppers are involved in transmission of stolbur phytoplasma. The third study provides the tools for discriminating between

Cacopsylla melanoneura Förster and *C. affinis* Löw, which are involved in apple proliferation (AP) transmission (11). Besides these, the molecular identification of *Cacopsylla picta* Förster, based on a species-specific PCR assay on the wingless (*wg*) gene, is also available (12).

3.3.1. Hyalesthes and Reptalus Species Identification

Molecular keys for the identification of three *Hyalesthes* species and four *Reptalus* species have been published (9, 10) (see Notes 19 and 20).

1. Species identification is based on PCR of the *COI* gene using the primer pair C1-J-2195/L2-N-3014 (Table 1) and on ITS2, using the primer pair 5.8Sfw/28Srv (Table 2).
2. For *Hyalesthes*, perform a PCR-RFLP assay using *Taq*I digestion of the 890-bp COI amplicon.
   - *H. obsoletus* produces two restriction fragments (~210, ~680 bp).
   - *H. luteipes* produces two restriction fragments (~390, ~500 bp).
   - *H. scotti* produces two restriction fragments (~280, ~610 bp).
3. For *Hyalesthes*, PCR amplification of ITS2 (Fig. 2) yields an amplicon of:
   - 1,130 bp for *H. luteipes*.
   - 880 bp for *H. obsoletus*.

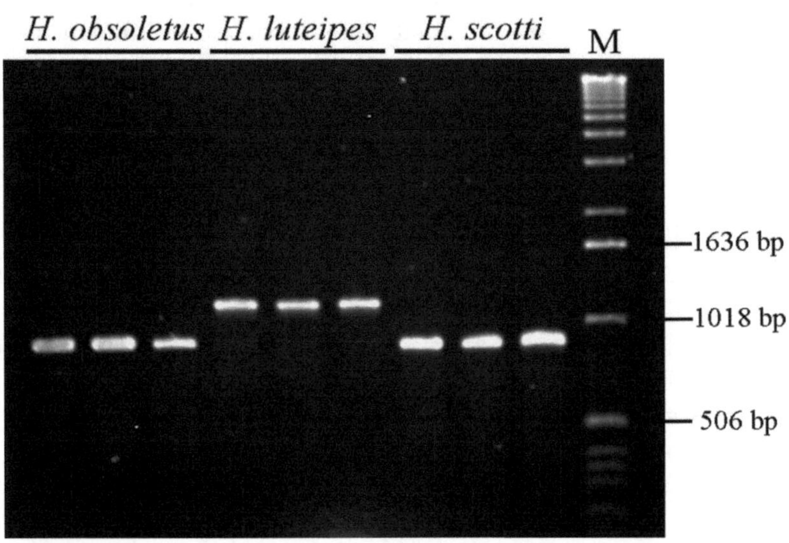

Fig. 2. Agarose gel electrophoresis of ITS2 polymerase chain reaction (PCR) products from three *Hyalesthes* species: *H. obsoletus*, *H. luteipes*, and *H. scotti*. *M* molecular weight marker (1 kb ladder). Figure reproduced from (9) with permission from John Wiley and Sons.

- 830 bp for *H. scotti*.
- To differentiate *H. obsoletus* and *H. scotti*, use an RFLP assay with *Taq*I.
  - *H. scotti* produces two restriction fragments (~130, ~700 bp).
  - *H. obsoletus* remains uncut (880 bp).

4. For *Reptalus*, perform a PCR-RFLP assay using *Alu*I digestion of the COI Amplicon (Fig. 3).
   - *R. quinquecostatus* produces three restriction fragments (~170, ~260, ~490 bp).
   - *R. cuspidatus* produces three restriction fragments (~80, ~330, ~510 bp).
   - *R. panzeri* produces two restriction fragments (~440, ~480 bp).
   - *R. melanochaetus* produces four restriction fragments (~60, ~120, ~170, ~570 bp).
5. For *Reptalus*, PCR amplification of ITS2 yields.
   - 1,450 bp for *R. quinquecostatus*.
   - 960 bp for *R. melanochaetus*.

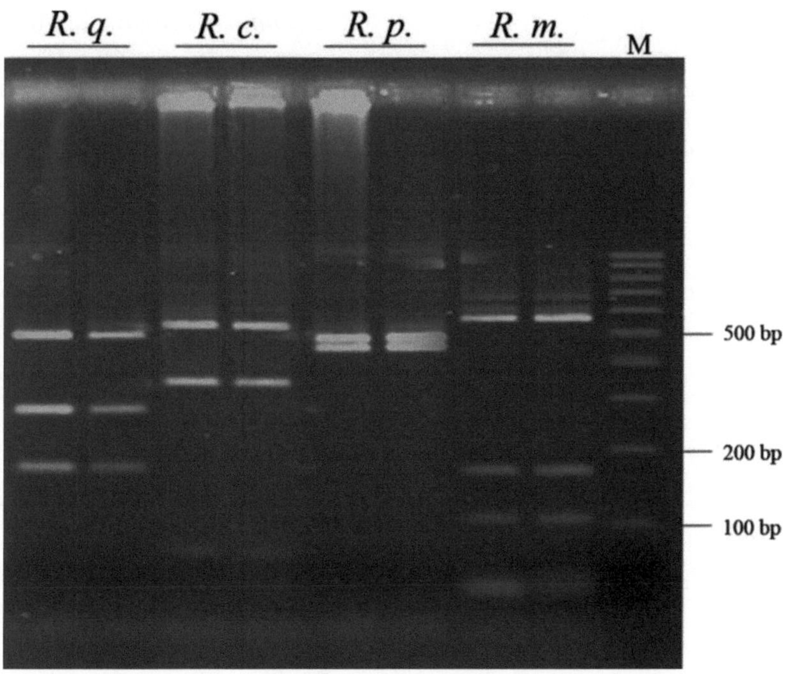

Fig. 3. RFLP profile after *Alu*I digestion of the cytochrome-*c* oxidase I (COI) PCR products of four *Reptalus* species. *R.q.*, *R. quinquecostatus*; *R.c.*, *R. cuspidatus*; *R.p.*, *R. panzeri*; *R.m.*, *R. melanochaetus*. *M* molecular weight marker (100 bp ladder). Figure reproduced from (10) with permission from Cambridge University Press.

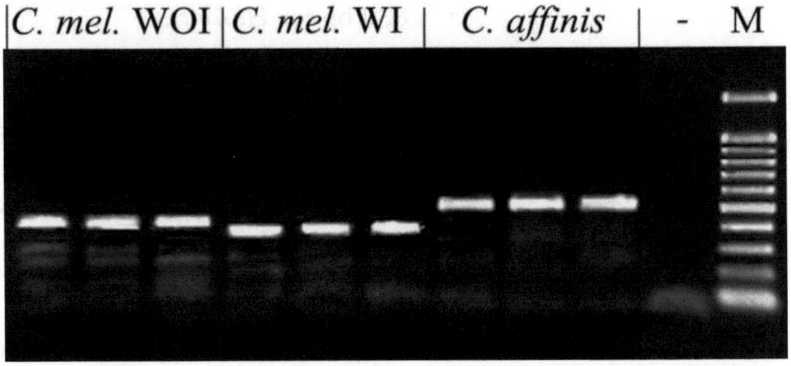

Fig. 4. Products of multiplex PCR assay with the four primers MEL_fw/MEL_rev and AFF_fw/AFF_rev from *Cacopsylla melanoneura* (*C. mel.*) and *C. affinis*. Two intraspecific variants of *C. melanoneura*, having (WI) or not having (WOI) a 56-bp indel, are shown. *Dash* water; *M* molecular weight marker (100 bp ladder). Figure reproduced from (11) with permission from Cambridge University Press.

- 850 bp for *R. cuspidatus* and *R. panzeri*.
- To differentiate *R. cuspidatus* and *R. panzer*, use an RFLP assay with *Taq*I.
  - *R. panzeri* produces four restriction fragments (~20, ~230, ~260, ~340 bp).
  - *R. cuspidatus* remains uncut (850 bp).

3.3.2. Discrimination Between C. melanoneura and C. affinis

The two species (see Note 21) can be distinguished using the control region (CR) of the mitochondrial DNA (11). Note that there are two haplotypes of *C. melanoneura* differing by the presence/absence of a 56-bp indel (Fig. 4).

1. For *C. affinis*, PCR with AFF_fw/AFF_rev (Table 1) produces a 499 bp amplicon.
2. For *C. melanoneura*, PCR with MEL_fw/MEL_rev (Table 1) produces 381 and 436 bp amplicons.
3. The amplification can be performed as multiplex PCR, using all the four primers in the same reaction tube.

4. Notes

1. The techniques are already in use in the laboratories that deal with phytoplasmas, so the same DNA preparations can serve as a PCR template for either insect species or phytoplasma identification. The versatility of the methods makes insect species identification affordable even for scientists working on phytoplasma epidemiology and lacking traditional insect

taxonomy skills. A particular emphasis has been placed on the importance of analyzing the widest range of specimens for each species, in order to assess the reliability of the molecular taxonomic keys. Samples from museum collections offer an important source of material, although in some cases the low quality of DNA can hinder the molecular analysis. A special procedure for analyzing museum samples has been proposed. The molecular approach cannot be used in isolation from traditional taxonomy: the molecular tools provide further information which can at most strengthen the identity of a known species, but the DNA itself is not enough to describe unknown specimens as a new species. Thus, the integration of molecular information with various types of data, mainly morphological, is essential for the discovery and description of new species. Similarly, taxonomic databases of planthoppers, leafhoppers, and psyllids (e.g., http://naturalhistory.museumwales.ac.uk/vectors, http://www.psyllidkey.eu/index.html, http://ctap.inhs.uiuc.edu/dmitriev) could be usefully supplemented with species-specific DNA sequences.

2. The identification of the vector species can be performed together with the assessment of their role in phytoplasma transmission. Indeed the insects collected in the vicinity of symptomatic plants during the disease monitoring can serve for the taxonomic classification too. The most common method to survey the insect populations in and near the target crops is the use of sticky traps. Alternatively, malaise traps can be adapted for capturing the most recalcitrant species (1). In order to make possible the molecular analyses, the collected insects should be rapidly removed from the traps before DNA degradation occurs, even though, in our experience, DNA purified from insects stuck on traps stored for some weeks at room temperature is suitable for PCR. Sweep netting and vacuum sampling are also effective techniques, especially in the field, and have the advantage of providing live insects.
3. Degradation of DNA in insects occurs over time and fresh starting material is preferable for molecular studies. Yield and quality of the DNA are optimized when live insects are killed by freezing and directly processed. However, the field-collected samples often cannot be immediately transferred to the laboratory and the DNA extraction is necessarily delayed. During that time, the specimens must be correctly stored in order to preserve the DNA. Among the different storage methods, ethanol is widely employed. The use of absolute ethanol is usually preferred, because the use of alcohol solutions mixed with water (e.g., 70% ethanol) may lead to DNA extraction results that are worse than when no preservative is used (13). Alternatively, the samples can be stored long-term by freezing

at −80°C. Quick and effective insect preservation has been reported for hemipterans using acetone solutions at room temperature, making this option cheap and easy to use in the insect archive (14). Furthermore, acetone appears to be the best preserving agent also for phytoplasma DNA, yielding the same diagnostic PCR efficiency as freshly extracted samples (15).

4. Other DNA-based approaches are widely employed at evolutionary levels other than species level, in both phylogeny and population genetic studies. Some of these have been successfully applied to phytoplasma vectors. SSRs and RAPD have been employed for population genetics of *H. obsoletus* Signoret, *Scaphoideus titanus* Ball, and *Cacopsylla pruni* Scopoli (16–20).

5. mtDNA is widely used in constructing phylogeny and inferring evolutionary history among closely related taxa. The suitability of mtDNA for taxonomic purposes is mainly due to the relatively high rate of evolution, the maternal inheritance, and the reduced chance of recombinations. Moreover the small size (16–20 kb in length) and the abundance in each cell make mtDNA relatively easy to obtain, even from degraded samples (3).

   The circular molecule of mtDNA contains 37 genes coding for small and large subunits of ribosomal RNA, for 13 proteins and for 22 transfer RNAs (tRNAs). The genes are compactly arranged on both DNA strands and intergenic sequences are small or absent. At least one noncoding region is present (control region) and contains regulatory elements for replication and transcription. With the exception of tRNAs, the gene order appears to be highly conserved within insects (21).

   The evolution of the mitochondrial genes occurred at different rates, making each region suitable for analyses at diverse taxonomic levels. Both ribosomal genes 12S and 16S, coding for small and large subunits respectively, proved to be informative for distantly related taxa such as genera, tribes, and families. The genes coding for tRNA have been used less frequently for phylogenetic analyses, because they are small and highly conserved. The protein-coding genes are the most successful taxonomic tools and provide the best performances at genus and species level. Indeed, when used to depict relationships among families and among orders, they repeatedly produced poorly supported results (22). Most species-level studies exploit the polymorphic sites at the genes coding for subunits I and II of the cytochrome-*c* oxidase (*COI* and *COII*), cytochrome b (*cyt*b), and the subunit 5 of the NADH dehydrogenase system (*ND5*) (2).

   In insects, a further advantage of using mitochondrial markers is that many of these loci can be readily amplified using universal primers designed from conserved regions. A list of primers amplifying all the mitochondrial regions in insects was first compiled by Simon et al. (1994) (22) and then updated by Simon et al. (2006) (23). A sample of those primers is reported in Table 1. We chose the primer pairs that amplify the most informative regions of the *COI*, *COII*, *cytb*, and *ND5* genes and worked successfully within the Hemiptera.

Among the mitochondrial sequences, the *COI* gene has been chosen as reference point for DNA barcoding, a universal system of animal species identification. DNA barcoding is based on the premise that sequence diversity within a short, standardized segment of the genome can provide a "biological barcode" that enables identification at the species level (4, 5). Since its introduction in 2003, the number of studies using the DNA barcode has rapidly increased. The extreme diversity of insects and their economical, epidemiological, and agricultural importance have made this group a major target of DNA bar-coding. An International Barcode of Life (iBOL) project was proposed (iBOL, 2010; http://ibol.org/) to develop a global standard barcoding protocol and to construct a comprehensive DNA barcode library. Within the iBOL project, the Barcode of Life Data Systems (BOLD; http://www.boldsystems.org/views/login.php) is the official informatics workbench, developed by the Canadian Center for DNA Barcoding (CCDB). BOLD provides a data repository for DNA barcodes, an identification support system based on them, and Web services for other system developers (24).

One of the key features of the barcoding project is the designation of a single 648-bp fragment at the 5′ end of *COI* as reference point (4, 5). This region is amplified with a universal primer pair, LCO/HCO, that was originally designed for marine invertebrates, but can be applied to all animal phyla (25) (Table 1). However, the DNA barcoding literature reveals that the majority of projects actually rely on other universal or taxon-specific primers, which provide better PCR performance. On the other hand, with current sequencing technology, up to 900 bp can be routinely obtained with accuracy from a single run. This widens the chance to identify genetic variability and to obtain a larger amount of phylogenetic information. In this regard, even the ~800-bp *COI* segment positioned downstream of the HCO/LCO amplicon is greatly informative in terms of mean percentage of divergence among diverse species of Lepidoptera and Diptera (26). This downstream region can be easily amplified by the conserved primers pairs C1-J-2183/L2-N-3014 or C1-J-2195/L2-N-3014 (22) (Table 1). Currently, a *COI* sequence of at least 500 bp is requested for recording a specimen in BOLD (24).

6. Among the nuclear markers, rDNA has been widely used owing to its universal presence in all the protein-synthesizing organisms and its abundance within a genome. The rDNA gene

family is well characterized in a broad range of eukaryotes, including many insect species (2). The gene organization conforms to a very highly conserved structure, which is based on multiple copies of tandemly repeated transcription units, separated from each other by intergenic spacers (IGS). The transcription unit includes regions coding for the three major ribosomal RNA subunits (18S, 5.8S, and 28S), a leader promoter region known as the external transcribed spacer (ETS), and two noncoding internal transcribed spacers (ITS1 and ITS2) whose functions are not precisely known. The basic rDNA repeat thus consists of the transcription unit plus an adjacent IGS (21).

The different regions of each transcription unit evolve at different rates in the nuclear genome. Thus, evolutionary studies employ analyses of different rDNA segments, depending on the taxonomic level being studied. In general, a higher degree of variability has been found in the noncoding segments IGS, ETS, and ITS. Notably, IGS and ITS have proved useful for detecting molecular differences between species by either length or sequence polymorphisms. In contrast, the coding regions change relatively slowly and can be used for systematic studies of higher taxa (21). The rDNA undergoes concerted evolution so that the sequence similarity is expected to be greater within a species than between species. However, it has been observed that the many copies of rDNA within a genome can differ at those sites that are relatively free to diverge, such as IGS, ITS, and even parts of the coding regions. The net effect of these polymorphisms in both spacers and functional domains is the rapid spread of new rDNA variants within the species and even within a single individual (27). If present, such variants would invalidate the reliability of rDNA as a species identifier. Therefore, the possible intraspecific and/or intra-individual variability must be carefully inspected when ribosomal identification tools are set up. This check is made by analyzing different populations from the same species as well as different clones from the same individual.

Intraspecific and intra-individual polymorphisms frequently occur in the IGS sequence. Indeed this region, despite the evolutionary rate that makes it an appealing target for identifying species, is often neglected. Moreover, the overall IGS length makes difficult its amplification in a single PCR run. These drawbacks lead to a preference for using the ITS sequences, which are shorter and generally less variable at the intraspecific level (27). The use of ITS as taxonomic tools is also encouraged by the availability in the literature of universal primers, which are complementary to the highly conserved flanking regions 18S, 5.8S, and 28S (Table 2) (27, 28). These primers enable ITS amplification across a wide range of insect orders, making easy the production of taxonomic information.

Although the coding rDNA regions tend to evolve at a slower rate than the spacer regions, polymorphisms in the 28S gene even between species have been reported. Indeed, the conserved core of the 28S gene is interspersed with 12 divergent domains (D1–D12) that are much more variable (29). These domains have proven to be informative for species identification and a set of primers that amplify the region spanning D2–D10 has been adapted to Hemiptera (Table 2) (30).

7. Phylogenetic studies based on nuclear protein-coding genes are fewer in number than studies of either mtDNA or nuclear rDNA. However a few loci have received attention for their taxonomic interest, such as alcohol dehydrogenase (*adh*), dopa decarboxylase (*ddc*), elongation factor-1-alpha (*EF-1α*), glycerol-6-phosphate dehydrogenase (*g6pdh*), Hunchback (*hb*), Period (*per*), phosphoenolpyruvate carboxykinase (*PEPCK*), and Wingless (*wg*) (2).

   Protein-coding genes exhibit a wide range of evolutionary rates and have been used for resolving recent divergences as well as conserved relationships (2). However, working with nuclear genes poses potential difficulties with heterozygosity and low copy number, which may hinder the amplification. In addition, many genes may contain large introns that make it difficult to amplify more than one exon unless reverse transcriptase PCR (RT-PCR) is carried out on mRNA (21). Among the taxonomically informative genes, *EF-1α* is the most popular. Its sequences have proven to be useful for studies among species and genera of different insect orders, including Hemiptera (2). A couple of primer pairs amplifying two different segments of *EF-1α* are suggested here (31).

8. Several commercial kits for insect DNA extraction are available (e.g., *prep*GEM™ Insect, ZyGEM, Solana Beach, CA, USA; E.Z.N.A.® Insect DNA Kit, Omega Bio-Tek, Norcross, GA, USA; EZgeneTM Insect gDNA Kit). These options are more expensive than traditional methods but yield greater DNA purity. Nevertheless, attention must be paid in verifying that the insect-specific kits are effective also for phytoplasma DNA.

9. Even if the identification of species mostly relies on fresh material collected in the field during the survey of phytoplasma diseases, the museum collections can represent an alternative source of material. Dried, pinned insects can be used for DNA analyses, although the percentage of successful samples after both DNA extraction and PCR is quite variable. Commercial DNA extraction kits specific for small sample volumes (e.g., QIAamp®DNA Investigator Kit, QIAGEN)

can ensure acceptable DNA quality. In all cases, the time of storage has a significant effect on the length of the sequence that can be amplified by PCR: an increase in the sample age corresponds to a reduction in amplicon length. However, even after years, it is possible to amplify fragments that are generally long enough (~600–700 bp) to serve as molecular markers (32). For the most problematic samples, post-PCR strategies such as nested PCR can improve the amplification yield. The drawback of using dried insects for molecular analyses is that the specimens will be destroyed, causing loss of museum materials. To overcome this limitation, semi-destructive methods have been developed: the DNA can be extracted with minimal, visible, external morphological damage to the specimens by using portions of insects, such as portion of one or two legs (33). An alternative chemical extraction method, that allows DNA sampling without conferring any external morphological damage, has been established recently for Coleoptera (8). This protocol is of considerable interest for different arthropods (34), and perhaps may lead to possible application even to Hemiptera.

10. Besides the technical tools, a few other points should be considered in planning the procedure. First, morphology-based identification cannot be completely replaced by a molecular taxonomic key. Indeed, the analysis of the external body traits is necessary for the identification of a sample at the family and/or genus level. This information allows the choice of the most appropriate molecular key among those which are already available for the identified taxa. Moreover, when a taxonomy key is developed for the first time, a univocal correlation between species-specific morphology and molecular profile must be defined. Secondly, the developed molecular keys must be always validated on the widest number of specimens collected at different sites and, when possible, on different host plants. This step is essential for ensuring the reliability and species-specificity of the marker.

    Concerning the molecular techniques, several alternatives are available. DNA sequencing has become the main method of choice for molecular systematics, given its potential of generating highly comparative data. However, DNA sequencing remains relatively expensive and time-consuming. Other molecular tools can be more practical, especially when a large number of individuals must be analyzed. Among these, the RFLP technique is versatile and has been widely applied in species differentiation (2). RFLP exploits sequence polymorphisms that occur at the recognition site of a specific restriction enzyme. The method consists in amplifying the divergent nuclear or mitochondrial region and then digesting the amplicons with the appropriate endonuclease. The fragments produced are sorted by length using agarose or acrylamide gel electrophoresis. Gel staining allows easy visualization of the diagnostic profiles. RFLP can be applied either without sequencing, to identify diagnostic sites by direct trials, or with a prior sequencing. In this case, sequencing can be performed first on a representative sample subset. The diagnostic endonuclease is then chosen by in silico-simulated digestion and finally the RFLP is applied to the whole sample set.

11. In order to test any potential variability at intraspecific level, specimens from multiple sites (e.g., from different geographical localities, different host plants, different ecological conditions, etc.) must be provided.

12. The purification can be performed directly from the PCR reaction mix using a variety of kits. If you experience difficulty obtaining a single specific PCR product, it is possible to purify the product of the expected size from an agarose gel using an appropriate kit.

13. Given the small expected size of the fragments (300–1,400 bp), single sequencing reactions, on a single DNA strand, or double, on both strands, should be adequate to obtain the whole region. This avoids the need to design internal primers and decreases the number of reactions.

14. Several alignment programs are available and most of them are freely downloaded, e.g., ClustalX, http://www.clustal.org/download/current/ (35); Opal, http://opal.cs.arizona.edu (36); Muscle, http://www.drive5.com/muscle/downloads.htm (37). Other programs are Web-interfaced, e.g., ClustalW, http://www.ebi.ac.uk/Tools/msa/clustalw2/ (35) and T-Coffee, http://tcoffee.crg.cat/ (38).

15. When working with ribosomal DNA, polymorphisms can be detected even at the intra-individual level. In order to exclude this level of variability, the ribosomal region must be cloned from each single individual. The clones are then sequenced and compared. If intra-individual polymorphisms are detected, the considered region is discarded.

16. One of the most versatile and easy-to-use programs for primer design is Primer3 (39), freely available at http://frodo.wi.mit.edu/primer3/input.htm. In this way different species-specific PCR assays can be attempted. Moreover, primer combinations for multiplex PCR may be used, so that the discrimination among species can be achieved in a single amplification reaction.

17. Discriminant restriction sites are found by in silico enzyme digestion of the amplicon sequences. Several programs are freely available on the Web, such as Webcutter (http://bio.lundberg.

gu.se/cutter2/), NEBcutter (http://tools.neb.com/NEBcutter2/) (40), and pDRAW32 (http://www.acaclone.com/). The virtual digestion allows enzymes that produce restriction fragments differing in length among the species to be identified. The enzyme producing the clearest and easiest-to-read RFLP profile is then chosen from within this selection. For this purpose, endonucleases that cut frequently (providing too many very light and faint bands) and those that produce bands with similar length (the bands will not be distinguishable in the gel) should be avoided. The restriction assay is then performed on the amplification products and the resulting fragments are separated on agarose or acrylamide gels, in order to produce species-specific RFLP patterns.

18. The DNA of dried specimens may be degraded and thus hinder efficient PCR analyses. Low DNA quality combined with the imperfect match between universal primers and target sequences may reduce the PCR yield, so that the amount of amplicon may not be enough for downstream applications. This trouble can be overcome by performing a nested PCR. New primers that specifically anneal within the amplified region are needed. The primers might be designed based on the sequences which have already been obtained from fresh specimens. The new primers should be chosen to amplify the longest internal sequence, in order to limit the loss of genetic information. PCR with the internal primers is performed by using the diluted products of the direct PCR (driven by universal primers) as template. As nested PCR is more sensitive than direct PCR, it should provide an acceptable amount of amplification product, enough for sequencing and RFLP assays. For example, successful application of nested PCR has been achieved in amplifying target DNA from specimens of *Empoasca* spp., stored at room temperature in an entomological collection for 20 years (Bertin, personal observations).

19. To date, *H. obsoletus* is the only established vector of the stolbur phytoplasma associated with a serious grapevine yellows disease known as bois noir (BN). However, other cixiids are suspected to be involved in BN transmission: *H. luteipes* and *H. scotti* have already been found to be PCR-positive to the phytoplasma and *R. quinquecostatus* can successfully inoculate BN into an artificial feeding medium. *R. panzeri*, besides acquiring BN, also transmits the stolbur phytoplasma responsible for the maize redness disease.

20. In both studies, the reliability at species level of the ITS2 and COI markers has been tested on different populations. For each species, the specimens were collected from different host plants from several sites in Italy and central and Eastern Europe.

No polymorphisms ascribable to the geographic origins and host plants were noted. Given their conservation and specificity, both ribosomal and mitochondrial regions have been accepted as reliable identification keys for species of the genera *Hyalesthes* and *Reptalus*.

21. *C. melanoneura* and *C. affinis* are two psyllid species which share a similar life cycle and morphology. *C. melanoneura* is the vector of '*Candidatus* Phytoplasma mali' in North-Western Italy, whereas *C. affinis* can acquire phytoplasmas of the AP group but its role as vector still needs to be assessed. The two species are often observed to form mixed populations and they are hard to distinguish by means of morphology.

Acknowledgement

The research carried out by the authors on molecular identification of insect species is supported by the Italian Ministry of Research (MIUR), under the grant PRIN 2008 "Identification of species and intra-specific variants of sap-sucking and parasitoid insects with molecular markers."

References

1. Weintraub PG, Beanland LeA (2006) Insect vectors of phytoplasmas. Annu Rev Entomol 51:91–111
2. Caterino MS, Cho S, Sperling FAH (2000) The current state of insect molecular systematics: a thriving tower of Babel. Annu Rev Entomol 45:1–54
3. Behura SK (2006) Molecular marker systems in insects: current trends and future avenues. Mol Ecol 15:3087–3113
4. Hebert PDN et al (2003) Biological identifications through DNA barcodes. Proc R Soc Lond B 270:313–321
5. Hebert PDN, Ratnasingham S, deWaard JR (2003) Barcoding animal life: cytochrome *c* oxidase subunit 1 divergences among closely related species. Proc R Soc Lond B 270:S96–S99
6. Marzachì C, Veratti F, Bosco D (1998) Direct PCR detection of phytoplasmas in experimentally infected insects. Ann Appl Biol 133: 45–54
7. Doyle JJ, Doyle JL (1990) Isolation of plant DNA from fresh tissue. Focus 12:13–15
8. Gilbert MTP et al (2007) DNA extraction from dry museum beetles without conferring external morphological damage. PLoS One 2:e272
9. Bertin S et al (2010) Molecular identification of the *Hyalesthes* species (Hemiptera: Cixiidae) occurring in vineyard agroecosystems. Ann Appl Biol 157:435–445
10. Bertin S et al (2010) Molecular differentiation of four *Reptalus* species (Hemiptera: Cixiidae). Bull Entomol Res 100:551–558
11. Tedeschi R, Nardi F (2010) DNA-based discrimination and frequency of phytoplasma infection in the two hawthorn-feeding species, *Cacopsylla melanoneura* and *Cacopsylla affinis*, in northwestern Italy. Bull Entomol Res 100:741–747
12. Jarausch W et al (2010) Development of specific primers for the molecular identification of *Cacopsylla picta*, the main vector of apple proliferation disease. In: Bertaccini A, Lavina A, Torres E (eds) Current status and perspectives of phytoplasma disease research and management. COST Action FA0807. http://costphytoplasma.eu/PDF%20files/WG%20BookwithiISBN.pdf
13. Mandrioli M (2008) Insect collections and DNA analyses: how to manage collections? Mus Manag Curatorship 23:193–199
14. Fukatsu T (1999) Acetone preservation: a practical technique for molecular analysis. Mol Ecol 8:1935–1945

15. Bosco D et al (2002) DNA-based methods for the detection and the identification of phytoplasmas in insect vector extracts. Mol Biotechnol 22:9–18
16. Johannesen J et al (2008) Invasion biology and host specificity of the grapevine yellows disease vector *Hyalesthes obsoletus* in Europe. Entomol Exp Appl 126:217–227
17. Imo M et al (2011) Highly polymorphic di- and trinucleotide microsatellite markers for the grapevine yellows disease vector *Hyalesthes obsoletus* (Auchenorrhyncha: Cixiidae). Eur J Entomol 108:161–163
18. Bertin S et al (2007) Diffusion of the Nearctic leafhopper *Scaphoideus titanus* Ball in Europe: a consequence of human trading activity. Genetica 131:275–285
19. Papura D et al (2009) Comparing the spatial genetic structures of the Flavescence dorée phytoplasma and its leafhopper vector *Scaphoideus titanus*. Infect Genet Evol 9:867–876
20. Sauvion N et al (2009) Nine polymorphic microsatellite loci from the psyllid *Cacopsylla pruni* (Scopoli), the vector of European stone fruit yellows. Mol Ecol Resour 9:1196–1199
21. Hoy MA (2003) Insect molecular genetics, 2nd edn. Academic Press/Elsevier, San Diego, CA
22. Simon C et al (1994) Evolution, weighting, and phylogenetic utility of mitochondrial gene sequences and a compilation of conserved polymerase chain reaction primers. Ann Entomol Soc Am 87:651–701
23. Simon C et al (2006) Incorporating molecular evolution into phylogenetic analysis, and a new compilation of conserved polymerase chain reaction primers for animal mitochondrial DNA. Annu Rev Ecol Evol Syst 37:545–579
24. Ratnasingham S, Hebert PDN (2007) Bold: the Barcode of Life Data System (http://www.barcodinglife.org). Mol Ecol Notes 7:355–364
25. Folmer O et al (1994) DNA primers for amplification of mitochondrial cytochrome *c* oxidase subunit I from diverse metazoan invertebrates. Mol Mar Biol Biotechnol 3:294–299
26. Roe AD, Sperling FAH (2007) Patterns of evolution of mitochondrial cytochrome *c* oxidase I and II and implications for DNA barcoding. Mol Phylogenet Evol 44:325–345
27. Collins FH, Paskewitz SM (1996) A review of the use of ribosomal DNA (rDNA) to differentiate among cryptic *Anopheles* species. Insect Mol Biol 5:1–9
28. White TJ et al (1990) Amplification and direct sequencing of fungal ribosomal RNA genes for phylogenetics. In: Innis MA, Gelfand DH, Shinsky JJ, White TJ (eds) PCR protocols: a guide to methods and applications. Academic, San Diego, pp 315–322
29. Hancock JM, Tautz D, Dover GA (1988) Evolution of the secondary structures and compensatory mutations of the ribosomal RNAs of *Drosophila melanogaster*. Mol Biol Evol 5:393–414
30. Dietrich CH et al (2001) Phylogeny of the major lineages of Membracoidea (Insecta: Hemiptera: Cicadomorpha) based on 28S rDNA sequences. Mol Phylogenet Evol 18:293–305
31. Cryan JR et al (2000) Phylogeny of the treehoppers (Insecta: Hemiptera: Membracidae): evidence from two nuclear genes. Mol Phylogenet Evol 17:317–334
32. Dean MD, Ballard JWO (2001) Factors affecting mitochondrial DNA quality from museum preserved *Drosophila simulans*. Entomol Exp Appl 98:279–283
33. Starks PT, Peters JM (2002) Semi-nondestructive genetic sampling from live eusocial wasps, *Polistes dominulus* and *Polistes fuscatus*. Insectes Soc 49:20–22
34. Rowley DL et al (2007) Vouchering DNA-barcoded specimens: test of a nondestructive extraction protocol for terrestrial arthropods. Mol Ecol Notes 7:915–924
35. Larkin MA et al (2007) Clustal W and Clustal X version 2.0. Bioinformatics 23:2947–2948
36. Wheeler TJ, Kececioglu JD (2007) Multiple alignment by aligning alignments. Bioinformatics 23:i559–i568
37. Edgar RC (2004) MUSCLE: multiple sequence alignment with high accuracy and high throughput. Nucleic Acids Res 32:1792–1797
38. Notredame C, Higgins DG, Heringa J (2000) T-Coffee: a novel method for fast and accurate multiple sequence alignment. J Mol Biol 302:205–217
39. Rozen S, Skaletsky HJ (2000) Primer3 on the WWW for general users and for biologist programmers. In: Krawetz S, Misener S (eds) Bioinformatics methods and protocols: methods in molecular biology. Humana Press, Totowa, NJ, pp 365–386
40. Vincze T, Posfai J, Roberts RJ (2003) NEBcutter: a program to cleave DNA with restriction enzymes. Nucleic Acids Res 31:3688–3691

Chapter 9

Dienes' Staining and Light Microscopy for Phytoplasma Visualization

Rita Musetti

Abstract

The chapter describes the main light microscopy techniques used for the diagnosis of phytoplasmas. Because the described techniques are generally only effective in cases where the concentration of the pathogen inside the host phloem tissue is high, they are useful as preliminary methods to assess the presence of phytoplasmas and analyze the distribution in symptomatic plants.

Key words: Detection, Dienes' stain, Light microscopy, Phytoplasma

1. Introduction

Phytoplasmas occur worldwide and have been associated with several hundred diseases affecting economically important crops. Phytoplasmas are wall-less prokaryotes, pleomorphic in shape, belonging to the Class Mollicutes; they are bound by a trilaminate unit membrane, contain ribosomes, and DNA and range up to 1.0 μm in diameter. Their shape may be filamentous, beaded, or simply spheroid. They are obligate parasites that have not been cultured in vitro, and grow and reproduce in the phloem of the host plants and in the vector insects.

The association of these pathogens with plants exhibiting "yellows" symptoms was demonstrated by Doi et al. (1) using transmission electron microscopy (TEM). Since then, many authors have used electron microscopy to reveal the presence of phytoplasmas in the phloem tissues and to study cytological interactions between these pathogens and their hosts (2–8). In addition, light microscopy (LM) techniques, even if not designed to directly visualize phytoplasmas inside the host tissues, have been used successfully as preliminary methods for diagnosis to verify the

Matt Dickinson and Jennifer Hodgetts (eds.), *Phytoplasma: Methods and Protocols*, Methods in Molecular Biology, vol. 938,
DOI 10.1007/978-1-62703-089-2_9,

presence of phytoplasmas in symptomatic plants, so they constitute the first steps towards understanding the possible association between phytoplasmas and the disease symptoms in the plants. Moreover, LM methods are fast and less expensive than electron microscopy techniques.

The aim of this chapter is to indicate some of the LM techniques used in the diagnosis of phytoplasmas. Several staining methods, applied to semi-thin sections of free-hand made and resin-embedded materials, have been used to localize and identify phytoplasmas in the infected tissues by means of LM. The LM detection of phytoplasmas in semi-thin sections is a good method only for cells containing high concentrations of the pathogen, so for example it can be successfully used on herbaceous plant materials.

Dienes' stain was first developed as a specific stain to visualize cultivable mycoplasma colonies on agar surfaces, but the stain has subsequently been used to visualize phytoplasmas directly in the sieve tubes of the host plants (9). Deeley et al. (10) applied this method to hand-cut or freezing-microtome sections of stem tissues of plants infected with phytoplasmas and spiroplasmas. Dienes' stain represents a quick and easy preliminary test for rapid assessment of plant samples from field collections. Following this staining method, colonized sieve tubes can be identified as irregular patches of intensely stained cells when observed under the light microscope. The procedure is of diagnostic value, as phloem tissues of samples infected by phytoplasmas will be stained dark blue, while xylem will be stained turquoise and the cortex light blue. The stain has been shown to be specific for diseases associated with mollicutes and does not stain healthy tissues and or samples infected by other plant pathogens. Sections obtained from infected plant materials and treated with the suggested dilution of Dienes' stain are shown in Fig. 1, with groups of phloem cells stained in blue (Fig. 1a, arrows) contrasting with unstained phloem in healthy plant sections (Fig. 1b).

Specific stains that associate with phytoplasma nucleic acid are also useful because phloem sieve tubes do not have a nucleus. Toluidine blue stains of semi-thin sections (11) and thionin/acridine orange (12) have also been used successfully. Other possible stains for the detection of phytoplasmas in the tissues of the host plants (not described here) include methyl green and Feulgen staining procedures (13, 14).

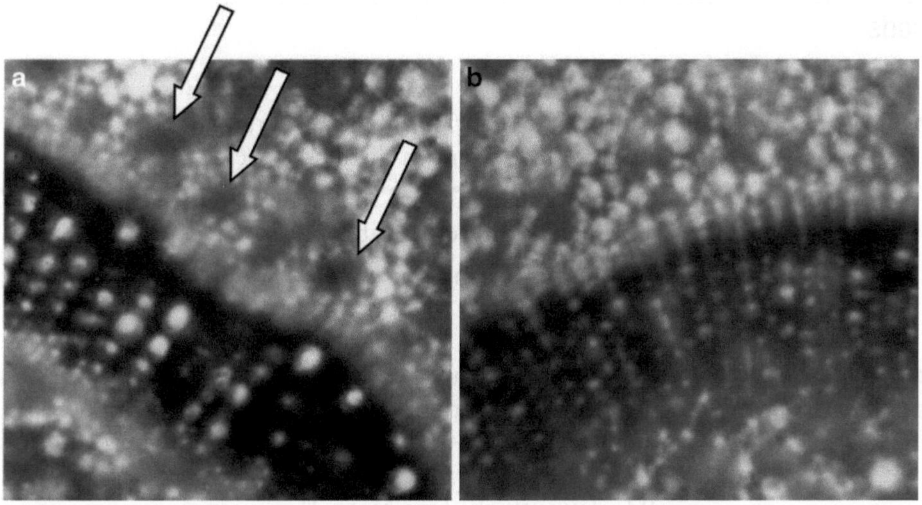

Fig. 1. Micrograph of hand-made cut stems of *Catharanthus roseus* stained with Dienes' stain. Areas corresponding to groups of phloem cells are stained in blue in sections obtained from infected plant material ((**a**) *arrows*), contrasting with sections from healthy plants with unstained phloem (**b**).

2. Materials

2.1. Dienes' Stain

1. Plant material: samples are taken from the main veins, shoots, and petioles of infected and healthy control plants.
2. Dienes' stain stock solution: 2.5 g methylene blue, 1.25 g Azure II, 10 g maltose, 0.25 g sodium carbonate (Na_2CO_3), 100 mL distilled water. Filter through filter paper and dilute to 0.2% (v/v) in distilled water.

2.2. Thionin and Acridine Orange Stain

1. Plant material: samples are taken from the main veins of infected and healthy control plants.
2. Thionin solution: 0.3 g thionin, 200 mL distilled water, 50 mL 0.1 M NaOH, 250 mL 90% ethanol.
3. Acridine orange solution: 6.25 g acridine orange, 200 mL distilled water, 50 mL 1 M NaOH.

2.3. Toluidine Blue Stain

1. Plant materials. Samples are taken from the main veins of infected and healthy control plants.
2. Toluidine blue solution: 1 g toluidine blue, 100 mL distilled water.

3. Methods

3.1. Dienes' Stain

1. Hand-made sections of healthy and infected main veins, shoots, and petioles are cut in distilled water.
2. Sections are incubated in Dienes' stain working solution for 10 min at room temperature.
3. Sections are washed in distilled water, mounted in water, and observed under LM (see Note 1).

3.2. Thionin and Acridine Orange Stain

1. Samples are prepared and fixed and embedded in resin according to a standard protocol for electron microscopy sample preparation (see Chapter 11). After polymerization, samples are cut by ultramicrotome to obtain semi-thin sections (0.5–1 μm thick) (see Note 2).
2. Semi-thin sections are collected on glass slides and covered with thionin solution for 10 min.
3. Sections are rinsed in distilled water and dried on a warming plate.
4. Cover the sections with the acridine orange solution for 2 min.
5. Mount the sections in the same resin as previously used for embedding the samples.
6. Observe under LM (see Note 3).

3.3. Toluidine Blue Stain

1. Samples are prepared and fixed and embedded in resin according to a standard protocol for electron microscopy sample preparation (see Chapter 11). After polymerization, samples are cut by ultramicrotome to obtain semi-thin sections (0.5–1 μm thick) (see Note 2).
2. Semi-thin sections are collected on glass slides, covered with the toluidine blue solution for 4–5 s.
3. Slides are dried briefly (few seconds) on a warming plate.
4. Rinse the slides with abundant distilled water and dry on the warming plate.
5. Observe under LM (see Note 4).

4. Notes

1. Sections can be left in distilled water for several hours before examination without any negative effect.
2. Polychromatic staining of semi-thin sections of in-resin embedding materials allows accurate phytoplasma detection.

3. Phytoplasmas can be observed as purple spots scattered along the walls of the sieve tubes. The cellulose cell walls are stained in yellow.
4. Phytoplasmas can be observed as blue/violet accumulations inside the sieve elements. Cell walls are stained in blue.

References

1. Doi Y et al (1967) Mycoplasma or PLT group-like microorganisms found in the phloem elements of plants infected with mulberry dwarf, potato witches' broom, aster yellows or Paulownia witches' broom. Ann Phytopathol Soc Jpn 33:259–266
2. Braun EJ, Sinclair WA (1976) Histopathology of phloem necrosis in *Ulmus americana*. Phytopathology 66:598–607
3. Favali MA et al (1990) A study on peach decline. II. Ultrastructural study. Riv Pat Veg S IV 26:11–22
4. Meignoz R et al (1992) Grapevine flavescence dorée. Presence of MLO and associated cytopathological effects in grapevine phloem. J Phytopathol 134:1–9
5. Musetti R et al (1992) An attempt to differentiate by microscopic methods two plant mycoplasma-like organisms. Cytobios 72:71–82
6. Musetti R et al (1994) Histological detection of mycoplasma-like organisms causing leptonecrosis in plum trees. Cytobios 78:81–90
7. Musetti R, Favali MA (1999) Histological and ultrastructural comparative study between *Prunus* varieties of different susceptibility to plum leptonecrosis. Cytobios 99:73–82
8. Musetti R, Favali MA, Pressacco L (2000) Histopathology and polyphenol content in plants infected by phytoplasmas. Cytobios 102:133–147
9. Dienes L et al (1948) The role of pleuropneumomia-like organisms in genitourinary and joint diseases. New Engl J Med 238:509–515
10. Deeley J, Stevens WA, Fox RTV (1979) Use of Dienes' stain to detect plant diseases induced by mycoplasmalike organisms. Phytopathology 69:1169–1171
11. Cousin MT, Sharma AK, Misra S (1986) Correlation between light and electron microscopic observations and identification of mycoplasmalike organisms using consecutive 350 nm thick sections. J Phytopathol 115:368–374
12. Hooper GR, Case JFW, Myers R (1971) Mycoplasma-like bodies associated with a flower greening disorder of a wild flower *Trillium grandiflorum*. Plant Dis Rep 55: 1108–1110
13. Giannotti J, Morvan G, Vago C (1968) Microorganismes de type mycoplasme dans les cellules liberiennes de *Malus sylvestris* L. atteint de la maladie des proliferation. C R Acad Sci Paris 267:76–77
14. Kartha KK, Cousin MT, Ruegg EF (1975) A light microscopic detection of plant mycoplasma infection by Feulgen staining procedure. Indian Phytopathol 28:51–56

Chapter 10

DAPI Staining and Fluorescence Microscopy Techniques for Phytoplasmas

Nancy M. Andrade and Nolberto L. Arismendi

Abstract

The 4′,6-diamidino-2-phenylindole (DAPI) stain technique is a simple method that was developed for confirming the presence of phytoplasmas in hand-cut or freezing microtome sections of infected tissues. DAPI binds AT-rich DNA preferentially, so that phytoplasmas, localized among phloem cells, can be visualized in a fluorescence microscope. The procedure is quick, easy to use, inexpensive, and can be used as a preliminary or quantitative method to detect or quantify phytoplasma-like bodies in infected plants.

Key words: 4′,6-diamidino-2-phenylindole, Phytoplasma, Fluorescence microscopy, Diagnosis, Preliminary techniques

1. Introduction

Phytoplasmas are cell wall-less, non-helical bacteria that are surrounded by a single-layer membrane, and are pleomorphic in shape with an average diameter of 200–800 nm (1). These prokaryotes live a parasitic lifestyle associated with diseases in more than a thousand plant species, and are transmitted by insect vectors, mainly leafhoppers, and psyllids (1–3). In infected plants, phytoplasmas inhabit the sieve cells of phloem tissue and induce disease symptoms related to disrupted physiological processes (such as hormonal balance, amino acid transport, and carbohydrate translocation), causing inhibition of photosynthesis and rapid senescence or morphological changes in infected tissues (4–7). Initially, these disorders were thought to be caused by viruses, but Doi and collaborators, in 1967, identified wall-less pleomorphic bodies in diseased plants which, due to their morphological similarity to mycoplasmas, were named mycoplasma-like organisms (MLOs) (2).

Matt Dickinson and Jennifer Hodgetts (eds.), *Phytoplasma: Methods and Protocols*, Methods in Molecular Biology, vol. 938, DOI 10.1007/978-1-62703-089-2_10, © Springer Science+Business Media, LLC 2013

Currently, PCR-based techniques are widely used to detect and identify phytoplasmas, but their major limitations are the high cost and the need for highly trained personnel to carry out the procedure, which are lacking in many laboratories in developing countries. In that sense, quick, inexpensive, and easy-to-use techniques are highly valued to detect phytoplasmas in a rapid diagnostic approach. For these reasons, techniques such as the 4′,6-diamidino-2-phenylindole (DAPI) and Dienes' stains are commonly used for quick and inexpensive phytoplasma diagnosis.

The Dienes' stain was developed as an indirect method for diagnosis of phytoplasma infection, but this method has relatively low sensitivity and is nonspecific, because the phloem of stem sections infected by phytoplasmas generally stains dark blue, while xylem can stain turquoise and the cortex pale purplish-blue (8). Thus, a stained tissue section indicates the possible presence of phytoplasmas, but there is no specific staining of phytoplasma-like bodies. On the other hand, the DAPI stain is considered more specific than the Dienes' stain. The sequenced genomes of some phytoplasmas proved to be AT-rich, with a correspondingly low GC content ranging from 21% (e.g., '*Candidatus* Phytoplasma mali') to 28% (e.g., '*Candidatus* Phytoplasma asteris') (9). These characteristics make phytoplasmas amenable to staining techniques such as the use of DAPI, a fluorescent stain that has the ability to pass through cell membranes and which binds strongly to AT-rich regions of DNA (10).

The interaction of DAPI with DNA has been the subject of numerous studies since this aromatic compound was first synthesized. Initially, several diamidine compounds were synthesized to be used as a trypanocide agent (11). However, the special spectral properties of DAPI have caused it to be used more as a DNA probe than a medication drug. It is currently used extensively in fluorescence microscopy. In phytopathology it is used especially to detect phytoplasmas and spiroplasmas in different plant species, as reported in many studies (12–18). Following DAPI staining, phloem cells of infected material show a strong fluorescence, brighter than that which is typical of the nuclei of parenchymal cells. Infected tissues generally show bright phytoplasma-like spots in the phloem sieve tubes that are not present in healthy tissues (18). The DAPI staining technique is considered to be a rapid and precise method of localization of phytoplasmas in the phloem sieve tubes of different tissues such as twigs, petioles, leaves, and roots (13, 14, 16, 18). Apart from fluorescence light microscopy for phytoplasmas, DAPI is also commonly used in the detection of DNA of mycoplasmas contaminating animal cell cultures: mycoplasma cells in the growth medium fluoresce when stained by DAPI, making them easily detectable (19). Other applications have also been explored: PCR electrophoresis (20), cytofluorometry (21), and staining chromosomes (22).

2. Materials

1. Petioles and stems (fresh or dried samples can be used).
2. Phosphate buffer, 0.1 M pH 6.8: KH_2PO_4. Add 6.8 g of KH_2PO_4 to 500 mL of sterile distilled water. Shake at 7 rpm for 10 min. Add 1.16 g of NaOH to 200 mL of sterile distilled water and stir at 7 rpm for 10 min. Adjust the pH of the phosphate solution to pH 6.8 by adding NaOH solution.
3. Phosphate buffer, 0.1 M pH 6.9: Add 6.8 g of KH_2PO_4 to 500 mL of sterile distilled water. Shake at 7 rpm for 10 min. Add 1.16 g of NaOH to 200 mL of sterile distilled water and stir at 7 rpm for 10 min. Adjust the pH of the phosphate solution to pH 6.9 by adding NaOH solution.
4. Glutaraldehyde (5% solution): 5 mL glutaraldehyde, 95 mL 0.1 M phosphate buffer pH 6.8.
5. DAPI stain (4′,6-diamidino-2-phenylindole dihydrochloride) (1,000× stock solution): add 1 mg of DAPI to 1 mL of sterile distilled water. Store at 4°C if not used immediately (see Note 1).
6. Sterile distilled water.
7. Stirring machine.
8. Freezing microtome.
9. Fluorescence microscope.
10. Electronic balance.
11. Conical flasks (1,000 mL).
12. Plastic bags and trays.
13. Graduated pipettes (5 and 10 mL).
14. Micropipettes (100 μL).
15. Microscope slides and coverslips.
16. Scalpel and tweezers.

3. Methods

The method described here was slightly modified from Romero (23) and can be used for determining the presence of, or quantifying, phytoplamas (see Fig. 1).

1. Dilute the stock solution to working concentration (1×): take 100 μL of stock solution (1,000×) and add 100 mL of sterile distilled water. Store the working solution at 4°C.
2. Take the young stems and petioles from the diseased plants (see Note 2).

Fig. 1. Schematic illustration of the different steps for 4′,6-diamidino-2-phenylindole (DAPI) stain diagnosis in samples affected by phytoplasmas. Drawings were prepared by Francisco Beluzan, Universidad Austral de Chile.

3. Cut the samples in 1 cm lengths and put them in sterile distilled water (see Note 3).
4. Immerse the samples in 5% glutaraldehyde and fix for 20 min.
5. Remove the samples from the glutaraldehyde solution and wash with 0.1 M phosphate buffer pH 6.9 for 5 min. Transfer the samples to fresh 0.1 M phosphate buffer pH 6.9.
6. With the freezing microtome, cut longitudinal sections of 10–15 μm from the fixed samples (see Note 4).
7. Place the sample sections in the DAPI stain (1× working solution) for 25 min.

8. Remove the sections from the stain and put them in sterile distilled water to remove some of the dye (must have at least 50% of the dye when mounted). Immediately mount sections on a slide and place the coverslip (see Note 5).
9. Observe the samples with a fluorescence microscope with a range of objective lenses (10×, 40×, 100×) (see Note 6).

4. Notes

1. Caution: DAPI is labeled as a nontoxic product, but the toxicological properties of this reagent have not been fully investigated according to its Material Safety Data Sheet (http://www.sigmaaldrich.com). We recommend using gloves while handling your samples.
2. We recommend taking samples which are less affected by disease, preferably from areas that do not show fully developed symptoms, with the aim of avoiding false positives due to secondary infections by other microorganisms in the affected tissues.
3. This step reduces the dehydration of the samples.
4. We recommend having one or two layers of cells for a better fit of the samples on the slide after the DAPI stain.
5. We recommend keeping the samples in dark to extend the stain's life in the prepared sample, because light reduces the quality of the samples and tends to remove the stain from them. For further observations, it is recommended to seal the samples, e.g., using a layer of vaseline. If you do not have a special slide sealer, nail polish works very well as a sealer.
6. If more information about the fluorescence microscopy procedure is required, we recommended reading various reviews, e.g., (24). Samples that contain phytoplasmas show high DAPI fluorescence in the sieve tubes due to the specific combination of this stain with DNA, whereas no fluorescence is seen in healthy plants (Fig. 2). Generally, small points and aggregations near the cell wall of the phloem tissue cells are observed. This same procedure can be used to quantify phytoplasma-like bodies in phloem cells of infected plants (Fig. 3).

Fig. 2. Symptomless plant tissues (control) of *Ugni molinae* (**a**) and *Gaultheria phillyreifolia* (**b**), and their phloem tissues stained with DAPI ((**c**, **d**) respectively); witches' broom symptoms in *U. molinae* (**e**) and *G. phillyreifolia* (**f**), and fluorescent phytoplasma-like bodies in their phloem (indicated with *white arrows*, (**g**, **h**) respectively). The sections were examined using a fluorescence microscope (BP 450–490, FT 510, and LP 520 filters; Carl Zeiss Axiolab, New York, USA) with UV light excitation of 450–490 nm, a 100× oil immersion lens and 10× ocular lenses to achieve a magnification of 1,000× (18).

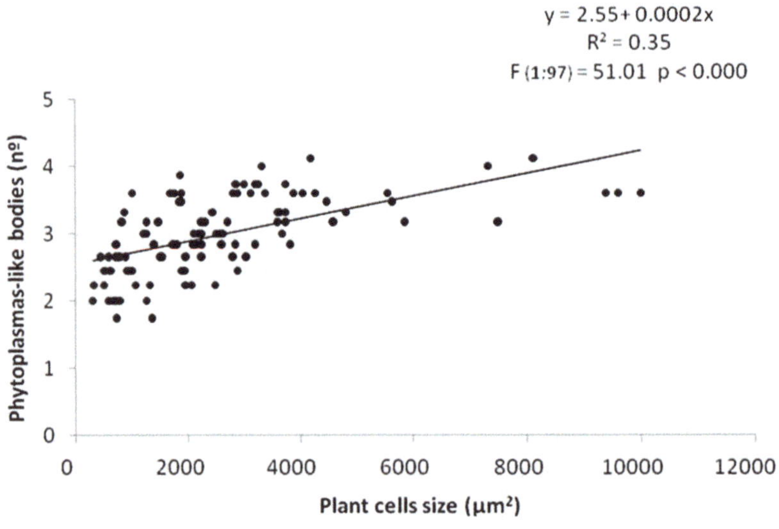

Fig. 3. Relationship between number of phytoplasma-like bodies and the size of phloem cells in petioles of *Hypochaeris radicata* L. Phloem tissues were stained with DAPI. Fluorescent phytoplasma-like bodies and phloem cells were quantified and measured with 10× grid eyepieces. The preparations were observed using a fluorescence microscope (BP 450–490, FT 510, and LP 520 filters; Carl Zeiss Axiolab, New York, USA) with a UV light excitation of 450–490 nm, a 100× oil immersion lens and total magnification of 1,000× (unpublished data).

References

1. Hogenhout SA et al (2008) Phytoplasmas: bacteria that manipulate plants and insects. Mol Plant Pathol 9:403–423
2. Lee I-M, Davis RE, Gundersen-Rindal DE (2000) Phytoplasma: phylogenetic mollicutes. Annu Rev Microbiol 54:221–255
3. Weintraub PG, Beanland L (2006) Insect vectors of phytoplasmas. Annu Rev Entomol 51: 91–111
4. Lepka P et al (1999) Effect of phytoplasmal infection on concentration and translocation of carbohydrates and amino acids in periwinkle and tobacco. Physiol Mol Plant Pathol 55: 59–68
5. Bertamini M, Grando MS, Nedunchezhian N (2003) Effects of phytoplasma infection on pigments, chlorophyll-protein complex and photosynthetic activities in field grown apple leaves. Biol Plant 47:237–242
6. Ambrožič-Dolinšek J et al (2008) Phytoplasma infection may affect morphology, regeneration and pyrethrin content in pyrethrum shoot culture. Sci Hortic 116:213–218
7. Bertaccini A, Duduk B (2009) Phytoplasma and phytoplasma diseases: a review of recent research. Phytopathol Mediterr 48:355–378
8. Deeley J, Stevens WA, Fox RTV (1979) Use of Dienes' stain to detect plant diseases induced by mycoplasmalike organisms. Phytopathology 69:1169–1171
9. Hogenhout SA, Šeruga MM (2010) Phytoplasma genomics, from sequencing to comparative and functional genomics—what have we learnt? In: Weintraub PG, Jones P (eds) Phytoplasmas: genomes, plant hosts and vectors. CABI Publishing, Oxfordshire, pp 19–36
10. Eriksson S et al (1993) Binding of 4′,6-diamidino-2-phenylindole (DAPI) to AT regions of DNA: evidence for an allosteric conformational change. Biochemistry 32:2987–2998
11. Jansen K, Nordén B, Kubista M (1993) Sequence dependence of 4′6-diamidino-2-phenylindole (DAPI)–DNA Interactions. J Am Chem Soc 115:10527–10530
12. Dale JI (1988) Rapid compression technique for detecting mycoplasmalike organisms in leaf midrib sieve tubes by fluorescence microscopy. Phytopathology 78:118–120
13. Sinclair WA et al (1992) Detection of ash yellows mycoplasmalike organisms in different tree organs and in chemically preserved specimens by a DNA probe vs. DAPI. Plant Dis 76: 154–158
14. Thomas S, Balasundaran M (1998) *In situ* detection of phytoplasma in spike-disease-affected sandal using DAPI stain. Curr Sci 74:989–993
15. Ghosh DK et al (1999) Association of a phytoplasma with witches' broom, a new disease of acid lime (*Citrus aurantifolia*). Curr Sci 77: 174–177
16. Bricker JS, Stutz JC (2005) Host range and distribution of the phytoplasma causing Arizona ash decline. J Arboric 3:257–262
17. Fletcher S, Melcher U, Wayadande A (2006) The phytopathogenic spiroplasmas. In: Falkow S, Dworkin M, Rosenberg E et al (eds) The prokaryotes: a handbook on the biology of bacteria: Firmicutes, Cyanobacteria, vol 4, 3rd edn. Springer, New York, pp 905–947
18. Arismendi AS et al (2010) Presence of a phytoplasma associated with witches'-broom disease in *Ugni molinae* Turcz. and *Gaultheria phillyreifolia* (Pers.) Sleumer determined by DAPI, PCR, and DNA sequencing. Chil J Agric Res 70:26–33
19. Jung H et al (2003) Detection and treatment of mycoplasma contamination in cultured cells. Chang Gung Med J 26:250–258
20. Mabuchi T, Nishikawa S (1991) Selective staining with two fluorochromes of DNA fragments on gels depending on their AT content. Nucl Acids Res 18:7461–7462
21. Hamada S, Fujita S (1983) DAPI staining improved for quantitative cytofluorometry. Histochem Cell Biol 79:219–226
22. Liu J-Y et al (2004) A new chromosome fluorescence banding technique combining DAPI staining with image analysis in plants. Chromosome 113:16–21
23. Romero J (2001) XI Curso Internacional Teórico-Práctico de Detección e Identificación de Virus, Viroides y Fitoplasmas. Instituto Nacional de Investigación y Tecnología Agraria y Alimentaria (INIA). Madrid, p 2–9
24. Lichtman JW, Conchelo JA (2005) Fluorescence microscopy. Nat Methods 2:910–919

Chapter 11

Visualization of Phytoplasmas Using Electron Microscopy

B. Jean Devonshire

Abstract

The use of electron microscopy, both transmission and scanning, provides reliable and accurate methods for detecting phytoplasmas in plants. Our understanding of these pathogens, their morphology, development, and intracellular location in plants and insect vectors has been greatly increased through the use of these instruments. Development of techniques such as immunolabeling, cryofixation with freeze substitution or plunge freezing with direct transfer to the microscope stage, together with advances in instrumentation is enabling us to study these pathogens under conditions close to their native state. The visualization of fine detail and ultrastructure, using modern and established techniques, can only be appreciated by the magnification and spatial resolution offered in the electron microscopes. Now that the full sequencing of four phytoplasma genomes (to date) has been achieved, electron microscopy can play an important role in identifying and understanding specific gene functions.

Key words: Cryofixation, Electron microscopy, Scanning electron microscopes, Transmission electron microscopes, Ultramicrotomy

1. Introduction

Phytoplasmas, known up to the early 1990s as mycoplasma-like organisms (MLOs), are prokaryotes lacking a cell wall, a nucleus, and any membrane-bound organelles. They are found in the sieve elements of plant phloem (Fig. 1), occasionally in parenchyma cells and in the salivary glands, mid-gut and ovaries (1, 2) of phloem-feeding insect vectors, namely leafhoppers (Cicadellidae), planthoppers (Fulgromorpha), and psyllids (Psyllidae). Electron microscopy has shown phytoplasmas to be pleiomorphic, having been observed in many shapes and sizes between 0.1 and 1 μm (3–5); this ability to modify their morphology enables phytoplasmas to pass through phloem sieve plates and plasmodesmata of the cell walls in plant vascular systems (Fig. 2). Although the mechanisms inducing disease in plants are still uncertain, the symptoms

Matt Dickinson and Jennifer Hodgetts (eds.), *Phytoplasma: Methods and Protocols*, Methods in Molecular Biology, vol. 938, DOI 10.1007/978-1-62703-089-2_11,

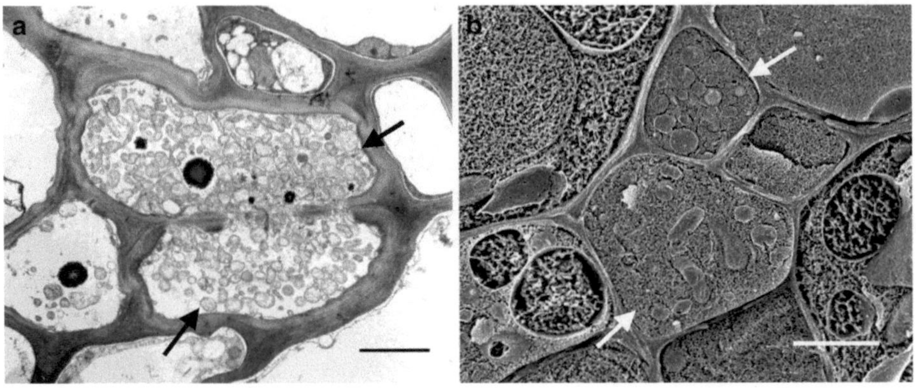

Fig. 1. Electron microscope images of (**a**) transmission electron microscopes (TEM) ultrathin cross-section of plant phloem and companion cell infected with sesame phyllody phytoplasma. Image courtesy P. Jones. (**b**) Scanning electron microscopes (SEM) of cryofractured cross-section of plant phloem and companion cell infected with Brinjal little leaf phytoplasma. *Arrows* indicate cells with an abundance of phytoplasmas. (**a**, **b**) Bar = 2 micron.

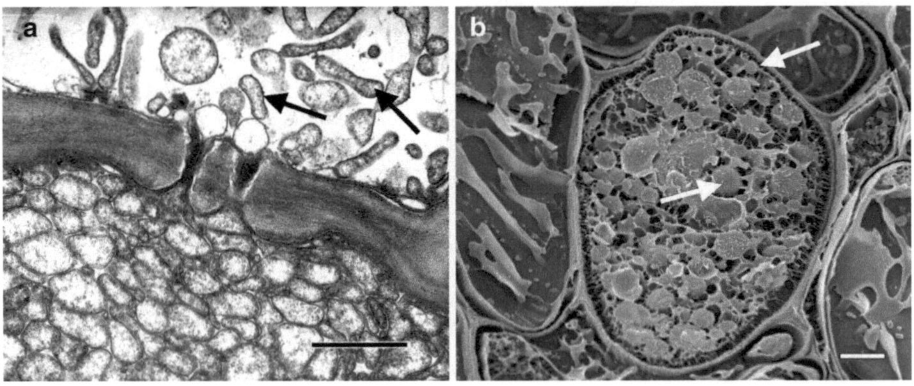

Fig. 2. A TEM image of phytoplasma bodies passing through the plasmadesmata of a densely occupied phloem cell into a less densely occupied companion cell. *Arrows* indicate changed morphology. Image courtesy P. Jones. (**b**) CryoSEM image of phloem cell packed with phytoplasma bodies of various sizes. (**a**) Bar = 1 micron, (**b**) bar = 2 micron.

are well documented. The deposition of callose in cells adjacent to sieve plates has been seen in longitudinal, thin sections under the electron microscope (6) and this appears to herald the onset of symptoms and necrosis.

To date phytoplasmas have not been cultured axenically (outside the plant cell) so, unlike other prokaryotes and fungi, researchers are presented with the task of obtaining or producing fresh infected tissue for their studies and for visualization. This must be taken directly from the infected plant or body of the insect vector.

Transmission (TEM) and scanning (SEM) electron microscopes offer opportunities to visualize ultrastructural detail, responses to environmental stimuli, and how gene manipulation could modify the ability of a phytoplasma to survive and thrive in plant cells. As early as 1967 phytoplasmas were detected in plant tissue using TEM (7); the pathogen was seen in the phloem of plants showing dwarfing, "'witches' broom" and yellowing-type disease symptoms.

Many preparation techniques have been used to visualize phytoplasmas in electron microscopes, the method of tissue fixation being the main area of variation. Two of the most widely used protocols for TEM and SEM will be described in this chapter.

1.1. Safety

Most of the chemicals used for fixation and embedding are hazardous and it is important that they are handled in accordance with the UN hazard number (e.g., acrylamide is UN2074), which is displayed on the container. Waste chemicals must be disposed of using the recommended procedures and all research laboratories should have clearly set out standard operating procedures (SOPs) or guidelines available. Protective disposable gloves, preferably nitrile, should be worn and all work should be performed in a fume cupboard unless otherwise stated.

Wherever possible the use of less dangerous chemicals can remove the associated problems, but in some circumstances it is difficult to find equivalent alternatives. Care should be taken when making and handling glass knives during ultramicrotomy. The cutting edge of the glass knife is of course extremely sharp and an appropriate method for disposal of used knives should be followed.

Guidelines on the safe use and handling of liquid nitrogen should be carefully followed. Particular attention should be given to suitable clothing including footwear. Eye and face protection should always be available for the handler. Serious burns can occur if liquid nitrogen is trapped against skin and clothing, and the wet surface of the eyes is a risk area for burns through splashing. The evaporation of liquid nitrogen increases the gaseous nitrogen content of the atmosphere very rapidly and asphyxiation occurs without warning. Low oxygen detectors should be installed or personal units should be available whenever liquid nitrogen is being used for cryopreparation particularly in an enclosed room.

1.2. Preparation Techniques for TEM

Excellent results for morphological studies using TEM have been obtained in laboratories around the world using established techniques which involve chemical fixation and dehydration, with final embedding in a resin followed by ultrathin sectioning (5, 8, 9). The cryofixation method using high-pressure freezing (HPF, see Subheading 3.7) is becoming more widely used and recognized as a method of choice. However, each technique has its drawbacks and the initial cost of the instrument, the accessibility of chemicals, and also sample size are important factors that have to be considered in order to make the best decision for a particular situation.

The visualization of phytoplasmas in plant and insect tissue using chemical fixation is still a popular technique. Although it requires more operator input than HPF, it is a method affordable in most laboratory setups and allows larger tissue sections to be processed.

The procedure starts with a primary fixation in glutaraldehyde and paraformaldehyde. Each component in the aldehyde mixture has a distinct role in host tissue preservation. The formaldehyde, released from the polymeric paraformaldehyde, penetrates rapidly and prevents autolysis by binding to proteases and lipases within the cell; the slightly larger glutaraldehyde molecules (and its oligomers also present in solution) penetrate more slowly but achieve more effective fixation by cross-linking proteins through its multiple aldehyde groups (10). After several washes in buffer, the samples are transferred to a secondary fixative/stain, osmium tetroxide. This binds preferentially to phospholipid bilayers, so enhancing the contrast of cell membranes and to a certain extent making the membranes less soluble in the solvents used during the dehydration routine (11). Placing the samples at 4°C for 1–2 h is recommended and double sealing the container is critical. Modifications of the fixative strengths and the complete omission of the secondary fixative/stain, osmium, are options often considered. The buffer affects the rate of protein cross-linking with glutaraldehyde and its role is fairly complex. Cacodylate buffer is sometimes preferred because the high phosphate concentration in Sørensen's buffer can precipitate the Ca^{2+} in plant tissue (12). However, phosphate buffer is less hazardous as it does not contain arsenic and, with careful preparation and choice of pH for the final solution, good results can be obtained.

The resins used for embedding are water-immiscible, so any water held in the fixed tissue must be removed in a controlled way using a solvent. The data sheet accompanying each resin kit should be followed carefully to ensure the correct choice and use of solvent. In the protocol outlined in this chapter, the Spurr's replacement resin from TAAB is chosen for embedding and acetone as the solvent for dehydration.

Embedding in resin is necessary to provide the tissue with support, allowing it to be oriented and held firmly during sectioning. Two of the main types of resins available are acrylic and epoxy, each offering specific advantages for different objectives. The acrylic resins such as LRW, LRG, and Lowicryl are suitable for immunolabeling with both electron and light microscopy (13). The epoxy resins, such as Epon and the Spurr's replacement from TAAB, are generally used for morphological studies for TEM (14). The original Spurr's resin is no longer available, but a less hazardous replacement is the TAAB low viscosity resin S032/D medium grade. It is sold as a premix kit and is chosen for this protocol because it has given good results and if the steps in the data sheet are followed, the mix will be reproducible every time.

1.3. Preparation Techniques for SEM

Previous attempts to visualize phytoplasmas using SEMs presented difficulties in identifying the phloem cells and keeping cell contents intact (15, 16). The samples were prepared using the critical-point drying technique (CPD), which causes some removal and modification of cell contents, leaving only the phytoplasma bodies firmly attached to the phloem cell walls and those trapped in sieve plate pores for visualization.

The low temperature cryofixation technique does not cause these problems and is being used increasingly in a bid to obtain the perfectly preserved specimen in its native state. The aim is to fix the tissue so rapidly that the redistribution of organelles is prevented. Likewise, changes to cell membranes, shrinkage, and the creation of artifacts are all minimized. Ideally the cell cytoplasm is converted to vitreous ice, eliminating the movement of cell components and labile elements (important in immunolabeling or microanalysis), without the formation of ice crystals that disrupt delicate membranes causing what is recognized as "crystal damage."

Cryoscanning electron microscopy (CryoSEM) is the low temperature electron microscopy of fully hydrated bulk samples. The samples are not modified by exposure to chemical fixatives or solvents, so preserving their integrity and external features such as wax coatings, extracellular mucilage, and fully hydrated cells with contents are visualized intact. CryoSEM offers in situ observations; it is easy to perform, allows much larger sample size, and is extremely quick. Together with freeze-fracturing it enables the visualization of cells, cell structure, and the general architecture beneath the surface (Fig. 3).

Plunging samples into liquid nitrogen slush (SLN_2) is a recognized method for fixing relatively large samples, e.g., ~2 × 5 × 5 mm. The tissue thickness does, however, influence the formation of vitreous ice, giving the best preservation in the outer layers and

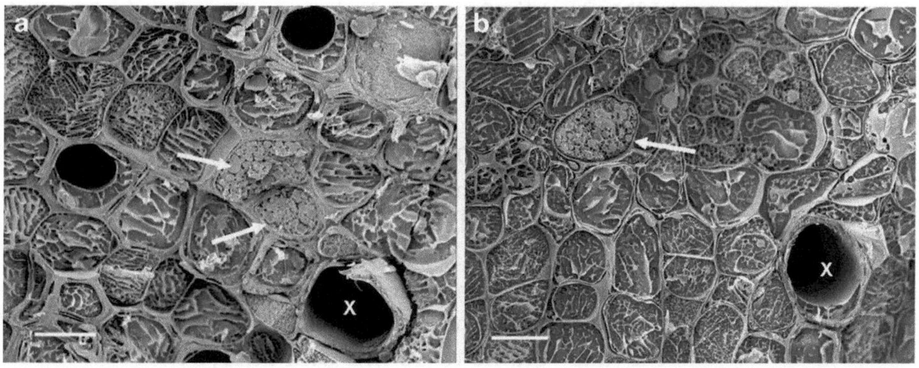

Fig. 3. Cryo SEM freeze fractured petioles from (**a**) Crotalaria phyllody and (**b**) brinjal little leaf. Both images of infected Madagascar periwinkle (*Catharanthus roseus*) vascular systems with *arrows* indicating phloem and companion cells packed with phytoplasmas. X = xylem. (**a**, **b**) Bar = 10 micron.

less perfect preservation towards the center of the tissue. The fixation process is not as fast as HPF (17) and sometimes ice contamination can be identified on surfaces. This can easily be removed by careful control of the stage temperature through sublimation. Provided the plunging is performed quickly and tissue stress has been avoided, successful imaging of a fractured surface can reveal information at the cellular level, i.e., the presence of phytoplasmas.

Sample fracturing is a random technique, very dependent on the nature of the sample. Often the fracture plane will reveal internal microstructure of particular interest whilst at other times it will be difficult to obtain a fracture plane of interest. A more controlled and uniform surface can be achieved by using a cryo-ultramicrotome and a technique called cryo-planing, which relies on the use of a liquid nitrogen-cooled cryo-ultramicrotome, although this technique is more time-consuming.

Whilst the presence of phytoplasmas can be identified by electron microscopy, their classification and taxonomy relies on molecular biology techniques.

2. Materials

Glutaraldehyde solution, paraformaldehyde, osmium tetroxide, and uranyl acetate are all electron microscopy (EM) grade; all other chemicals are analytical grade.

2.1. Buffers and Fixatives

1. Sørensen's phosphate buffer (PB) pH 7.2: Prepare a stock solution (0.2 M) solution *A* (dissolve 2.40 g NaH_2PO_4 *or* 2.76 g $NaH_2PO_4 \cdot H_2O$ *or* 3.12 g $NaH_2PO_4 \cdot 2H_2O$ in 100 mL double distilled water, ddH_2O) solution *B* (dissolve 5.36 g $Na_2HPO_4 \cdot 7H_2O$ *or* 7.16 g $Na_2HPO_4 \cdot 12H_2O$ in 100 mL ddH_2O). Mix 28 mL solution *A* + 72 mL solution *B*. Confirm pH 7.2 (see Note 1), store at 4°C. Dilute to 0.05 M for washing and the preparation of the osmium stain.
2. Primary fixative: 2.5% (w/v) glutaraldehyde, 4% (w/v) paraformaldehyde. Dissolve 1.6 g paraformaldehyde in 20 mL ddH_2O in a 45 mL screw cap Falcon tube. Place this in a glass beaker half filled with water and heat to 70°C, but do not allow it to boil (see Note 2). Add 1–2 drops of 1 M NaOH until the solution clears then allow it to cool. Add 4 mL 25% glutaraldehyde and 10 mL 0.2 M PB (stock solution). Make up the final volume to 40 mL with ddH_2O.
3. Secondary fixative/stain: 1% (w/v) osmium tetroxide (OsO_4). Pour 25 mL 0.05 M PB into a 45 mL screw cap Falcon tube. Hold the pre-scored glass ampoule of 0.25 g OsO_4 under hot

water to melt the crystals, at the same time remove the paper label. Under the fume hood and wearing nitrile gloves, check the contents have accumulated in the lower section of the ampoule, then snap open and immediately pour the contents into the buffer. The lower section can be dropped into the buffer to ensure all the osmium is dissolved. Cap the tube and gently invert to mix. Seal with Parafilm and foil secured with tape and label. Place the tube in a sealed plastic bag at 4°C and use as soon as possible (see Note 3).

2.2. Tissue Dehydration Resin Infiltration

1. Sample rotator with speed control 2–4 rpm or platform agitator.
2. Oven for polymerization set to 70°C situated in fume cupboard.
3. Glass 7 mL vials with snap-on polythene caps.
4. Acetone: 100%, dried over molecular sieve.
5. Acetone/water series: 30%, 50%, 70%, 90% prepared using ddH_2O.
6. TAAB low viscosity resin kit S032/D medium grade (Spurr replacement resin) (see Note 4).
7. BEEM® capsules or flat silicone rubber embedding moulds for samples.

2.3. Ultramicrotomy

1. Sectioning is performed using a high-precision ultramicrotome such as the Reichert Jung Ultracut or the Leica EMUC6 (Leica Microsystems UK) (see Note 5).
2. Glass knife maker EMKMR2 or similar and glass (Leica Microsystems UK) (see Note 6).
3. Fresh glass knives are prepared using the EMKMR2 and following the operating instructions for the individual model.
4. Plastic troughs: LKB Trufs which are attached to the back of the glass knife using hot wax produced on the LKB/Leica Multiplate hotplate (Fig. 4).
5. Formvar/carbon-coated slotted Cu grids.
6. Fine forceps and manipulating tool (see Note 7).
7. Double sided razor blades.

2.4. Stains

1. Uranyl acetate stain: 2% (w/v) uranyl acetate (UA). Add 0.5 g UA to 25 mL ddH_2O in a dark-colored glass bottle. Use a magnetic stirrer until no solid residue remains visible (see Note 8).
2. Sodium hydroxide solution: 1 M. Add 4.0 g to NAOH 100 mL ddH_2O. Use pellets from freshly opened container if possible and freshly boiled and cooled ddH_2O (CO_2-free).
3. Reynolds' lead citrate stain (18). Add 1.33 g lead nitrate $Pb(NO_3)_2$ and 1.76 g sodium citrate ($Na_3C_6H_5O_7{\cdot}2H_2O$) to

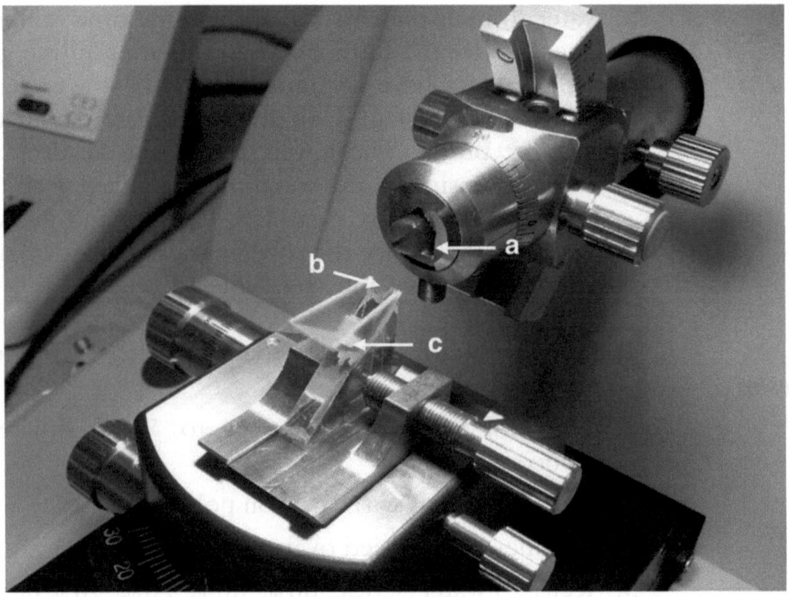

Fig. 4. The resin embedded sample block mounted in the sample holder of the Leica EMUC6 and a glass knife with trough attached mounted on the ultramicrotome stage. a = sample block, b = glass knife edge, and c = plastic trough filled with water.

30 mL of ddH_2O in a 50 mL flask and stir or shake continuously for 1 min then intermittently for 30 min. Add 8 mL 1 M NaOH and make the volume up to 50 mL with ddH_2O and mix by inversion. Check pH 12.0 ± 0.1 (see Note 9).

2.5. CryoSEM Equipment

1. A cryo-system such as the Alto2500 (Gatan UK). The preparation chamber is attached directly to the electron microscope chamber and is independently pumped to high vacuum using a vibration-free turbo molecular pump (see Note 10).
2. There is an option of coating targets, but gold/palladium (Au/Pd) is used for most routine imaging in the Gatan system.
3. A supply of about 25 L/day of liquid nitrogen is needed, preferably in a swing dewar for ease of pouring. A small 2 L liquid nitrogen dewar and two plastic funnels (see Note 11).
4. Aluminum cryo-stubs (Gatan or Agar Scientific) (see Note 12).
5. Tissue-Tek OCT compound (Gatan or Agar Scientific).
6. Sterile razor blades for removing tissue from the plant and scalpel blades either flat or hooked size 11 or 12 mounted on the cold knife—handle size 3 (Agar Scientific) (see Note 13).
7. Dry nitrogen gas.
8. Ultra-pure argon gas canister.

3. Methods

Perform this procedure under the fume hood wearing nitrile gloves. Use disposable pipettes to aspirate unwanted solutions ensuring the tissue remains immersed in liquid at all times. Follow the recommended method for disposing of all waste.

3.1. Fixation TEM

1. Using a sterile blade, select tissue from young shoots, leaf mid-vein, petiole, and/or stem, keeping about 1 mm below any necrotic region that may be visible on the infected plant.
2. In a Petri dish and under a pool of primary fixative, cut the tissue into small sections ~1 × 1 × 3 mm (see Note 14).
3. Transfer sections to glass vials with ~3 mL of primary fixative, include pencil written labels, cap and gently agitate for 1–2 h (see Note 15).
4. Wash 3× using 0.05 M PB with 10 min intervals and gentle rotation or agitation.
5. Replace the final wash with the secondary fixative/stain (OsO_4). Double-seal the samples in a taped box and sealed plastic bag and place in a fridge at 4°C for 4 h, or overnight (see Note 16).
6. Wash 3× as before using 0.05 M PB with 10 min intervals and gentle agitation.

3.2. Dehydration

Use disposable fine tip pipettes and gently agitate at each stage.

1. Remove most of the PB wash and replace sequentially with ~3 mL of the following solutions.
2. 30% acetone for 10 min then aspirate.
3. 50% acetone for 10 min.
4. 70% acetone 10 min or overnight at 4°C (see Note 17).
5. 90% acetone for 10 min.
6. 100% dry acetone, 3 × 30 min. Hold in final solution until ready to infiltrate with resin (see Note 18).

3.3. Resin Infiltration Embedding

Prepare the resin as per the manufacturer's instructions and store as recommended (see Note 4). If the resin is stored in the fridge overnight ensure that it is allowed to reach room temperature before using. Infiltrate the samples through a resin/acetone series using gentle agitation or rotation at each stage.

1. 1:2 resin:acetone 1 h.
2. 1:1 resin:acetone 1 h.
3. 2:1 resin:acetone 1 h.
4. 100% resin overnight at room temperature.

5. Change to fresh resin for 1 h and gently agitate.
6. Use wooden cocktail sticks to place the tissue in the moulds or capsules, fill with fresh resin, orientate the tissue, and then polymerize in the oven at 70°C for 12–15 h (see Note 19).
7. After checking the resin has hardened, remove the samples from the oven and allow them to cool down before releasing from the moulds using a single-sided razor blade (see Note 20).
8. Individually label each block and store in paper envelopes ready for ultrathin sectioning.

3.4. Ultramicrotomy

Sectioning is performed on an ultramicrotome using glass knives, freshly made as required. Within a matter of days and regardless of use, glass knives tend to lose their sharpness. A diamond knife can be used, alleviating the need to make fresh glass knives. However, diamond knives are very expensive and can be easily damaged; they should be handled carefully and only used by experienced operators.

1. Mount the sample block in the ultramicrotome holder provided and secure the holder firmly on the ultramicrotome stage.
2. Using a razor blade, trim away excess resin to form a trapezoid shape as small as possible, keeping your tissue in the center of the block face. Ensure the top and bottom edges are parallel to allow ribbon formation. (Trimming can also be done semi-automatically using a glass knife with the sample mounted in the specimen arm).
3. Transfer the sample to the specimen arm and plane the block face with a glass knife until the tissue is just visible at the surface.
4. Mount a fresh glass knife with a trough attached filled with ddH_2O (Fig. 4).
5. Set the appropriate thickness, (silver colored 60 nm) and cut sections using the manipulating tool to guide the sections as they float away from the knife edge.
6. Dip a folded Whatman No. 1 filter paper into chloroform and hold it just above the sections for a few seconds (see Note 21).
7. Sections can now be separated from the ribbon into sets using the manipulation tool and picked up on the grids.
8. Draw off excess water by touching the side of the forceps with a piece of filter paper, then place the grid section side up on a silicone rubber Grip-a-grid mat in a covered Petri dish to dry.

3.5. Staining

Sections are double-stained using 2% (w/v) uranyl acetate and Reynolds lead citrate.

1. Float the grids section side down on 30 μL drops of uranyl acetate stain placed on dental wax in a closed Petri dish for 5–10 min at room temperature.
2. Hold the grids in forceps and wash with a gentle stream of fresh ddH_2O, remove excess water by careful blotting with filter paper.
3. Float the grids section side down on 30 μL drops of lead citrate stain placed on a section of dental wax in the Petri dish for 2–10 min at room temperature. Place 4–5 pellets of NaOH inside the dish to absorb any carbon dioxide (see Note 22).
4. Wash the grids as before and again carefully remove excess water with a filter paper.
5. Place the grids on a silicone rubber Grip-a-grid mat in a covered Petri dish ready for TEM examination (see Note 23).

3.6. CryoSEM

1. Cool down the cryo-unit and electron microscope using liquid nitrogen (LN_2). Proceed when the temperature has reached about −170°C. The anticontaminators should always record a temperature ~10°C colder than the stage (see Notes 11 and 24).
2. Petiole or leaf mid-vein tissue ~5 × 1–2 mm wide is excised from the plant using a sterile blade.
3. This is mounted on a modified cryo-stub using Tissue-Tek, a low temperature cryoprotectant and adhesive. Slotted stubs are used for leaf fractures and stubs with holes are more successful for stem or petiole fractures (see Note 12).
4. Fill the container in the slush unit of the Gatan Alto 2500 with LN_2, cover the unit with the lid, and activate the vacuum pump (see Note 25) to convert LN_2 into semi-solid nitrogen.
5. Nitrogen gas is admitted into the unit to allow the lid to be removed and the sample is plunged rapidly into slushy liquid nitrogen.
6. When boiling ceases, the vacuum pump is activated again and, as the LN_2 starts to solidify, the sample is withdrawn into the vacuum transfer capsule. It is isolated in the capsule by closing the valve and a puff of N_2 gas into the unit releases the capsule and rod allowing transfer under vacuum to the prep-chamber.
7. Once the air lock is pumped out double valves are opened allowing access to the prep-chamber and the sample is loaded on the stage (see Note 26).
8. A single deft touch using the cold knife fitted in the chamber will fracture the specimen. Return the knife to the cold block when not in use (see Note 27).

9. The stage temperature is controlled by a heater which allows sublimation of the sample. This is set typically to about −95°C for 1–2 min to remove any ice on the surface. Etching will also reveal features, such as cell organelles and thin delicate membranes on the fractured face (see Note 28).
10. The stage heater is switched off and the temperature of the stage is allowed to recover to about −140°C before coating the sample with Au/Pd in an atmosphere of argon gas (see Note 29). Set the timer for coating: 60 s is usually adequate.
11. When the argon gas has been pumped away, the sample can be transferred to the electron microscope stage with the temperature maintained at −150°C for examination (see Note 30) and imaging.

3.7. High-Pressure Freezing

In a high-pressure freezer (HPF), the operating pressure and sample size, particularly thickness, are critical (17). These are carefully controlled to avoid crystal damage, especially at the center of the sample where the cold front does not penetrate so quickly. Cryoprotectants, for example yeast paste or 20% bovine albumin serum (BSA), are used to lower the freezing point of water, and fillers such as hexadecene can help to achieve more even freezing. After HPF the samples are brought to room temperature very slowly, typically over 3–4 days, using a freeze substitution unit where the frozen water in the tissue is replaced with acetone containing a fixative, usually osmium tetroxide. When room temperature is reached the samples are washed in dry acetone and the dehydration routine (Subheading 3.2) is completed followed by resin embedding and sectioning.

1. As briefly mentioned, HPF does have drawbacks, especially regarding limits to the sample size. The tissue has to be trimmed to fit into planchettes 1.5 mm diameter and 0.2 mm thick. This size restriction is a problem when trying to identify pathogens in situ, but it is being addressed by manufacturers and instruments using slightly larger planchettes, and other alternatives are becoming available. The value of this technique lies in its outstanding ability to preserve morphological detail, and its importance will increase should the axenic culture of phytoplasmas become possible.

4. Notes

1. Preparation of 0.2 M Sørensen's buffer giving pH 7.2 as indicated in the table in *Plant Microtechnique and Microscopy*, 1999, S. Ruzin, Oxford University Press, Oxford.

2. The fumes are very harmful so this must be performed under a fume hood.
3. Dispose of osmium solution after 1 week or sooner if it turns a purple/black color, remembering to use the recommended procedure for this hazardous material. Neutralizing OsO_4 waste by mixing it with twice its volume of corn oil in a glass bottle is a common method used. For extra protection two pairs of gloves can be worn.
4. Follow instructions in the data sheet for S032D premix resin kit. Add the contents of one 50 mL bottle to the contents of one 100 mL bottle then add the contents of one ampoule of accelerator and mix thoroughly. Approximately 100 mL of resin is ready for use and can be stored at 4°C for a few days, or at −20°C for a few months, but bring back to room temperature before use. No weighing of components is required.
5. Ensure that the instructions have been read and training has been received before operating the ultramicrotome.
6. Operating instructions should be read and practiced taking care during the handling of glass strips.
7. A single eye lash attached to the end of a cocktail stick with wax or tape works very well as a tool for manipulating sections.
8. If a dark bottle is not available cover the container with aluminum foil to exclude light. Pipette from below the surface and it is recommended the solution is passed through a 0.2 micron filter.
9. The stain can be kept for ~6 months in a screw top bottle at 4°C. It is advisable to pass through a 0.2 micron fifilter before use and if a white precipitate is observed dispose of and make up a fresh solution.
10. Fracturing, sublimation, and coating are all performed in the prep chamber. Control of the temperatures of the stages in the prep and the microscope chambers is necessary and anticontaminators are fitted just above the sample position in both chambers.
11. Tissue paper is placed in each funnel to absorb any moisture every time LN_2 is dispensed.
12. Aluminum stubs can be drilled or machined with holes or slots to accommodate stems, leaves, or petioles or a vice clamp sledge is another option and can be purchased (Gatan UK).
13. The flat blades are easier to use for leaf or stem fractures, remembering to fit this before cooling down the prep chamber and to park the blade in the cold block ready for use.

14. Record the orientation as this will aid alignment when embedding. A piece of dental wax can be used in the Petri dish as a base on which to cut the sections.
15. If after 5 min the sections remain floating, place the uncapped vials in a desiccator adding 1 drop of Tween 20 detergent (0.05% diluted in ddH_2O) to each vial. Carefully evacuate using a vacuum pump until the tissue sections sink.
16. Failure to seal the container will allow osmium vapor to fix any grease etc. inside the fridge such as finger prints on doors and shelving turning it black.
17. This is a convenient stage to hold the samples avoiding a harsh and rapid dehydration routine that could result in damaged tissue.
18. Holding tissue in 100% acetone can cause it to become very brittle and too delicate to handle through the resin infiltration routine so keep this time to a minimum.
19. Any spare resin can be placed in blank capsules and used to check on the progress of polymerization. These blanks can also be used as mounts for flat embedded samples.
20. If the resin feels tacky, the polymerization has not been complete and samples must be returned to the oven until properly cured and dry to touch.
21. The chloroform vapor stretches the silver colored sections helping to remove any wrinkles.
22. When using the stain, avoid breathing directly over the Petri dish as the CO_2 in your breath will cause formation of lead carbonate and this will contaminate the grids. Holding the Petri dish lid at an angle opening away from your body will help to prevent this.
23. Sections should be examined as soon as possible after staining as some fine detail can be lost on storage.
24. The cryo-stage usually fits over the dry stage in many electron microscopes and can be removed to perform microscopy at room temperature. If this is the case always check that the cryo-stage is in the operating position before starting the cool down procedure. Purge the system with nitrogen gas for about 10 min before starting to cool down, this will clear any moisture in the tubing and is an opportunity to check that there are no gas leaks.
25. The LN_2 in the slush unit is frozen just prior to plunging the sample, any moisture dissolved in it is removed thus reducing the possibility of ice contamination on the sample.
26. The stage is cooled directly by LN_2 maintaining a base temperature of about −180°C with the temperature of the anticontaminator

at −190°C. These temperatures are recorded on the control pad of the Alto unit.

27. Do not use a sawing or scraping movement, as this will produce knife marks. A precise touch or gentle flicking action is all that is required. The frozen sample will cleave very easily along a line of least resistance.
28. The timing of this procedure is very sample dependent with no sample being exactly the same, a trial run is advised to achieve optimal timing.
29. The plasma is formed in an atmosphere of argon gas which is attached to the chamber and connected via a solenoid switch. A timer is wired to the switch and will close the flow of gas once the set time, usually about 60 s, has been reached. Coating thickness is ~5 nm.
30. Transfer should be done from the cryoprep chamber to the microscope stage once the temperatures are within about 10°C of each other. Sample contamination can occur if the difference is significantly greater.

Acknowledgements

I would like to thank Prof. Phil Jones, former head of Bioimaging at Rothamsted, for his valued advice and from whom a version of the TEM protocol originated and the use of the TEM images. Also to Prof. John Lucas and Dr Allison van de Meene for helpful comments on the manuscript. Rothamsted Research receives grant-aided support from the Biotechnology and Biological Sciences Research Council (BBSRC) of the UK.

References

1. Weintraub PG, Beanland LeA (2006) Insect vectors of phytoplasmas. Annu Rev Entomol 51:91–111
2. Kawakita H et al (2000) Identification of mulberry dwarf phytoplasmas in the genital organs and eggs of leafhopper *Hishimonoides sellatiformis*. Am Phytopathol Soc 90:909–914
3. Hull R, Horne RW, Nayar RM (1969) Mycoplasma-like bodies associated with Sandal Spike Disease. Nature 224:1121–1122
4. Horne RW (1970) The ultrastructure of mycoplasma and mycoplasma-like organisms. Micron 2:19–38
5. Waters H, Hunt P (1980) The *in vivo* three dimensional form of a plant mycoplasma-like organism by analysis of serial ultrathin sections. J Gen Microbiol 116:111–131
6. Musetti R (2010) Biochemical changes in plants infected by phytoplasmas. In: Weintraub PG, Jones P (eds) Phytoplasmas: genomes, plant hosts and vectors. CAB International, Wallingford, pp 132–146
7. Doi Y et al (1967) Mycoplasma—or PLT group-like microorganisms found in the phloem elements of plants infected with mulberry dwarf, potato witches' broom, aster yellows, or paulownias witches' broom. Ann Phytopathol Soc Jpn 33:259–266
8. Siddique ABM et al (2001) Electron microscopy and molecular characterization of phytoplasmas associated with little leaf disease of brinjal (*Solanum melongena* L.) and periwinkle (*Catharanthus roseus*) in Bangladesh. J Phytopathol 149:237–244

9. Franova J et al (2003) Electron microscopy and molecular characterization of phytoplasmas associated with strawflower yellows in the Czech Republic. Eur J Plant Pathol 109: 883–887
10. Kiernan JA (2000) Formaldehyde, formalin, paraformaldehyde and glutaraldehyde: what they are and what they do. Microsc Today 00–1:8–12
11. Sabatini DD, Bensch K, Barrnett MD (1963) The preservation of cellular ultrastructure and enzymatic activity by aldehyde fixation. J Cell Biol 17:19–58
12. Dykstra MJ, Reuss LE (2003) Biological electron microscopy: theory, techniques and troubleshooting, vol 2. Springer Verlag GmbH, New York, pp 27–28
13. Musetti R et al (2002) Application of immunoelectron microscopy techniques in the diagnosis of phytoplasma diseases. Microsc Res Tech 56:462–464
14. Arocha Y et al (2009) Detection of phytoplasma and potyvirus pathogens in papaya (*Carica papaya* L.) affected with 'Bunchy Top Symptom' (BTS) in eastern Cuba. Crop Prot 28:640–646
15. Haggis GH, Sinha RC (1977) Scanning electron microscopy of mycoplasma like organisms after freeze fracture of plant tissues affected with Clover Phyllody and Aster Yellows. J Phytopathol 68:677–680
16. Poghosyan AV et al (2004) Possible phytoplasma disease in papaya (*Carica papaya* L.) from Baja California Sur: diagnosis by scanning electron microscopy. J Phytopathol 152: 376–380
17. Studer D et al (2001) A new approach for cryofixation by high-pressure freezing. J Microsc 203:285–294
18. Reynolds ES (1963) The use of lead citrate at high pH as an electron-opaque stain in electron microscopy. J Cell Biol 17:208–212

Chapter 12

Automated DNA Extraction for Large Numbers of Plant Samples

Nataša Mehle, Petra Nikolić, Matevž Rupar, Jana Boben, Maja Ravnikar, and Marina Dermastia

Abstract

The method described here is a rapid, total DNA extraction procedure applicable to a large number of plant samples requiring pathogen detection. The procedure combines a simple and quick homogenization step of crude extracts with DNA extraction based upon the binding of DNA to magnetic beads. DNA is purified in an automated process in which the magnetic beads are transferred through a series of washing buffers. The eluted DNA is suitable for efficient amplification in PCR reactions.

Key words: Phytoplasma, Homogenization, DNA extraction, Automation, Magnetic processor

1. Introduction

DNA extraction is usually one of the most laborious and time-consuming steps in the sample analysis process. This chapter describes an automated DNA extraction method, which has proven suitable for processing many grapevine and fruit tree samples.

The cell disruption process is achieved using a FastPrep®24 (MP Biomedicals) instrument, in which collisions of a matrix and the sample in the presence of buffer release the DNA from the cells. The QuickPick™ SML Plant DNA kit (Bio-Nobile) is used for total DNA extraction: DNA in the plant tissue sample is released using a proteinase K solution and lysis buffer. The lysate is centrifuged to remove debris, and the cleared lysate is mixed with magnetic beads and loaded into the wells of a KingFisher® extraction machine (Thermo Scientific). DNA binds to the surface of the magnetic beads and is purified as the KingFisher® machine passes the beads through a series of washing buffers. Finally, purified DNA is eluted into elution buffer.

Matt Dickinson and Jennifer Hodgetts (eds.), *Phytoplasma: Methods and Protocols*, Methods in Molecular Biology, vol. 938, DOI 10.1007/978-1-62703-089-2_12, © Springer Science+Business Media, LLC 2013

The KingFisher® mL system handles 15 samples in 1 h (not including the time needed for homogenization of plant material). However, greater numbers of samples can be processed simultaneously using other instruments, e.g., the KingFisher® Flex processes 96 samples together. The technique does not require any organic solvents and eliminates the need for repeated centrifugation, vacuum filtration, and column separation. The risk of sample cross-contamination is also reduced due to the use of disposable plasticware in each step and the simple protocol. The efficiency of the automated extraction procedure, including quality and quantity of extracted DNA, is comparable with CTAB extraction methods (1). Therefore, the described automated procedure of homogenization and DNA extraction has been successfully used in our laboratories in the Slovenian grapevine and fruit tree phytoplasma monitoring programme (1, 2), as well as for diagnosis of other bacteria (3). Extracted DNA can be used for downstream applications such as PCR, real-time PCR, and isothermal amplifications such as LAMP (see Note 1).

2. Materials

1. Samples: fresh plant tissue (e.g., leaves, roots, phloem).
2. Tissue homogenizer, e.g., FastPrep®-24 with TN 12×15-TeenPrep™ Adapter (MP Biomedicals) (see Note 2).
3. 15 mL sterile polypropylene disposable centrifuge tubes, and sterile 2 mL and 1.5 mL microfuge tubes (see Note 3).
4. Matrix, e.g., Matrix A: garnet matrix and a minimum of 5 1/4-in. ceramic spheres (MP Biomedicals).
5. Centrifuge.
6. Vortexer.
7. Heating block or waterbath (65°C).
8. Set of pipettes (dedicated set for DNA extraction).
9. Disposable filter pipette tips.
10. QuickPick™ SML Plant DNA kit (Bio-Nobile): magnetic particles (MagaZorb™), proteinase K solution, lysis buffer, binding buffer, wash buffer, elution buffer (see Note 4).
11. Magnetic particle processor (e.g., KingFisher® mL, Thermo Scientific; including disposable 5-well tube strips and disposable tip combs designed for the instrument) (see Note 5).

3. Methods

Various homogenization methods can be used. The method described here works well; two other methods that were used in our laboratory in the past were much more time-consuming (see Note 6).

3.1. Homogenization of Plant Material

1. Prepare about 1 g of leaf mid-vein tissue or root phloem tissue for each sample (see Note 7). Cut the sample into small pieces (see Note 8).
2. Transfer samples into labeled 15 mL centrifuge tubes containing Matrix A (see Note 9).
3. Add 2 mL of lysis buffer to each tube. Include at least one extraction blank control in the process (see Note 10).
4. Place the tubes in the FastPrep® adapter and operate the FastPrep® homogenizer for at least 40 s at 5 m/s (see instrument manual for detailed instructions) (see Note 11).
5. Centrifuge samples for 10 min at 15,000 × *g*.
6. Carefully pipette 800 μL of supernatant into new 2 mL tubes and proceed immediately with the extraction protocol.

3.2. Extraction of Total DNA

1. Pipette 25 μL of proteinase K solution into each supernatant and mix thoroughly by pulse-vortexing.
2. Lyse the samples for 30 min at 65°C.
3. During the lysis step, place 5-well tube strips into the stainless steel rack supplied with the KingFisher® (see Note 12) and pipette reagents into wells: 20 μL magnetic particles and 500 μL binding buffer (well A), 800 μL wash buffer (wells B and C), 200 μL elution buffer (well D), and 100 μL water (well E). Set up a new tube strip for each sample. Load new tip combs into their slots (see Note 13).
4. Remove the samples from the 65°C incubator, and centrifuge the tubes for 1 min at 6,000 × *g*.
5. Gently transfer 220 μL supernatant into the first wells (A) of strips containing magnetic particles and binding buffer (see Note 14).
6. Insert the strips into the KingFisher® instrument, placing them onto the strip carrier in the correct position (see instrument manual for detailed instructions about using the instrument).
7. The following instrument program in KingFisher® mL should be used: binding time in well A, 3 × 1 min release plus 2 min binding; wash in well B, 15 s; wash in well C, 15 s; elution in

well D, 10 min; releasing the magnetic particles in well E (see Notes 15 and 16). The program in detail is:

(a) Collect beads: Well = A, Premix = No, Collect count = 1.

(b) Bind: Well = A, Release = Yes, time = 1 min 0 s, speed = Fast dual mix; Bind time = 2 min 0 s, speed = Slow; Collect beads = No; Release = Yes, time = 1 min 0 s, speed = Fast; Bind time = 2 min 0 s, speed = Slow; Collect beads = No; Release = Yes, time = 1 min 0 s, speed = Fast dual mix; Bind time = 2 min 0 s, speed = Slow; Collect beads = Yes, count = 4.

(c) Wash: Well = B, Release = Yes, time = 0 s, speed = Fast; Wash time = 15 s, speed = Fast dual mix; Collect beads = Yes, count = 3.

(d) Wash: Well = C, Release = Yes, time = 0 s, speed = Fast; Wash time = 15 s, speed = Fast dual mix; Collect beads = Yes, count = 3.

(e) Elution: Well = D, Release = Yes, time = 10 s, speed = Fast; Elution time = 10 min 0 s, speed = Bottom slow; Pause for manual handling = No.

(f) Remove beads = Yes, collect count = 4, disposal well = E.

8. Once the program has run to completion, remove the rack and transfer the eluted DNA from well D into a sterile 1.5 mL microcentrifuge tube. Store at −20 ± 5°C until required.

4. Notes

1. Use of names of chemicals or equipment in this chapter implies no approval of them to the exclusion of others. This information is given for the convenience of users and does not constitute an endorsement of the products or services. Equivalent products may be used if they can be shown to lead to the same results.
2. A few other homogenizers and matrixes were tested in our laboratory, but until now we have not found any other homogenizer and/or matrix suitable for homogenization of leaf mid-vein tissue or root phloem tissue.
3. At all stages clean, DNAase-free, disposable plasticware should be used.
4. All solutions should be stored at room temperature and should be clear when used. If precipitates have formed, warm the solutions gently until the precipitates have dissolved. Magnetic particles should be mixed thoroughly just before pipetting: vortexing of the magnetic particles is not recommended.

5. With the KingFisher® mL instrument, up to 15 samples per run can be processed. Each run uses one tube strip containing five wells. One tip comb containing five tips is used for processing five samples at a time. Therefore, if 15 samples are processed at once, 15 tube strips and 3 tip combs are needed. If more samples are processed at once, the use of a high-throughput instrument, e.g., KingFisher® Flex, is recommended. With KingFisher® Flex, up to 96 samples per run can be processed and five 96-deep-well plates and one 96 format tip comb are needed for each run.
6. Homogenization with liquid nitrogen using mortar and pestle.
   (a) Freeze plant tissue sample in liquid nitrogen immediately after harvesting. Do not let the sample thaw at any time during homogenization.
   (b) Pre-cool equipment by pouring liquid nitrogen into mortar and placing the pestle's grinding end in the liquid nitrogen.
   (c) Place the frozen plant tissue sample in mortar and grind until a fine whitish powder results.
   (d) Add liquid nitrogen as necessary but be careful not to spill the sample out of the mortar.
   (e) Using a pre-cooled spatula, transfer about 200 mg of the powdered plant tissue sample into a pre-cooled 2 mL tube. All liquid nitrogen has to be evaporated before closing the tube.
   (f) If plant tissue sample is not processed immediately, the tube should be kept on dry ice or liquid nitrogen or stored at –80°C to prevent the sample from thawing. To extract DNA, add 600 μL of lysis buffer and proceed immediately with the extraction protocol.

   Homogenization in extraction bags.
   (a) Weigh approximately 1 g of plant tissue samples into bags (e.g., Universal-U-form, BIOREBA). If plant tissue sample is not processed immediately the bags should be kept at –20°C for a maximum of 1 month.
   (b) Add 1 mL of lysis buffer and homogenize using semi-automated Homex 6 homogenizer (BIOREBA) or hand-held roller grinder until the plant tissue samples are homogenous.
   (c) Add 3 mL of lysis buffer and gently rub over the bag surface (buffer should not foam).
   (d) Using a disposable pipette, transfer about 800 μL of homogenate into a new 2 mL tube and proceed immediately with the extraction protocol.

7. The recommended quantity of sample material to use is 1 g. This amount allows for a good cross-section of sample to be taken because uneven distribution of the phytoplasma is suspected.
8. The sample preparation step should be done in a separate location, away from those used for homogenization of the samples, DNA extractions, and DNA processing. Clean all surfaces (using DNA remover and clean water) before starting and at the end of each step: sample preparation, homogenization, and DNA extraction.
9. The tube should be kept on ice and the sample should be transferred to the tube immediately after harvesting. Do not freeze the tubes in liquid nitrogen, as this may lead to breakage of the tubes. The tube should be kept at –20°C for a maximum of 1 month if the plant tissue sample is not to be processed immediately.
10. For each extraction series, you should perform an extraction blank control. The extraction blank control is used to demonstrate the absence of sample contamination during extraction and should always be the last sample in the extraction series. The extraction blank control is the negative control of extraction. Lysis buffer alone is used in all steps of extraction in parallel with samples where DNA is extracted. The extraction blank control is tested in parallel with the samples in any further procedures.
11. The duration of disruption and homogenization depends on the tissue being processed and can be extended until no tissue debris is visible. With many years of experience we found that 100% homogenization is difficult to obtain, especially if root samples are processed, but if the described protocol is followed, the yield of extracted DNA is satisfactory.
12. The tube strip will only fit one way: there is a small ridge on the flanged end of the strip that locates into a small slot in the rack. The well closest to the flanged end is well A, followed by well B, C, D and the furthest is well E.
13. Push tip combs in just until a dull "click" is felt. This ensures that the comb is in the correct position. If the combs are not aligned with the magnetic rods, the machine will not be able to operate properly which may cause permanent damage to the machine.
14. Clearly mark each strip with the number/label of the specific sample to be extracted in that strip. It is easy to put the number on the flanged end of the strip. A fine-tip permanent marker pen should be used. If the KingFisher® Flex instrument is used, make a note of the scheme of sample positions in the 96-well plate.

15. During the collection (binding) of the magnetic particles, the magnetic rod is fully inside the tip. The magnetic rods together with the tip comb(s) move slowly up and down in the wells and the magnetic particles are collected onto the wall of the tips. The magnetic rods together with the tip comb(s), having collected the magnetic particles, can be lifted out of the wells and transferred into the next wells containing a reagent. Magnetic particles are released by moving the tip comb(s) up and down several times with considerably high speed until all the particles have been mixed with the substance in the next reaction. Washing is a combination of the release and collection processes in wells filled with washing solution.
16. Extraction can also be processed manually using, e.g., PickPen® manual tools (Bio-Nobile) or other systems.

References

1. Boben J, Mehle N, Ravnikar M (2007) Optimization of extraction procedure can improve phytoplasma diagnostics. Bull Insectol 60:249–250
2. Nikolić P et al (2010) A panel of real-time PCR assays for specific detection of three phytoplasmas from the apple proliferation group. Mol Cell Probes 24:303–309
3. Pirc M et al (2009) Improved fireblight diagnostics using quantitative real-time PCR detection of *Erwinia amylovora* chromosomal DNA. Plant Pathol 58:872–881

Chapter 13

DNA Extraction from Arborescent Monocots and How to Deal with Other Challenging Hosts

Nigel A. Harrison, Robert E. Davis, and Ericka E. Helmick

Abstract

Detection of pathogen DNA by polymerase chain reaction (PCR) assays is the most widely used method for diagnosing phytoplasma diseases. Reliable and efficient detection of phytoplasmas, especially in woody perennial plants, is challenging due to the unusually low abundance and sporadic distribution of phytoplasmas within infected host tissues. Detection success depends largely upon the host species and sampling procedures and, to a lesser extent, on the protocol used for DNA extraction. Here we describe a simple, straightforward, nondestructive stem sampling protocol to confirm phytoplasma infection of palms and other arborescent monocots of large stature. The protocol requires minimal processing of excised tissues and yields phytoplasma DNA preparations in suitable quantity for reliable detection by nested PCR assays.

Key words: DNA extraction, Palms, Phytoplasmas, Stem tissue, Woody plants

1. Introduction

Detection and diagnosis of phytoplasma diseases is based primarily upon selective amplification of phytoplasma DNA by polymerase chain reaction (PCR) assays that require nucleic acids or DNA preparations from infected plants to perform these analyses. Phytoplasmas that infect woody perennial plants are most difficult to detect consistently due to their low concentrations and irregular distribution within host tissues (1–4). Detection success may also be influenced by other factors such as seasonal fluctuations in phytoplasma populations within affected plants as well as the particular host species, organ, or tissues selected for analysis (4–8). Thus decisions about how, where, and when to sample plants are vital considerations during etiological investigation of suspected cases of phytoplasma disease.

Matt Dickinson and Jennifer Hodgetts (eds.), *Phytoplasma: Methods and Protocols*, Methods in Molecular Biology, vol. 938,
DOI 10.1007/978-1-62703-089-2_13, © Springer Science+Business Media, LLC 2013

No universal sampling protocol has been devised that can be applied to all plant species with equal efficiency (1, 9–13). Typically, phytoplasmas are found most reliably in tissues rich in functional phloem. These include immature leaf bases, developing inflorescences or young roots of arborescent monocots (3, 14–16), and buds, young shoots, leaf petioles and midveins, peduncles, sapwood scrapings, or roots of woody dicots (1, 5, 7, 9, 17, 18). Pooling multiple samples of each tissue type from the same affected host plant at different times during the growing season is a recommended strategy to compensate for fluctuations in pathogen concentrations and distribution (1, 10, 13).

Several different protocols for obtaining DNA from phytoplasma-infected tissues suitable for analysis by PCR have been devised (17–22). For many plant species, pathogen detection is often performed on total DNA extracts derived directly from fresh, frozen, or otherwise preserved host tissues after preparatory grinding of tissues in extraction buffer containing a strong detergent, antioxidants, and polyvinylpyrollidone (PVP) to complex plant polyphenols (20, 21). For challenging hosts for which detection success is not assured, DNA extraction is usually preceded by an enrichment step (17, 21, 23) to increase the proportion of phytoplasma DNA while reducing or eliminating unwanted host DNA and other plant constituents (i.e., polysaccharides and secondary metabolites) in final preparations. Enrichment processing is recommended for fresh tissues only and may require inclusion of further host species-dependent modifications such as an incubation step with proteinase K (8) or by performing enrichment processing at 25°C (24) rather than 4°C as recommended (17) to acquire DNA preparations of acceptable quality. A final purification of DNA preparations using a commercial kit designed for this purpose is a prudent step prior to quantification and use of the DNA as template in PCRs.

Diagnosis of lethal yellowing and similar fatal phytoplasma diseases of palms has traditionally relied upon sampling and analysis of immature tissue of the stem apex (4). As young palms rarely contract these diseases, affected palms are often large in stature and must be felled in order to readily access and remove these tissues. This approach to sampling is rarely feasible in most urban and suburban landscape settings and is impractical in situations where multiple diseased palms are involved. The following protocol for pathogen detection in stem tissues by nested PCR assay offers an efficient and reliable alternative strategy that we (25, 26), and other investigators (3, 16, 27), have used successfully for confirming palm phytoplasma diseases.

2. Materials

2.1. Sampling Kit

1. Hand-held GPS unit, e.g., Garmin *etrex* Legend.
2. Digital camera, e.g., Olympus Stylus Tough, 10 megapixel.
3. Cordless drill driver with rechargeable batteries, e.g., DeWalt DC720, 18 V.
4. 5/16th in. (8 mm) diameter, 6.5 in. (16.5 cm) long multipurpose drill bits.
5. Portable propane torch.
6. 2–4 in. (5–10 cm) lengths of 3/8 in. (1 cm) wooden dowel.
7. Small hammer.
8. Clean resealable plastic bags (1 L size).
9. Polypropylene wash bottle(s) filled with deionized water.
10. Portable cooler with crushed ice.

2.2. DNA Extraction Buffer Components

Prepare all solutions with deionized or distilled water unless otherwise specified. Diligently follow all regulations for safe disposal of waste materials or solutions.

1. DNA extraction buffer: 2% CTAB, 100 mM Tris–HCl, pH 8.0, 1.4 M NaCl, 20 mM EDTA, pH 8.0, 1% PVP-40, 1% 2-mercaptoethanol. Begin by preparing a stock 0.5 M EDTA buffer by adding 186.1 g of disodium EDTA·$2H_2O$ to 800 mL of deionized water. Stir vigorously on a magnetic stirrer. Adjust the pH to 8.0 with NaOH (~20 g of NaOH pellets). Adjust the volume to 1 L with water. Sterilize the buffer by autoclaving 121°C for 20 min. Prepare a stock 1 M solution of Tris buffer by dissolving 121.1 g of Tris base in 800 mL of deionized water. Adjust the pH of the solution to 8.0 by adding ~42 mL of concentrated HCl. Adjust the volume of the buffer to 1 L with water and sterilize by autoclaving for 20 min. Store both stock buffer solutions at room temperature. Prepare the DNA extraction buffer by adding the following ingredients to 600 mL of deionized water with vigorous stirring: add 10 g of PVP-40 (soluble form), 100 mL of 1 M Tris pH 8.0 buffer, 40 mL of 0.5 M EDTA pH 8.0, and 81.8 g of NaCl. Pour the extraction buffer into a graduated glass (Pyrex) screw cap bottle. Add 20 g of CTAB (hexadecyltrimethylammonium bromide). Adjust volume of the buffer to 1 L with water. Cap the bottle, briefly shake the contents then sterilize by autoclaving for 20 min. Once the buffer has cooled to room temperature, place the bottle in a fume hood and add 10 mL of 2-mercaptoethanol. Store the buffer at room temperature.

2. Chloroform:isoamyl alcohol (24:1 v/v): 192 mL chloroform, 8 mL isoamyl alcohol. Store the mixture in a clean brown glass bottle in a fume hood at 25°C.
3. 2-propanol (also known as isopropanol).
4. 95% ethanol.
5. 70% ethanol.
6. TE buffer, pH 8.0: 50 mM Tris, 5 mM EDTA, pH 8.0. To 94 mL of deionized water add 5 mL of 1 M Tris buffer pH 8.0 and 1 mL of 0.5 M EDTA buffer pH 8.0. Sterilize by autoclaving for 20 min. Store at 25°C.
7. TES buffer: 10 mM Tris, 1 mM EDTA, 1 M NaCl, pH 8.0. Dissolve 29.22 g of NaCl in 250 mL of deionized water. Add 5 mL of 1 M Tris buffer, pH 8.0 and 1 mL of 0.5 M EDTA buffer pH 8.0. Adjust the volume to 500 mL with water and then sterilize by autoclaving for 20 min. Store at 25°C.
8. 50 mL polypropylene Oak Ridge centrifuge tubes.
9. Sterile 15 mL Falcon polypropylene tubes.
10. DNA-OFF solution (Clontech Laboratories, Inc.).
11. Water bath at a constant temperature setting of 65°C.
12. Bench top centrifuge with 6 × 100 mL fixed angle rotor, e.g., Sorvall Legend Mach 1.6.
13. Wizard DNA Clean-Up System (Promega).

2.3. Gel Electrophoresis Components

1. 50× Tris-acetate (TAE) buffer: Dissolve 242 g of Tris base in 500 mL by vigorous stirring. Add 100 mL of 1 M of 0.5 M EDTA buffer, pH 8.0, and 57.1 mL of glacial acetic acid. Adjust the volume to 1 L with water. Store at 25°C.
2. 1% agarose gel: Add 1 g of molecular biology grade (low EEO/multipurpose) agarose and 100 mL of a 1:50 dilution of the 50× TAE buffer stock to a 250 mL glass flask. Weigh the flask and contents. Loosely cap the flask with an inverted 150 mL glass beaker and heat in a microwave until the agarose is fully dissolved. Use protective gloves to remove the flask. Weigh the flask and contents again. Adjust any weight loss by adding deionized water.
3. Gel loading buffer: 0.25% bromophenol blue, 0.25% xylene cyanol FF, 15% Ficoll in water. Store at 25°C.
4. Ethidium bromide stock solution (EtBr, 10 mg/mL): 1 g ethidium bromide in 100 mL of deionized water. Store the solution in a 250 mL dark-colored polypropylene screw cap bottle at 25°C (see Note 1).
5. Agarose gel electrophoresis tank, e.g., Wide Mini-Sub Cell GT System (Bio-Rad).

6. Double-stranded DNA molecular weight markers, e.g., *Hin*dIII-digested phage lambda DNA.
7. Polyacrylamide gel vertical electrophoresis system, e.g., for gels 16 × 14 × 0.08 cm.
8. 5× Tris-borate EDTA (TBE) buffer: 54 g Tris base, 27.5 g boric acid and 20 mL of 0.5 M EDTA, pH 8.0 added to 700 mL of deionized water. Stir vigorously until all components are dissolved. Adjust the final volume to 1 L with deionized water.
9. Polyacrylamide stock solution: 29% acrylamide, 1% *N,N′*-methylene bisacrylamide. Add 29 g of acrylamide and 1 g of *N,N′*-methylene bisacrylamide to 50 mL of deionized water. Heat the solution to 37°C with continuous stirring to dissolve the chemicals. Adjust the volume to 100 mL with deionized water. Store the solution in a dark-colored polypropylene bottle at 4°C (see Note 2).
10. pGEM double-stranded DNA markers (Promega).
11. UV transilluminator.
12. Gel documentation system, e.g., Kodak EDAS 290.

2.4. PCR Analytical Components

Prepare all solutions with filter sterile molecular biology grade water unless indicated otherwise.

1. *Taq* DNA polymerase, 5 U/μL with 10× polymerase reaction buffer.
2. 100 mM dNTP stock solutions.
3. Sterile 0.2 mL thin-walled PCR tubes.
4. Programmable thermal cycler with hot lid, e.g., PTC-100 (MJ Research, Inc.).
5. DNA fluorometer, e.g., Qubit™ (Invitrogen).
6. Wizard PCR Preps Purification system (Promega).

3. Methods

3.1. Stem Sampling (Fig. 1)

1. Attach a multipurpose drill bit to a cordless electric drill (see Note 3). Flame sterilize the bit using a propane torch and then cool it with a stream of deionized water.
2. Begin sampling by drilling a shallow pilot hole in the lower stem to remove the outermost layer of pseudobark. Discard these tissues (see Note 4).
3. Resume drilling incrementally through the pilot hole into the interior stem to a final depth of ~15 cm while using a

Fig. 1. Stem sampling for DNA-based diagnosis phytoplasma diseases in large arborescent monocots. (**a**) An inventory of basic equipment used for removing samples of interior stem tissues from palms and other arborescent monocots; (**b**) pseudobark shavings from drilling a superficial pilot hole are a potential source of nontarget microbes and should be discarded; (**c**) removal and collection of interior stem shavings into a clean sealable plastic bag, should be labeled with the sample identity and stored on ice in a cooler to reduce tissue discoloration during transport to the laboratory. (**d**) At least 3 g of shavings can be obtained by drilling a 6 in. deep hole using a 6.5 in. long (5/16th dia.) bit; (**e**) residual stem tissues usually cling to the bit grooves after removing a palm stem sample; (**f**) tissues are easily washed from the bit using a stream of water applied from a polypropylene squirt bottle. (**g**) Flame sterilization with a propane torch followed by cooling the bit with an additional application of water is essential before sampling the next palm to avoid carry over of DNA from the previous sample. (**h**) Tightly sealing the sampling hole with a piece of wooden dowel prevents copious stem bleeding and provides a visual indicator that a sample has been removed from the palm; (**i**) for palm stems tightly encased by old leaf bases or a thick layer of pseudobark, a 9 in. long wood boring bit is recommended for sampling.

Fig. 1. (continued).

back-and-forth motion to dislodge shavings. Passively collect shavings into a clean sealable plastic freezer bag positioned at a safe distance below the rim of the drill hole (see Note 5).

4. Once sampling is completed (see Note 6), the stem can be sealed, if necessary, by tapping a piece of wooden dowel into the hole. This will prevent copious sap bleeding and provide a barrier against invasion by pests.
5. Use water to thoroughly wash residual tissues from the drill bit. Flame sterilize and cool the bit again before sampling the next palm.
6. Document the location (GPS coordinates) and image of each sampled palm species. Store packaged samples on ice during field collection and at 4°C for no longer than 24–48 h prior to laboratory processing (see Note 7).

3.2. Sample DNA Extraction

1. Add 3 g of tissue sample in a sterile polypropylene Oak Ridge tube (see Note 8) with 15 mL of CTAB extraction buffer. Cap and mix the tube contents. Place the tube in a 65°C water bath for 30 min. Shake the tube contents a few times during the course of incubation (see Note 9).

2. Remove tube from water bath and allow the tube to cool to 25°C before adding 15 mL of chloroform:isoamyl alcohol. Cap the tube tightly and shake it vigorously for 5–10 s to emulsify the contents.
3. Centrifuge the tube for 10,000 × *g* for 15 min at 25°C to pellet tissue debris. Transfer the upper aqueous phase to a second sterile Oak Ridge tube using a sterile disposable polypropylene transfer pipette.
4. Repeat steps 2 and 3 once more.
5. Transfer the aqueous layer to another sterile Oak Ridge tube. Add 10 mL of 2-propanol. Cap the tube and invert several times to mix the contents. Incubate the tube at 25°C for 30 min (see Note 10).
6. Centrifuge the tube at 10,000 × *g* for 10 min to pellet the nucleic acids. Carefully decant the 2-propanol from the visible nucleic acid pellet. Add 1–3 mL of TES buffer to resuspend the pellet.
7. Transfer the nucleic acid solution to a sterile 15 mL disposable polypropylene centrifuge tube. Add 2.5 volumes of 95% ethanol. Cap and invert the tube several times to gently mix the contents. Pellet the precipitated nucleic acid by centrifugation for 10 min at 10,000 × *g*.
8. Carefully remove the liquid and briefly wash the visible pellet by adding 0.5 mL of 70% ethanol to the tube. Centrifuge the tube at 10,000 × *g* for 2 min, decant the ethanol, invert the tube to air dry the pellet for 15 min. Resuspend the pellet in 200 μL of TE buffer pH 8.0 (see Note 11).
9. Purify 50 μL aliquots of the crude nucleic acid sample on spin columns. Elute DNA from each column in 50 μL of warm (65°C), sterile ultrapure water or TE buffer pH 8.0.
10. Each purified DNA sample should be quantified by comparison with known amounts of genomic DNA (e.g., lambda phage DNA) using standard agarose gel electrophoresis, by spectrophotometry (A_{260}), or by fluorometry.

3.3. PCR Analysis

Refer to Chapters 14, 15, 16, and 17 for more specific detail.

1. For each 50 μL PCR mixture, add 50–100 μg of sample DNA template.
2. Mix appropriate stock solutions of PCR reagents to a final concentration of 10 mM Tris–HCl, 50 mM KCl, 1.5 mM $MgCl_2$, 200 μM each NTP, 0.15 μM each primer, 1.25 U *Taq* DNA polymerase, and sterile ultrapure water (see Note 12).
3. After 35 cycles of PCR, electrophorese 10 μL of each final reaction mixture through a 1% agarose gel using 1× TAE as running buffer.

4. Stain PCR products by immersing the gel in a 5 μg/mL EtBr solution for 5 min. Destain the gel by immersion in 1 L of deionized water for 15 min.
5. Visualize PCR products by UV transillumination of the gel and document the results.
6. For nested PCR assays, prepare a 1:10 dilution of each primary PCR product with sterile ultrapure water. Use 2 μL of each dilution as template during 35 cycles of re-amplification with nested primers.
7. Repeat steps 3–5 with products resulting from nested PCR amplification.

3.4. Phytoplasma Identification

Refer to Chapters 14, 27, and 28 for more specific detail.

1. For identification of rDNA amplified from palm-associated phytoplasmas, mix 5 μL of each nested PCR product with 8 μL of sterile deionized water, 0.5 μL (5 U) of restriction endonuclease (see Note 13), and 1.5 μL of 10× endonuclease buffer. Incubate overnight at an appropriate temperature to suit the restriction enzyme (generally 37°C).
2. Add an additional 0.5 μL (5 U) of restriction endonuclease with 0.5 μL of 10× endonuclease buffer and 4 μL of sterile water to each digest. Continue incubation for 4 h before adding 2 μL of 6× gel loading buffer.
3. Load half (11 μL) of each digest to a well of an 8% non-denaturing polyacrylamide gel. Separate the products of each digest by electrophoresis at 250 V using 1× TBE as running buffer.
4. Stain DNA fragments by immersion of the gel in an aqueous 0.5 μg/mL EtBr solution for 15 min. Destain in water for 15 min. Visualize DNA fragments by UV transillumination and document the restriction fragment profiles.
5. Purify residual PCR products using a suitable commercial kit (e.g., Wizard PCR Preps) designed for this purpose. Quantify and sequence the purified products (see Note 14).

4. Notes

1. EtBr is a powerful mutagen and is moderately toxic. Gloves should be worn when working with solutions that contain this dye.
2. Acrylamide is a potent neurotoxin and is absorbed through the skin. The effects of acrylamide are cumulative. Wear suitable gloves and a mask when weighing powdered acrylamide and methylene bisacrylamide or handling solutions of these chemicals. Polyacrylamide is considered nontoxic but should

be handled with care because it might contain small quantities of unpolymerized acrylamide.

3. A carpenter's brace or increment borer can be used in place of a cordless electric drill.
4. The dead outermost layer of pseudobark varies in thickness according to the age, condition, and species of palms. These dead tissues may contribute DNA sources from nontarget Gram-positive bacteria that may interfere with the specificity of PCR assays that incorporate phytoplasma universal rRNA gene primer pairs.
5. For coconut or other host palms with minimal pseudobark, a hole drilled to a depth of 15 cm provides at least 3 g of stem shavings. Slow-growing palms such as *Sabal palmetto* retain remnant leaf bases tightly attached to stems for several years once lowermost leaves are shed. This species may require use of an auger-style drill bit of at least 20 cm length to extract adequate samples of interior stem tissues. Samples yield sufficient DNA for diagnostic and archival purposes.
6. The vasculature of monocots, combined with enhanced sensitivity of nested PCR assays, makes single-point sampling of palms possible. Reliable pathogen detection is unaffected by the location of stem sampling sites. Excision of tissues from the basal stem is cosmetically desirable for palms in ornamental or amenity plantings with questionable cases of disease.
7. Stem samples shipped within 24 h after harvest by overnight courier to laboratories for diagnostic testing show no apparent deterioration in quality due to growth of secondary microbial contaminants. Samples from remote locations requiring extended travel times to reach diagnostic laboratories should be preserved for shipment by immersion in CTAB extraction buffer.
8. Oak Ridge tubes are recycled for DNA extractions. Between extractions treat each tube with DNA-OFF solution according to the manufacturer's recommendations to remove any traces of contaminating DNAs. Wash all treated tubes and caps with two to three changes of deionized water. Drain, dry, and reassemble tubes then sterilize by autoclaving for 20 min.
9. Preparatory grinding of fresh tissues shavings with a mortar and pestle is unnecessary as it does not result in any appreciable increase in DNA yields.
10. Strands of nucleic acid may not visible in solution at this stage although the mixture will be noticeably turbid.
11. Store tubes at 4°C for at least 2 h or overnight to passively resuspend nucleic acid pellets. Gentle aspiration of tube

contents with a sterile, disposable polypropylene pipette may be required to fully resuspend the pellet.

12. Detection of phytoplasmas associated with lethal yellows-type diseases of palms is most reliably accomplished by PCR assays incorporating phytoplasma group-specific, rather than phytoplasma "universal," rRNA gene primer pairs. The latter primers may amplify rDNA products from nontarget bacteria present in or on palm stem tissues. Sequencing of rDNA product(s) derived solely from diseased palms is mandatory for unequivocal disease diagnosis.
13. A comparative analysis of restriction fragment profiles resulting from separate digestion of each PCR product with *Alu*I, *Dde*I, *Hha*I, *Hin*fI, and *Mse*I endonucleases is usually sufficient to collectively identify each of the various palm phytoplasma groups and subgroup strains.
14. Direct sequencing of purified products may be achieved with the same primers used to amplify the products during PCR assay. Resulting sequence data may be compared for homology with other phytoplasma nucleotide sequences archived in the GenBank database.

References

1. Boudon-Padieu E et al (2003) Grapevine yellows: comparison of different procedures for DNA extraction and amplification with PCR for routine diagnosis of phytoplasmas in grapevine. Vitis 42:141–149
2. Christensen NM et al (2004) Distribution of phytoplasmas in infected plants as revealed by real-time PCR and bioimaging. Mol Plant Microbe Interact 17:1175–1184
3. Oropeza C et al (2011) Phytoplasma distribution in coconut palms affected by lethal yellowing disease. Ann Appl Biol 159:109–117
4. Thomas DL, Norris RC (1980) The use of the electron microscope for lethal yellowing diagnosis. Proc Fla State Hortic Soc 93:196–199
5. Baric S et al (2011) Seasonal colonisation of apple trees by '*Candidatus* Phytoplasma mali' revealed by a new quantitative TaqMan real-time PCR approach. Eur J Plant Pathol 129:455–467
6. Berges R, Rott M, Seemüller E (2000) Range of phytoplasma concentrations in various plant hosts as determined by competitive polymerase chain reaction. Phytopathology 90:1145–1152
7. Jarausch W, Lansac M, Dosba F (1999) Seasonal colonization pattern of European stone fruit yellows phytoplasma in different *Prunus* species detected by specific PCR. J Phytopathol 147:47–54
8. Marzachì C (2006) Molecular diagnosis of phytoplasmas. Arab J Plant Prot 24:139–142
9. Gibb K, Padovan A (1994) A DNA extraction method that allows reliable PCR amplification of MLO DNA from "difficult" plant host species. Genome Res 4:56–58
10. Green MJ, Thompson DA, MacKenzie DJ (1999) Easy and efficient DNA extraction from woody plants for the detection of phytoplasmas by polymerase chain reaction. Plant Dis 83:482–485
11. Heinrich M et al (2001) Improved detection methods for fruit tree phytoplasmas. Plant Mol Biol Rep 19:169–179
12. Lee I-M, Davis RE (1983) Phloem-limited prokaryotes in sieve elements isolated by enzyme treatment of diseased plant tissues. Phytopathology 73:1540–1543
13. Palmano S (2001) A comparison of different phytoplasma DNA extraction methods using competitive PCR. Phytopathol Mediterr 40:99–107
14. Cordova I et al (2000) First report of a phytoplasma-associated leaf yellowing syndrome of Palma Jipi plants in southern Mexico. Plant Dis 84:807

15. Thomas DL, Donselman HM (1979) Mycoplasmalike bodies and phloem degeneration associated with declining *Pandanus* in Florida. Plant Dis Rep 63:911–916
16. Vázquez-Euán R et al (2011) Occurrence of a 16SrIV group phytoplasma not previously associated with palm species in Yucutan, Mexico. Plant Dis 95:256–262
17. Kirkpatrick BC et al (1995) Isolation of mycoplasma-like organism DNA from plant and insect hosts. In: Razin S, Tully JG (eds) Molecular and diagnostic procedures in mycoplasmology Vol. 1, molecular characterization. Academic, New York, pp 105–117
18. Prince JP et al (1993) Molecular detection of diverse mycoplasmalike organisms (MLOs) associated with grapevine yellows and their classification with aster yellows, X-disease, and elm yellows MLOs. Phytopathology 83:1130–1137
19. Ahrens U, Seemüller E (1992) Detection of DNA of plant pathogenic mycoplasmalike organisms by a polymerase chain reaction that amplifies a sequence of the 16S rRNA gene. Phytopathology 82:828–832
20. Doyle JJ, Doyle JL (1990) Isolation of plant DNA from fresh tissue. Focus 12:13–15
21. Kollar A et al (1990) Isolation of the DNA of various plant pathogenic mycoplasmalike organisms from infected plants. Phytopathology 80:233–237
22. Zhang Y-P, Uyemoto JK, Kirkpatrick BC (1998) A small-scale procedure for extracting nucleic acids from woody plants infected with various phytopathogens for PCR assay. J Virol Methods 71:45–50
23. Jiang YP, Chen TA (1987) Purification of mycoplasma-like organisms from lettuce with aster yellows disease. Phytopathology 77:949–953
24. Harrison NA et al (2001) Detection and characterization of an elm yellows (16SrV) group phytoplasma infecting Virginia creeper plants in southern Florida. Plant Dis 85: 1055–1062
25. Harrison NA, Helmick EE, Elliott ML (2008) Lethal yellowing-type diseases of palms associated with phytoplasmas newly identified in Florida, USA. Ann Appl Biol 153:85–94
26. Harrison NA, Womack M, Carpio ML (2002) Detection and characterization of a lethal yellowing (16SrIV) group phytoplasma in Canary Island date palms affected by lethal decline in Texas. Plant Dis 86:676–681
27. Nejat N et al (2009) Phytoplasmas associated with disease of coconut in Malaysia: phylogenetic groups and host plant species. Plant Pathol 58:1152–1160

Chapter 14

Nested PCR and RFLP Analysis Based on the 16S rRNA Gene

Bojan Duduk, Samanta Paltrinieri, Ing-Ming Lee, and Assunta Bertaccini

Abstract

Current phytoplasma detection and identification methods are primarily based on nested polymerase chain reaction followed by restriction fragment length polymorphism analysis and gel electrophoresis. These methods can potentially detect and differentiate all phytoplasmas including those previously not described. The present protocol describes the application of this method for identification of phytoplasmas at 16S rRNA (16Sr) group and 16Sr subgroup levels.

Key words: Detection, Identification, Ribosomal group, Ribosomal subgroup

1. Introduction

Phytoplasma identification at 16Sr group level is currently performed by polymerase chain reaction (PCR)/restriction fragment length polymorphism (RFLP) analyses of 16S rDNA gene. The first comprehensive phytoplasma classification scheme was based on RFLP analysis of PCR-amplified 16S rRNA (1, 2), providing a reliable means for the differentiation of a broad array of phytoplasmas. This system allowed classification of phytoplasmas into 19 groups and more than 40 subgroups and has become the most comprehensive and widely accepted phytoplasma classification system so far (3–9). Sensitive and accurate detection of these microorganisms is a prerequisite for the management of phytoplasma-associated diseases. Following their discovery, phytoplasmas have been difficult to detect due to their low concentration in hosts, especially woody hosts, and their erratic distribution in the sieve tubes of the infected plants (10). The establishment of electron microscopy (EM)-based techniques represented an alternative approach to the traditional indexing procedure for phytoplasmas based on graft transmission of the pathogen to healthy indicator

Matt Dickinson and Jennifer Hodgetts (eds.), *Phytoplasma: Methods and Protocols*, Methods in Molecular Biology, vol. 938, DOI 10.1007/978-1-62703-089-2_14, © Springer Science+Business Media, LLC 2013

plants. EM observation (11, 12) and less frequently scanning EM (13) were the only diagnostic techniques available until staining with DNA-specific dyes such as DAPI was developed (14). Later, protocols for the production of enriched phytoplasma-specific antigens were developed, introducing serological-based detection techniques for the study of these pathogens in plants or insect vectors (15).

The first phytoplasma detections based on 16S rDNA were performed in the 1990s (16–20), and now are conducted by different nucleic acid-based techniques, primarily the PCR (21–24). The procedures developed in the last 20 years detect the presence of phytoplasmas in plant material and insect vectors, helping to preventing the spread of the diseases and lessening their economic impact. Several primers designed based on the 16S rDNA sequence that are universal for phytoplasmas or group specific have been developed (Table 1); they can be used in different combinations in direct, nested or semi-nested systems for routine detection of phytoplasmas as well as for identification of new phytoplasmas or new hosts (25–31).

Efforts to improve diagnostic procedures aim to develop quicker, more economic and robust methods. Sensitivity is not an issue *per se*, as the current nested PCR protocols are extremely sensitive, but the achievement of high levels of sensitivity without the risk of false positive results that can be associated with nested PCR is highly desirable. However, because of the highly conserved nature of the 16S rRNA gene and of the not uncommon presence of 16S rDNA interoperon sequence heterogeneity (1, 32–37), some additional tools for phylogenetic analyses and finer strain differentiation have been developed. More variable single copy genes, such as ribosomal proteins (*rpl22* and *rpS3*), *SecY*, *SecA, tuf*, and *GroEL* have been reported to be suitable for differentiation of some phytoplasmas (6, 8, 38–46).

Phytoplasma differentiation is routinely based on 16S rRNA gene sequences, and is carried out by RFLP analysis of PCR amplified DNA sequences using 17 restriction endonuclease enzymes (1, 2) according to the scheme presented in Fig. 1. This procedure allows the identification of the majority of phytoplasmas at the 16Sr group/subgroup level, which is relevant for epidemiology and quarantine purposes. Because the collective RFLP pattern characteristics of each phytoplasma are conserved (for both R16F2/R16R2 and P1/P7 amplicons), unknown phytoplasmas can be identified by comparing the patterns with those of known phytoplasmas, without having to analyze all reference phytoplasmas simultaneously (1, 2, 47–51). In some cases, the sample may contain phytoplasma(s) with the presence of ribosomal RNA interoperon heterogeneity or mixed phytoplasmas.

Table 1
Primers for amplification of the 16Sr DNA of phytoplasmas

Primer[a]	Specificity 16S group	Sequence 5′-3′	Position[b]	References
P1	Universal	AAGAATTTGATCCTGGCTCAGGATT	16S rDNA	(13)
P7	Universal	CGTCCTTCATCGGCTCTT	23S rDNA	(51)
R0	Universal	GAATACCTTGTTACGACTTAACCCC	16S rDNA	(29)
P3	Universal	GGATGGATCACCTCCTT	16S rDNA	(51)
P4	Universal	GAAGTCTGCAACTCGACTTC	16S rDNA	(51)
P5	Universal	CGGCAATGGAGGAAACT	16S rDNA	(51)
R16F2n	Universal	GAAACGACTGCTAAGACTGG	16S rDNA	(29)
R16R2	Universal	TGACGGGCGGTGTGTACAAACCCCG	16S rDNA	(29)
F1	Universal	AAGACGAGGATAACAGTTGG	16S rDNA	(11)
B6	Universal	TAGTGCCAAGGCATCCACTGTG	IS	(47)
R16mF2	Universal	CATGCAAGTCGAACGGA	16S rDNA	(21)
R16mR2	Universal	CTTAACCCCAATCATCGA	16S rDNA	(21)
P1A	Universal	AACGCTGGCGGCGCGCCTAATAC	16S rDNA	(33)
16Sr-SR	Universal	GGTCTGTCAAAACTGAAGATG	IS	(34)
P7A	Universal	CCTTCATCGGCTCTTAGTGC	23S rDNA	(33)
Pc399	Universal	AACGCCGCGTGAACGATGAA	16S rDNA	(53)
Pc1694	Universal	ATCAGGCGTGTGCTCTAACC	IS	(53)
fU5	Universal	CGGCAATGGAGGAAACT	16S rDNA	(39)
rU3	Universal	TTCAGCTACTCTTTGTAACA	16S rDNA	(39)

(continued)

Table 1 (continued)

Primer[a]	Specificity 16S group	Sequence 5′-3′	Position[b]	References
PA2f	Universal	GCCCCGGCTAACTATGTGC	16S rDNA	(23)
PA2r	Universal	TTGGTGGGCCTAAATGGACTC	IS	(23)
M1(758F)	Universal	GTCTTTACTGACGC	16S rDNA	(19)
M2(1232R)	Universal	CTTCAGCTACCCTTTGTAAC	16S rDNA	(19)
SN910601	Universal	GTTTGATCCTGGCTCAGGATT	16S rDNA	(46)
SN910502	Universal	AACCCCGAGAACGTATTCACC	16S rDNA	(46)
R16(I)F1	I, II, XII, XV	TAAAAGACCTAGCAATAGG	16S rDNA	(28)
R16(I)R1	I, II, XII, XV	CAATCCGAACTAAGACTCT	16S rDNA	(28)
R16(III)F2	III	AAGAGTGGAAAAACTCCC	16S rDNA	(28)
R16(III)R1	III	TTCGAACTGAGATTGA	16S rDNA	(28)
R16(V)F1	V	TTAAAAGACCTTCTTCGG	16S rDNA	(28)
R16(V)R1	V	TTCAATCCGTACTGAGACTACC	16S rDNA	(28)
R16(X)F1	X	GACCCGCAAGTATGCTGAGAGATG	16S rDNA	(29)
R16(X)R1	X	CAATCCGAACTGAGAGTCT	16S rDNA	(29)
fO1	X	CGGAAACTTTTAGTTTCAGT	16S rDNA	(39)
rO1	X	AAGTGCCCAACTAAATGAT	16S rDNA	(39)

[a]Protocol is provided here for primers in bold italics
[b]IS, spacer region between 16S and 23S rDNA

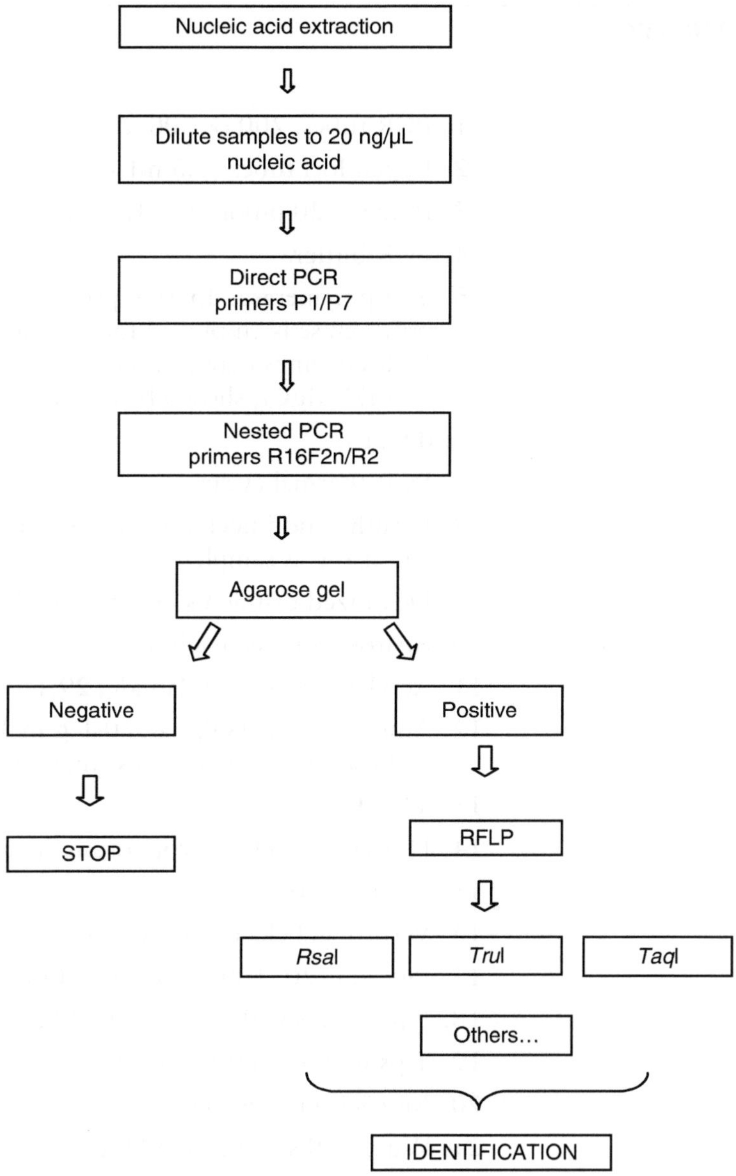

Fig. 1. Phytoplasma identification procedure.

The RFLP patterns will appear unusual, typically with extra DNA fragments, and the sum of total fragment sizes will be greater than that of the expected amplicon. In the case of mixed infections, the use of specific primers, when available, is recommended (Table 1); however cloning the amplicons is also helpful in solving the problem, but this is beyond the scope of this chapter.

2. Materials

1. PCR tubes, 200 or 500 μL.
2. Eppendorf tubes (1.5 mL).
3. Primers: 20 pmol/μL (Table 1).
4. PCR buffer.
5. *Taq* polymerase. In this protocol, Sigma-Aldrich *Taq* DNA polymerase is employed for PCR amplification; however other PCR enzymes may be used. For other *Taq* polymerases usually 25 mM $MgCl_2$ should be added (3 μL for 50 μL mix).
6. dNTPs.
7. PCR thermal cycler.
8. Healthy and infected plant controls preferably of same species and tissue as samples.
9. Deionized distilled sterile water (ddH_2O).
10. Agarose for electrophoresis.
11. Acrylamide/*bis*-acrylamide, 29:1.
12. Ammonium persulphate: 0.1 g in 1 mL ddH_2O, keep at 4°C and use fresh (solution less than 10 days old).
13. TEMED.
14. Horizontal gel electrophoresis apparatus.
15. Power supply.
16. Vertical gel electrophoresis apparatus.
17. Pipettes (20, 100, 200, and 1,000 μL).
18. Pipette tips with filters (20, 100, 200, and 1,000 μL).
19. Tips without filters (20 μL).
20. Restriction enzymes.
21. Water baths or heating blocks.
22. TAE buffer (10×): Trizma base 48.44 g, ethylenediainetetraacetic acid (EDTA) 7.44 g, ddH_2O 500 mL, adjust pH to 8.0 with glacial acetic acid, adjust volume to 1 L with ddH_2O, store at 4°C. Dilute to 1× in ddH_2O (if instructed) prior to use.
23. TBE buffer (10×): Trizma base 108 g, boric acid 55 g, 0.5 M EDTA 40 mL, adjust volume to 1 L with ddH_2O, store at 4°C. Dilute to 1× in ddH_2O (if instructed) prior to use.
24. Floating racks.
25. Loading dye: In this protocol, the Sigma-Aldrich *Taq* DNA polymerase used for PCR amplification has loading dye in the 10× PCR buffer. However, if the PCR buffer does not contain a loading dye, add 1.5 μL of any suitable DNA loading dye per 6 μL of PCR product before running the gel.

26. DNA size ladder: GeneRuler 1 kb DNA ladder (Fermentas) is suitable; however other DNA ladders may be used.
27. Φx174 DNA/*Bsu*RI (*Hae*III) marker (Fermentas). Other DNA ladders of appropriate size may be used.
28. Ethidium bromide: 10 mg/mL stock solution. Keep stock solution in a dark place. Make a 0.5 μg/mL working solution: add 50 μL of stock solution to 1 L distilled H_2O.

3. Methods

3.1. DNA Extraction

Extraction of phytoplasma DNA is explained in details in Chapters 12 and 13. Healthy samples of the analyzed species and negative controls must always be included.

3.2. Reaction Mix Preparation and PCR (See Note 1)

1. Before preparing the PCR reaction mix, label PCR tubes (number of tubes is number of samples + 2, one for negative and one for positive control) (see Note 1). Set aside the last tube for the negative control.
2. The PCR reaction mix should be prepared for all tubes together, following amounts shown in Table 2. Each 25 μL PCR reaction mix contains 20 ng template DNA, 2.5 μL 10× PCR buffer, 0.5 U *Taq* polymerase, 0.2 mM dNTPs (200 μM each dNTP), and 0.4 μM each primer. Samples lacking DNA must be run as negative controls in each PCR reaction (see Note 1).

Table 2
Volumes of PCR reagents, without template DNA, reduced to one half (to reduce the cost per reaction)

Loading order	Reagents	One sample (μL)	Two samples (μL)	Ten samples (μL)
1	ddH_2O	16.35	32.7	163.5
2	Primer # 1 (20 μM)	0.5	1.0	5.0
3	Primer # 2 (20 μM)	0.5	1.0	5.0
4	10× PCR buffer	2.5	5.0	25.0
5	$MgCl_2$ (25 mM)	1.5	3.0	15.0
6	dNTPs (10 mM)	2.5	5.0	25.0
7	*Taq* DNA polymerase (5 U/μL)	0.15	0.3	1.5
	Total	24.0	48.0	240.0

3. Pipette all reagents (excluding the DNA) into an Eppendorf tube and mix well. Aliquot 24 μL of the PCR reaction mix into all the PCR tubes.
4. Add 1 μL of DNA samples (diluted to 20 ng/μL or 1:100 with ddH_2O), then add a mineral oil layer if the thermal cycler does not have a heated lid.
5. Load the tubes into the PCR thermocycler and run the program (see Subheadings 3.2.1 and 3.2.2). For other primers listed in Table 1 use the conditions from the original publication.

3.2.1. Direct Generic PCR

1. The P1/P7 primers are recommended for generic PCR (16, 52). These primers amplify virtually the whole length of the 16S, intergenic 16S–23S, and a small part of the 23S rRNA gene.
2. The reaction mixture contains: 5 μL PCR buffer (10×), 2.5 μL dNTP (10 mM), 1 μL primer P1 (20 μM), 1 μL primer P7 (20 μM), 2 μL *Taq* polymerase (5 U/μL), 2 μL DNA extract (20 ng/μL), ddH_2O to 50 μL (Table 2). It is possible to use a volume of 25 μL, reducing the amount of reagents by half for economy.
3. PCR conditions are as follows: 1 cycle at 95°C for 3 min; 35 cycles of 94°C for 1 min, 50°C for 2 min, 72°C for 3 min; then a final extension of 72°C for 10 min.
4. If the samples are from a herbaceous host, an agarose gel can be run as described below (Subheading 3.3); if from a woody host or insect proceed straight to nested PCR below (Subheading 3.2.2).

3.2.2. Nested Generic PCR

1. If negative results are obtained with P1/P7 primers, or the original material is from wood or insect samples, nested PCR with R16F2n/R16R2 primers (28) must be carried out.
2. Reaction mixtures and PCR cycles are as for the direct PCR (Subheading 3.2.1), but substitute the R16F2n/R16R2 primers in place of P1/P7 primers, using the same amounts. Use 1.0 μL of the product of the direct PCR (P1/P7), diluted 1:30 in ddH_2O in place of the DNA.

3.3. Visualization of PCR Products (Agarose Gels)

1. To prepare 100 mL of a 1% agarose gel (see Note 2): Weigh out 1 g of agarose. Add 100 mL of 1× TAE buffer.
2. Carefully dissolve agarose using a microwave oven.
3. Cool agarose solution to about 40°C. Pour the agarose into a tray, insert comb and remove any air bubbles. Let it solidify (approximately 15 min).
4. Place the gel into the electrophoresis chamber and cover with 1× TAE buffer.

5. Load 6 μL of PCR product (mix PCR product and loading dye if not already present) into the wells. Always add the DNA ladder (2.5 μL) to a well at the edge of the gel.
6. Connect the electrodes to the power source (5 V/cm) and run the gel, stopping it before the loading dye leaves the gel.
7. Stain the gel by soaking it in ethidium bromide working solution for 10–15 min, then wash in water for 10–15 min.
8. Observe the gel on a transilluminator under UV light. Take a picture and store the print or electronic file in a laboratory book or computer.
9. Only PCR products visible on the agarose gel should undergo RFLP analysis with the informative restriction enzymes (Subheading 3.4).

3.4. RFLP Analyses for Group/Subgroup Identification (See Note 3)

1. Use 2–15 μL of PCR product (see Note 4) in the selected enzyme mix made according to the manufacturers' instructions.
2. Incubate at the recommended temperature in a water bath or heating blocks (commonest temperatures are 37°C or 65°C) for at least 16 h (see Note 5).

3.5. Visualization of RFLP Products

The choice between agarose and polyacrylamide gel for RFLP analyses depends on the length of expected fragments; for a clear separation of short fragments (less than 300 bp), a polyacrylamide gel is necessary.

3.5.1. Agarose Gels

1. To prepare 100 mL of a 3% agarose gel (see Note 2), weigh out 3 g of agarose. Add 100 mL of 1× TBE buffer.
2. Carefully dissolve the agarose in a microwave oven.
3. Cool the agarose solution to about 40°C. Pour the agarose into a tray, insert comb and remove any air bubbles. Let it solidify (approximately 15 min).
4. Place the gel into the electrophoresis chamber and cover it with 1× TBE buffer.
5. Load the total volume of digested product into the wells. Add 1.5 μL of ϕx174 DNA/*Bsu*RI (*Hae*III) marker in the outside lane of the gel.
6. Connect the electrodes to the power source (5 V/cm) and run the gel until the loading dye is about to leave the gel.
7. Stain the gel by soaking it in ethidium bromide working solution for 20–25 min, then wash in water for 20–25 min.
8. Observe the gel on a transilluminator under UV light. Take a picture and store the print or electronic file in a laboratory book or computer.

3.5.2. Polyacrylamide Gels

1. To prepare 25 mL of 7% polyacrylamide gel (see Note 6) combine the reagents in a clean glass beaker and in the following order:

ddH_2O	17.83 mL
10× TBE	2.5 mL
Acrylamide/bis-acrylamide solution	4.37 mL
Ammonium persulphate	310 μL
TEMED	15.6 μL

2. Gently mix and pour the mix between the glass plates. Add the comb and leave for 20 min to polymerize.
3. Take out the comb. Fix the glass plates containing the gel in the vertical electrophoresis apparatus.
4. Cover with 1× TBE. Gently clean the wells by rinsing with 1× TBE with an appropriate plastic pipette. Remove air bubbles at the bottom between the glass plates.
5. Load the complete volume of digested mix into the wells.
6. Load 1.5 μL of ϕx174 DNA/*Bsu*RI (*Hae*III) marker in an outside lane of the gel.
7. Connect the electrodes to the power source (7 V/cm) and run the gel until the loading dye is about to leave the gel.
8. Stain the gel by soaking it in ethidium bromide working solution for 10–15 min, then wash in water for 1–2 min.
9. Observe the gel on a transilluminator under UV light. Take a picture and store the print or electronic file in a laboratory book or computer.

4. Notes

1. At all stages while setting up PCR reactions the following precautions should be taken to avoid contamination of samples and reagents:
   - Use only dedicated pipettes for preparing PCR reagents which are never used for DNA manipulations.
   - Use tips with filters (for agarose gel electrophoresis tips with filters are not necessary unless processing samples to be employed in nested PCR afterwards) and use a new tip for each pipetting step.
   - Close reagent and sample tubes immediately after an aliquot has been removed.

- Wash hands carefully with soap any time they become contaminated (i.e., when you leave the room, pick something up off the floor).
- Use only clean, sterile plasticware.
- Each of the following procedures must be performed in separate locations with dedicated laboratory equipment (pipettes, tips, tubes, racks for tubes):

  Reaction mix preparation (without DNA)

  Addition of DNA (in a UV-sterilizable chamber)

  Amplification of the target sequence

  Analysis of PCR products

2. Correctly measure the amount of agarose gel needed in relation to the capacity of the tray; for a small tray, 80 mL should be adequate, but for larger trays approximately 120 mL or more will be needed.
3. The 17 restriction enzymes that are used for differentiation of ribosomal groups and subgroups are: *Alu*I, *Bam*HI, *Bfa*I, *Dra*I, *Eco*RI, *Hae*III, *Hha*I, *Hin*fI, *Hpa*I, *Hpa*II, *Kpn*I, *Mse*I (*Tru*1I), *Rsa*I, *Sau*3AI, *Ssp*I, *Taq*I, and *Tha*I. However, some of the most informative restriction enzymes (*Alu*I, *Hha*I, *Hpa*II, *Mse*I, *Rsa*I, *Taq*I) can be used for preliminary classification, because they can distinguish most of the ribosomal groups, and if needed additional enzymes can be employed (following figures and tables in appropriate literature listed in the introduction).
4. To determine the amount of PCR product to use for restriction enzyme digestion, consider the agarose gel band intensity. Very strong bands only need 1–2 μL in RFLP, medium intensity bands about 3–6 μL, low intensity bands 10–15 μL or more. Total volume should not exceed 20 μL or the well capacity.
5. Fast restriction enzymes are also available from some manufactures which reduce the digestion time to a few minutes. Beware that the enzymes can become inactivated with frequent defrosting so it may be prudent to prepare aliquots of smaller volumes.
6. The amount of the reagents needed for the polyacrylamide gel preparation depends on the size of the glass plates and thickness of the gel.

References

1. Lee I-M et al (1998) Revised classification scheme of phytoplasmas based an RFLP analyses of 16S rRNA and ribosomal protein gene sequences. Int J Syst Bacteriol 48:1153–1169
2. Lee I-M, Gundersen-Rindal DE, Bertaccini A (1998) Phytoplasma: ecology and genomic diversity. Phytopathology 88:1359–1366
3. Al-Saady NA et al (2008) '*Candidatus* Phytoplasma omanense', a phytoplasma associated with witches' broom of *Cassia italica* (Mill.) Lam. in Oman. Int J Syst Evol Microbiol 58:461–466
4. Arocha Y et al (2005) '*Candidatus* Phytoplasma graminis' and '*Candidatus* Phytoplasma caricae', two novel phytoplasmas associated with diseases of sugarcane, weeds and papaya in Cuba. Int J Syst Evol Microbiol 55: 2451–2463
5. Bertaccini A, Duduk B (2009) Phytoplasma and phytoplasma diseases: a review of recent research. Phytopathol Mediterr 48:355–378
6. Lee I-M et al (2004) '*Candidatus* Phytoplasma asteris', a novel taxon associated with aster yellows and related diseases. Int J Syst Bacteriol 54:1037–1048
7. Lee I-M et al (2004) Classification of phytoplasma strains in the elm yellows group (16SrV) and proposal of '*Candidatus* Phytoplasma ulmi' for the phytoplasma associated with elm yellows. Int J Syst Evol Microbiol 54: 337–347
8. Lee I-M, Zhao Y, Bottner KD (2006) *SecY* gene sequence analysis for finer differentiation of diverse strains in the aster yellows phytoplasma group. Mol Cell Probes 20:87–91
9. Montano HG et al (2001) '*Candidatus* Phytoplasma brasiliense', a new phytoplasma taxon associated with hibiscus witches' broom disease. Int J Syst Evol Microbiol 51: 1109–1118
10. Berges R, Rott M, Seemüller E (2000) Range of phytoplasma concentration in various plant hosts as determined by competitive polymerase chain reaction. Phytopathology 90: 1145–1152
11. Bertaccini A, Marani F (1982) Electron microscopy of two viruses and mycoplasma-like organisms in lilies with deformed flowers. Phytopathol Mediterr 21:8–14
12. Cousin MT, Sharma AK, Isra S (1986) Correlation between light and electron microscopic observations and identification of mycoplasma-like organisms using consecutive 350 nm think sections. J Phytopathol 115: 368–374
13. Haggis GH, Sinha RC (1978) Scanning electron microscopy of mycoplasma-like organisms after freeze fracture of plant tissues affected with clover phyllody and aster yellows. Phytopathology 68:677–680
14. Seemüller E (1976) Investigation to demonstrate mycoplasma-like organisms in diseases plants by fluorescence microscopy. Acta Hortic 67:109–112
15. Hobbs HA, Reddy DVR, Reddy AS (1987) Detection of a mycoplasma-like organism in peanut plants with witches' broom using indirect enzyme-linked immunosorbent assay (ELISA). Plant Pathol 36:164–167
16. Deng SJ, Hiruki C (1991) Amplification of 16S ribosomal-RNA genes from culturable and nonculturable mollicutes. J Microbiol Methods 14:53–61
17. Firrao G, Gobbi E, Locci R (1993) Use of polymerase chain reaction to produce oligonucleotide probes for mycoplasma-like organism. Phytopathology 83:602–607
18. Lee I-M et al (1993) Universal amplification and analysis of pathogen 16S rDNA for classification and identification of mycoplasma-like organisms. Phytopathology 83:834–842
19. Namba S et al (1993) Detection and differentiation of plant-pathogenic mycoplasma-like organisms using polymerase chain reaction. Phytopathology 83:786–791
20. Schneider B et al (1993) Classification of plant-pathogenic mycoplasma-like organisms using restriction site analysis of PCR-amplified 16S rDNA. J Gen Microbiol 139:519–527
21. Baric S, Dalla VJ (2004) A new approach to apple proliferation detection: a highly sensitive real-time PCR assay. J Microbiol Meth 57:135–145
22. Green MJ, Thompson DA, MacKenzie DJ (1999) Easy and efficient DNA extraction from woody plants for the detection of phytoplasmas by polymerase chain reaction. Plant Dis 83:482–485
23. Schaff DA, Lee I-M, Davis RE (1992) Sensitive detection and identification of mycoplasma-like organisms by polymerase chain reactions. Biochem Biophys Res Comm 186:1503–1509
24. Zhang Y, Uyemoto JK, Kirkpatrick BC (1998) A small-scale procedure for extracting nucleic acids from woody plants infected with various phytopathogens for PCR assay. J Virol Meth 71:45–50
25. Alvarez E et al (2009) Characterization of a phytoplasma associated with frogskin disease in cassava. Plant Dis 93:1139–1145
26. Cozza R et al (2008) Molecular identification of '*Candidatus* Phytoplasma asteris' inducing

histological anomalies in *Silene nicaeensis*. Phytoparasitica 36:290–293

27. Duduk B et al (2004) Identification of phytoplasmas associated with grapevine yellows in Serbia. J Phytopathol 152:575–579
28. Lee I-M et al (1995) Detection of multiple phytoplasmas in perennial fruit trees with decline symptoms in Italy. Phytopathology 85: 728–735
29. Lorenz KH et al (1995) Detection of the apple proliferation and pear decline phytoplasmas by PCR amplification of ribosomal and nonribosomal DNA. Phytopathology 85:771–776
30. Montano HG et al (2011) Hibiscus witches' broom disease associated with different phytoplasmas taxa in Brazil. Bull Insectol 64:S249–S250
31. Tolu G et al (2006) Identification of 16SrII-E phytoplasmas in *Calendula arvensis* L., *Solanum nigrum* L. and *Chenopodium* spp. Plant Dis 90:325–330
32. Davis RE et al (2003) Differential amplification of sequence heterogenous ribosomal RNA genes and classification of the '*Fragaria multicipita*' phytoplasma. Microbiol Res 158: 229–236
33. Duduk B et al (2007) Identification of phytoplasmas belonging to aster yellows ribosomal group in vegetables in Serbia. Bull Insectol 60:341–342
34. Duduk B et al (2009) Multigene analysis for differentiation of aster yellows phytoplasmas infecting carrots in Serbia. Ann Appl Biol 154:219–229
35. Jomantiene R et al (2002) New group 16SrIII phytoplasma lineages in Lithuania exhibit rRNA interoperon sequence heterogeneity. Eur J Plant Pathol 108:507–517
36. Liefting LW et al (1996) Sequence heterogeneity in the two 16S rRNA genes of *Phormium* yellow leaf phytoplasma. Appl Environ Microbiol 62:3133–3139
37. Schneider B, Seemüller E (1994) Presence of two sets of ribosomal genes in phytopathogenic mollicutes. Appl Environ Microbiol 60:3409–3412
38. Gundersen DE, Lee I-M (1996) Ultrasensitive detection of phytoplasmas by nested-PCR assays using two universal primer pairs. Phytopathol Mediterr 35:144–151
39. Lee I-M, Davis RE, Gundersen-Rindal DE (2000) Phytoplasmas: phytopathogenic mollicutes. Annu Rev Microbiol 56:1593–1597
40. Lee I-M et al (2010) Phylogenetic analysis and delineation of phytoplasmas based on *SecY* gene sequences. Int J Syst Evol Microbiol 60: 2887–2897
41. Marcone C et al (2000) Classification of aster yellows-group phytoplasmas based on combined analyses of rRNA and *tuf* gene sequences. Int J Syst Evol Microbiol 50:1703–1713
42. Martini M et al (2002) Genetic variability among Flavescence dorée phytoplasmas from different origins in Italy and France. Mol Cell Probes 16:197–208
43. Martini M et al (2007) Ribosomal protein gene-based phylogeny for finer differentiation and classification of phytoplasmas. Int J Syst Evol Microbiol 57:2037–2051
44. Mitrović J et al (2011) The *groEL* gene as an additional marker for finer differentiation of '*Candidatus* Phytoplasma asteris'-related strains. Ann Appl Biol 159:41–48
45. Schneider B, Gibb KS, Seemüller E (1997) Sequence and RFLP analysis of the elongation factor *Tu* gene used in differentiation and classification of phytoplasmas. Microbiol 143:3381–3389
46. Wei W et al (2004) An antibody against the SecA membrane protein of one phytoplasma reacts with those of phylogenetically different phytoplasmas. Phytopathology 94: 683–686
47. Cai H et al (2008) Genetic diversity among phytoplasmas infecting *Opuntia* species: virtual RFLP analysis identifies new subgroups in the peanut witches'-broom phytoplasma group. Int J Syst Evol Microbiol 58: 1448–1457
48. Duduk B, Bertaccini A (2006) Corn with symptoms of reddening: new host of stolbur phytoplasma. Plant Dis 90:1313–1319
49. Khan AJ et al (2002) Molecular identification of a new phytoplasma associated with alfalfa witches'-broom in Oman. Phytopathology 92:1038–1047
50. Wei W et al (2007) Computer-simulated RFLP analysis of 16S rRNA genes: identification of ten new phytoplasma groups. Int J Syst Evol Microbiol 57:1855–1867
51. Wei W et al (2008) Automated RFLP pattern comparison and similarity coefficient calculation for rapid delineation of new and distinct phytoplasma 16Sr subgroup lineages. Int J Syst Evol Microbiol 58:2368–2377
52. Schneider B et al (1995) Phylogenetic classification of plant pathogenic mycoplasma-like organisms or phytoplasmas. In: Razin S, Tully JG (eds) Molecular and diagnostic procedures in mycoplasmology. Academic, San Diego, pp 369–380
53. Skrzeczkowski L et al (2001) Bacterial sequences interferring in detection of phytoplasma by PCR using primers derived from the ribosomal RNA operon. Acta Hortic 550:417–424

Chapter 15

PCR and RFLP Analyses Based on the Ribosomal Protein Operon

Marta Martini and Ing-Ming Lee

Abstract

Differentiation and classification of phytoplasmas have been primarily based on the highly conserved 16S rRNA gene. RFLP analysis of 16S rRNA gene sequences has identified 31 16S rRNA (16Sr) groups and more than 100 16Sr subgroups. Classification of phytoplasma strains can, however, become more refined and specific if moderately conserved genes, such as the ribosomal protein (*rp*) genes, are used as genetic markers. The use of additional genetic markers enhances the resolving power of phytoplasma classification. This chapter describes the methodology of detection, differentiation, and classification of phytoplasma strains based on *rp* gene sequences. RFLP analysis of amplicons obtained by group- or subgroup-specific *rp* gene-based primers is used for finer differentiation of phytoplasma strains within a given group or subgroup. The *rp* gene-based classification not only readily resolves 16Sr subgroups within a given 16Sr group, but also provides finer differentiation of closely related phytoplasma strains. Many individual 16Sr subgroups can be further differentiated into two or more distinct *rp* subgroups.

Key words: *rplV* (*rpl22*), *rpsC* (*rps3*), Semiuniversal primers, 16Sr group-specific primers, 16Sr subgroup-specific primers, Restriction enzymes, Phytoplasma strain differentiation

1. Introduction

Over the last decade, epidemiological studies have revealed that diverse phytoplasma strains are involved in similar diseases associated with different cultivars of a given plant species grown in the same or different geographical regions. Often, the strains are closely related based on analysis of the 16S rRNA gene sequences, but it is not uncommon that closely related phytoplasma strains have unique ecological niches encompassing both plant host range and insect vectors. For epidemiological studies and quarantine purposes, it is relevant and essential to differentiate and identify these closely related but ecologically distinct strains.

Matt Dickinson and Jennifer Hodgetts (eds.), *Phytoplasma: Methods and Protocols*, Methods in Molecular Biology, vol. 938, DOI 10.1007/978-1-62703-089-2_15, © Springer Science+Business Media, LLC 2013

Currently, the primary genetic marker for phylogenetic studies and phytoplasma classification is the highly conserved 16S rRNA gene. Thus far, 31 16Sr groups and more than 100 16Sr subgroups have been identified (1, 2). However, because of its conserved nature, the 16S rRNA gene alone often cannot be used to differentiate closely related but ecologically distinct strains (3). This underscores the need to seek additional markers that permit finer differentiation of closely related strains. Several less-conserved genes, including ribosomal protein (*rp*) genes, are suitable as supplemental molecular markers for finer strain differentiation.

Ribosomal protein (*rp*) genes are more variable than 16S rRNA genes and have more phylogenetically informative characters, which substantially enhances the resolving power of classifying distinct phytoplasma strains within a given 16Sr group. Our earlier studies on differentiation of phytoplasmas in groups 16SrI and 16SrV indicated that analysis of *rp* gene sequences not only readily delineated subgroups that are consistent with 16Sr subgroups but also identified, within some subgroups, additional distinct strains that could not be resolved by analysis of highly conserved 16S rRNA gene sequences (4–7). Ten and twelve RFLP subgroups were differentiated on the basis of ribosomal protein gene sequences of 16SrI and 16SrV phytoplasma strains, respectively. Most of the additional strains identified have distinct biological and ecological properties. For example, maize bushy stunt (MBS) phytoplasma, a member of subgroup 16SrI-B, whose specific hosts are distinct from other members of the 16SrI group, is readily separated from other strains of subgroup 16SrI-B based on RFLP analysis of the *rp* gene sequence (5). Likewise, subgroup 16SrV-C can be further differentiated into several *rp* subgroups based on RFLP analyses with several selected key restriction enzymes (6, 7). The efficacy of *rp* gene sequences for finer differentiation among phytoplasma strains in a given 16Sr group has also been evaluated by putative restriction sites analysis of *rp* gene sequences of phytoplasmas of 16SrVI and X groups (8). For example, the three '*Candidatus* Phytoplasma' species '*Ca.* Phytoplasma mali', '*Ca.* Phytoplasma pyri', and '*Ca.* Phytoplasma prunorum' of 16SrX group were readily delineated by RFLP analysis of *rp* gene sequences (8).

This chapter describes the protocols for phytoplasma detection using semiuniversal or selective, 16Sr-group and subgroup-specific, *rp* gene-based primers, and for finer differentiation and classification of phytoplasma strains based on RFLP analysis of amplicons obtained by group- or subgroup-specific primers.

2. Materials

2.1. PCR

All PCR reagents and template DNAs are stored at –20°C unless otherwise stated.

1. Nuclease-free water.
2. 10× amplification buffer (Applied Biosystems).
3. 25 mM $MgCl_2$.
4. dNTPs solution containing all four dNTPs, 2.5 mM (see Note 1).
5. Forward and reverse primers (20 μM) (see Note 2).
6. AmpliTaq Gold polymerase (Applied Biosystems).
7. Template DNA: total plant genomic DNA 20 ng/μL, obtained from test plant samples and phytoplasma reference strains maintained in periwinkle.
8. Micropipettes and barrier or filter tips (see Note 3).
9. Microfuge tubes (0.2–0.5 mL thin-walled for PCR reactions, sterile 1.5 mL).
10. Thermal cycler, programmed with desired amplification protocol.

2.2. Agarose Gel Electrophoresis

1. Agarose.
2. 50× TAE electrophoresis buffer: 242 g of Tris base, 57.1 mL of glacial acetic acid, 100 mL of 0.5 M EDTA (pH 8.0). Use diluted to 1× TAE in ddH_2O.
3. 6× gel loading buffer type I: 0.25% bromophenol blue, 0.25% xylene cyanol FF, 40% (w/v) sucrose in water.
4. Staining solution: ethidium bromide 0.5 μg/mL stock solution, GelRed™ 10,000× stock solution (Biotium), or SYBR®Safe DNA Gel stain 10,000× stock solution in DMSO (Invitrogen) (see Note 4).
5. DNA size standard.
6. PCR products (see Note 5).
7. Micropipettes and tips.
8. Gel sealing tape.
9. Horizontal electrophoresis apparatus with chamber and comb.
10. Power supply device for electrophoresis apparatus.
11. Microwave oven.
12. Magnetic heating stirrer.
13. Gel documentation system, e.g., DigiDoc-It (UVP).

2.3. RFLP Analysis of PCR Products

1. Nuclease-free water.
2. Selected key restriction enzyme stored at –20°C.
3. 10× buffer supplied with the restriction enzyme stored at –20°C.
4. PCR products obtained from test samples and phytoplasma reference strains.
5. Mineral oil.
6. Micropipettes and tips.
7. Microfuge tubes (0.5 mL).
8. Water bath.

2.4. Polyacrylamide Gel Electrophoresis

1. Sterile distilled water.
2. 10× TBE electrophoresis buffer: 108 g of Tris base, 55 g of boric acid, 40 mL of 0.5 M EDTA (pH 8.0). Use diluted to 1× TBE.
3. Acrylamide/bisacrylamide (29:1) (30% w/v) solution, stored at 4°C (see Note 6).
4. Ammonium persulphate (APS): 10% w/v solution in water (see Note 7).
5. *N*, *N*, *N*, *N*′-tetramethyl-ethylenediamine (TEMED).
6. Gel loading buffer type I (6× stock): 0.25% bromophenol blue, 0.25% xylene cyanol FF, 40% (w/v) sucrose in water.
7. DNA size standard.
8. Agarose: 1% solution.
9. Vertical electrophoresis gel system, glass plates, comb and spacers, binder or "bulldog" paper clips.
10. Power supply device for electrophoresis system.
11. Micropipettes and gel-loading pipette tips.
12. Microwave oven or magnetic heating stirrer.
13. Gel documentation system, e.g., DigiDoc-It (UVP).
14. 100% Ethanol.

3. Methods

3.1. PCR Analysis Using Semiuniversal, 16Sr Group-Specific or Subgroup-Specific Primers

1. Choose the *rp* primer pair according to your requirements. For amplification of a partial *rp* operon, several semiuniversal, 16Sr group-specific and subgroup-specific primer pairs have been designed (Table 1). The majority of them can also be used in nested or seminested PCR (see Note 8).

Table 1
***Rp* gene-based, semiuniversal, 16Sr group-specific and 16Sr subgroup-specific primers designed for amplification of a portion of the phytoplasma *rp* operon**

Primer pair	Sequence (5′-3′)	Location	Expected size of PCR product (bp)	Specificity	Reference
rpF1/rpR1	GGACATAAGTTAGGTGAATTT ACGATATTTAGTTCTTTTTGG	*rpsS* *rplP*	1,245–1,389[a]	16SrI, III, IV, V, VII, VIII, IX, XIII	(9)
rp(I)F1A/rp(I)R1A	TTTTCCCCTACACGTACTTA GTTCTTTTTGGCATTAACAT	*rpsS* *rpsC/rplP*	1,200	16SrI	(10)
rp(I-A)F1/rp(I-A)R1	CAAGAGCTAAAGGTTCTGGC GAGGGCGCCTGTTAGGGTTA	*rplV* *rpsC*	800	16SrI-A	(10)
rp(I-B)F1/rp(I-B)R2	AAGAGCTAAAGGTTCTGGT GAGGGCGTCTGTTAGGAGTG	*rplV* *rpsC*	800	16SrI-B	(10)
rpL2F3/rp(I)R1A	WCCTTGGGGYAAAAAAGCTC GTTCTTTTTGGCATTAACAT	*rplB* *rpsC/rplP*	1,600	16SrI, III, IV, V, VI, VII, IX, X, XII, XIII, XVIII	(8)
rpF1C/rp(I)R1A	ATGGTDGGDCAYAARTTAGG GTTCTTTTTGGCATTAACAT	*rpsS* *rpsC/rplP*	1,212–1,386[a]	16SrI, II, III, IV, V, VI, VII, IX, X, XII, XIII, XVIII	(8)
rp(II)F1/rp(II)R1	GCTCTTACTCGTAAAYATGTAGT TTACTTGATTTTCTGGTTTTGA	*rpsS* *rpsC*	1,200	16SrII	(8)
rp(II)F/rp(I)R1A	ACTTATTCTCGTGATACTAG GTTCTTTTTGGCATTAACAT	*rpsS* *rpsC/rplP*	1,390	16SrII	(11)
rp(II)F2/rp(I)R1A	ATGGTAGGTTATAAATTAGG GTTCTTTTTGGCATTAACAT	*rpsS* *rpsC/rplP*	1,290	16SrII	(11)
rp(III)F1/rp(III)R1	TTAGAGAAGGCATTAAAC CTCTTTCCCCATCTAGGACG	*rpsS* *rpsC*	1,200	16SrIII	(8)

(continued)

Table 1 (continued)

Primer pair	Sequence (5′-3′)	Location	Expected size of PCR product (bp)	Specificity	Reference
rp(V)F1/rpR1	TCGCGGTCATGCAAAAGGCG ACGATATTTAGTTCTTTTTGG	*rpsS* *rplP*	1,200	16SrV	(4, 9)
rp(V)F2/rpR1	TTGCCTCGTTTATTTCCGAGAGCTA ACGATATTTAGTTCTTTTTGG	*rplV* *rplP*	950	16SrV	(4, 9)
rp(V)F1A/rp(V)R1A	AGGCGATAAAAAAGTTTCAAAA GGCATTAACATAATATATTATG	*rpsS* *rplP*	1,200	16SrV	(6)
rp(VI)F2/rp(VI)R2	GGTTGTTGATTTAATTCGTGGTC CCAGATATTCGTCTAGTATCAGAA	*rplV* *rpsC*	1,000	16SrVI	(8)
rp(VIII)F2/rp(VIII)R2	AGTTGTCGATTTAATTCGTGGCA CAGCAGATATTTGTCTAGTATCTGCG	*rplV* *rpsC*	1,000	16SrVII, VIII	(8)
rp(IX)F2/rp(IX)R2	GCACAAGCTATTTTAATGTTTACACCC CAAAGGGACTAAACCTAAAG	*rplV* *rpsC*	800	16SrIX	(8)
rpAP15f2/rp(I)R1A	CTCCTAAATCAGCTTCAAGT TTCTTTTTGGCATTAACAT	*rplV* *rpsC*/*rplP*	1,036	16SrX-A	(12)
rpAP15f/rpAP15r	AGTGCTGAAGCTAATTTGG TGCTTTTTATAGCAAAAGGTT	*rplV* *rpsC*	920	16SrX-A	(12)
rpStolF/rpStolR	CGTACAAAATAATCGGGAGA CGAAACAAAAGGTTTACGAG	*rplB* *rpsC*	1,372	16SrXII-A	(8)
rpStolF2/rpStolR	AAACTTGGTCACGTAGTTCC CGAAACAAAAGGTTTACGAG	*rplP* *rpsC*	1,253	16SrXII-A	(8)

2. Thaw the PCR reagents on ice. In a sterile 1.5 mL microfuge tube, make a master mix using the reagents in the following order (volumes are given for one reaction, multiply these as necessary for the number of samples):

Component	Final volume (μL)	Final concentration
Nuclease-free water	16.875	–
10× amplification buffer	2.5	1×
$MgCl_2$ solution, 25 mM	1.5	1.5 mM
dNTPs, 2.5 mM	2	200 μM
Forward primer 20 μM	0.5	0.4 μM
Reverse primer 20 μM	0.5	0.4 μM
AmpliTaq Gold polymerase (5 U/μL) (see Note 9)	0.125	0.625 U
Total volume	24	

3. Mix thoroughly and briefly spin down the reaction mixture, then aliquot 24 μL of the reaction mixture in each PCR tube.
4. Add 1 μL of template DNA (≥20 ng/μL). Include PCR positive and negative controls in each set (see Note 10).
5. Mix thoroughly, spin down, and place the tubes in the thermal cycler to amplify the nucleic acids using number of cycles, denaturation, annealing, and polymerization times and temperatures listed (Table 2) for the given primer pair.
6. To perform nested PCR, make a 1:30 dilution (in ddH_2O) of the direct PCR product in 0.5 mL tubes and then use 1 μL of the diluted product as the template DNA in the nested PCR.

3.2. Agarose Gel Electrophoresis

1. Seal the open ends of the plastic tray supplied with the electrophoretic apparatus with tape to form a mold. Place the mold on a horizontal section of bench. Choose an appropriate comb and position it 0.5–1.0 mm above the plate so that a complete well is formed when the agarose is added to the mold.
2. Prepare sufficient 1× TAE electrophoresis buffer to fill the tank and cast the gel (see Note 11).
3. Prepare agarose 1% (w/v) solution in 1× TAE in a flask or a glass bottle (see Note 12).
4. Heat the slurry in a microwave oven until the agarose dissolves (see Note 13).
5. Add a magnetic stirring bar to the flask or bottle and then transfer it to a stirrer to let the melted agarose solution to cool to about 60°C.

Table 2
PCR cycling conditions used with *rp* gene-based PCR primers

Primer pair	Initial denaturation	Cycles	Denaturation	Anneal	Extension	Final extension
rpF1/rpR1 rpL2F3/rp(I)R1A rpF1c/rp(I)R1A rp(II)F1/rp(II)R1 rp(II)F/rp(I)R1A rp(II)F2/rp(I)R1A rp(III)F1/rp(III)R1 rp(V)F1/R1 rp(V)F1A/rp(V)R1A rp(VI)F2/rp(VI)R2 rp(VIII)F2/rp(VIII)R2 rp(IX)F2/rp(IX)R2	94°C 11 min	38	94°C 1 min	50°C 2 min	72°C 3 min	72°C 7 min
rp(I)F1A/rp(I)R1A	94°C 11 min	38	94°C 1 min	55°C 2 min	72°C 3 min	72°C 7 min
rp(I-A)F1/rp(I-A)R1	94°C 11 min	38	94°C 1 min	64°C 2 min	72°C 3 min	72°C 7 min
rp(I-B)F1/rp(I-B)R2	94°C 11 min	38	94°C 1 min	60°C 2 min	72°C 3 min	72°C 7 min
rpAP15f2/rp(I)R1A	94°C 2 min	35 40[a]	94°C 1 min	54°C 45 s 52°C 80 s	72°C 1.5 min 72°C 2 min[a]	72°C 8 min
rpAP15f/rpAP15r	94°C 2 min	35/40[b] 36[a]	94°C 1 min	55°C 45 s 53°C 1 min[a]	72°C 1.5 min 72°C 2 min[a]	72°C 8 min
rpStolF/rpStolR rpStolF2/rpStolR	94°C 2 min	40	94°C 1 min	53°C 45 s	72°C 1.5 min	72°C 8 min

[a]PCR conditions used with insect total DNA template
[b]Number of cycles used in direct PCR with primers rpAP15f/rpAP15r

6. (Optional) Add 1 μL of GelRed™ (10,000× stock solution) or SYBR®Safe DNA Gel stain (10,000× stock solution) to 100 mL of agarose solution and mix the gel solution thoroughly by gentle swirling.
7. Pour the warm agarose solution into the mold (see Note 14). Allow the gel to set completely for 20–30 min at room temperature, and then pour a small amount of 1× TAE buffer on the top of the gel, and carefully remove the comb. Pour the electrophoresis buffer off and carefully remove the tape. Mount the gel in the electrophoresis tank and add enough 1× TAE buffer to cover the gel completely.
8. Withdraw 5 μL from the sample and the control PCR reaction mixtures. Add 1 μL of 6× gel loading buffer (see Note 15) and slowly load the sample mixture into the wells of the gel using a P10 or P20 micropipette. Load 5 μL of a DNA size standard such as GeneRuler™ 1 kb DNA Ladder (Fermentas) into wells on both the right and left sides of the gel.
9. Close the lid of the gel tank and apply a voltage of 1–5 V/cm (see Note 16). Run the gel until the bromophenol blue and xylene cyanol FF have migrated an appropriate distance through the gel.
10. If GelRed™ or SYBR®Safe DNA Gel stain is present in the gel (see Note 17), examine the gel by UV light on a transilluminator and photograph the gel by a fixed-focus digital camera using a gel documentation system. Otherwise, stain the gel by immersion in ethidium bromide (0.5 μg/mL in water) for 10 min at room temperature followed by destaining for 10 min in water.

3.3. RFLP Analysis of PCR Products

1. Choose the appropriate key restriction enzymes to use according to the 16Sr group phytoplasma strains and 16Sr groups you are studying (Table 3).
2. Set the water bath(s) to the appropriate temperature as indicated by the enzyme manufacturer.
3. Thaw the enzyme reagents. In a sterile 0.5 mL microfuge tube, combine the reagents in the following order (volumes are given for one sample reaction):

Component	Final volume (μL)	Final concentration
Nuclease-free water	9.5–11.5	–
10× restriction buffer	2	1×
Restriction enzyme	0.5	0.5–1U
PCR product	6–8	300–500 ng/20 μL
Total volume	20	

Table 3
Set of key restriction enzymes which are useful in differentiating phytoplasma strains within different 16Sr groups or subgroups

16Sr group	Restriction enzymes	Reference
I	*Alu*I, *Mse*I, *Tsp*509I	(4, 5, 10)
II	*Alu*I, *Dra*I, *Hha*I, *Mse*I, *Rsa*I, *Ssp*I, *Taq*I, *Tsp*509I	(11)
III	*Alu*I, *Dra*I, *Mse*I	(4, 13)
V	*Alu*I, *Dra*I, *Hae*III, *Hha*I, *Hpa*II, *Mse*I, *Ssp*I, *Taq*I, *Tsp*509I	(6, 7)
VI	*Alu*I, *Dra*I, *Taq*I, *Tsp*509I	(8)
VII	*Mse*I	(14)
X	*Alu*I, *Dra*I	(8, 12)
XII-A	*Mse*I	(15)
XII-B	*Alu*I, *Dra*I, *Mse*I	(16)

4. Mix thoroughly, spin down, and add a drop of mineral oil to any restriction enzyme reactions which need incubation at 60–65°C.
5. Place the tubes in the water bath(s) at the temperature indicated by the manufacturer and leave overnight.

3.4. Polyacrylamide Gel Electrophoresis

1. Clean the glass plates, spacers, and comb with 100% ethanol.
2. Assemble the glass plates with spacers: lay the notched plate on the bench and arrange the spacers on each side, lay the unnotched plate in position, resting on the spacer bars, clamp the plates together with binder or "bulldog" paper clips.
3. Dissolve the 1% agarose solution in a microwave oven and use it to seal the entire length of the two sides and the bottom of the plates (see Note 18).
4. Taking the size of the glass plates and the thickness of the spacers (see Note 19) into account, calculate the volume of gel required. Prepare the gel solution with the desired polyacrylamide percentage (Table 4) in a clean beaker or flask (see Note 20).
5. Immediately after the addition of TEMED, mix the solution by swirling gently.
6. Pour the solution into the space between the two glass plates filling the space almost to the top (see Note 21).

7. Immediately insert the appropriate comb into the gel and do not allow air bubbles to become trapped under the teeth. The tops of the teeth should be slightly higher than the top of the glass. Make sure that no acrylamide solution is leaking from the gel mold.
8. Allow the acrylamide to polymerize for 30–40 min at room temperature (see Note 22).
9. Remove the spacer at the bottom of the gel and attach the gel to the electrophoresis tank using large bulldog clips on the sides (see Note 23).
10. Fill the reservoirs of the electrophoresis tank with 1× TBE (see Note 24).
11. Carefully pull the comb from the polymerized gel and wash the wells out with 1× TBE thoroughly as soon as the comb is removed (see Note 25).
12. Use a syringe with a bent needle to remove any air bubbles trapped beneath the bottom of the gel.
13. Mix the digested DNA sample (20 μL) with 4 μL of 6× gel loading buffer. Load about 12 μL of the mixture with a P20 micropipette and gel-loading pipette tips (see Note 26). Load 10 μL of a DNA size standard such as *Hae*III-digested ΦX174 DNA into wells on both the right and left sides of the gel.
14. Connect the electrodes to a power supply and apply a voltage of 1–8 V/cm. Run the gel until the bromophenol blue and xylene cyanol FF have migrated an appropriate distance through the gel.
15. Detach the glass plates and lay them on the bench. Lift a corner of the upper glass using a spatula, pull away the upper plate and remove the spacers.
16. Gently submerge the gel attached to the lower plate in ethidium bromide (0.5 μg/mL in water) for 10 min at room temperature then destain for 10 min in water (see Note 27).
17. Invert the gel, remove the glass plate, and place the gel on the transilluminator.
18. Examine the gel by UV light on the transilluminator and photograph the gel by a fixed-focus digital camera using a gel documentation system.
19. Compare the RFLP patterns obtained with phytoplasma reference strains or previously published patterns (Fig. 1) (4–6, 10, 12–14, 16).

Table 4
Volume of reagents used to cast polyacrylamide gels (25 mL)

Final polyacrylamide gel (%)	30% acrylamide/bisacrylamide (mL)	H_2O (mL)	10× TBE (mL)	10% APS (μL)	TEMED (μL)
5	4.2	18	2.5	310	15.6
12	10	12.2	2.5	310	15.6

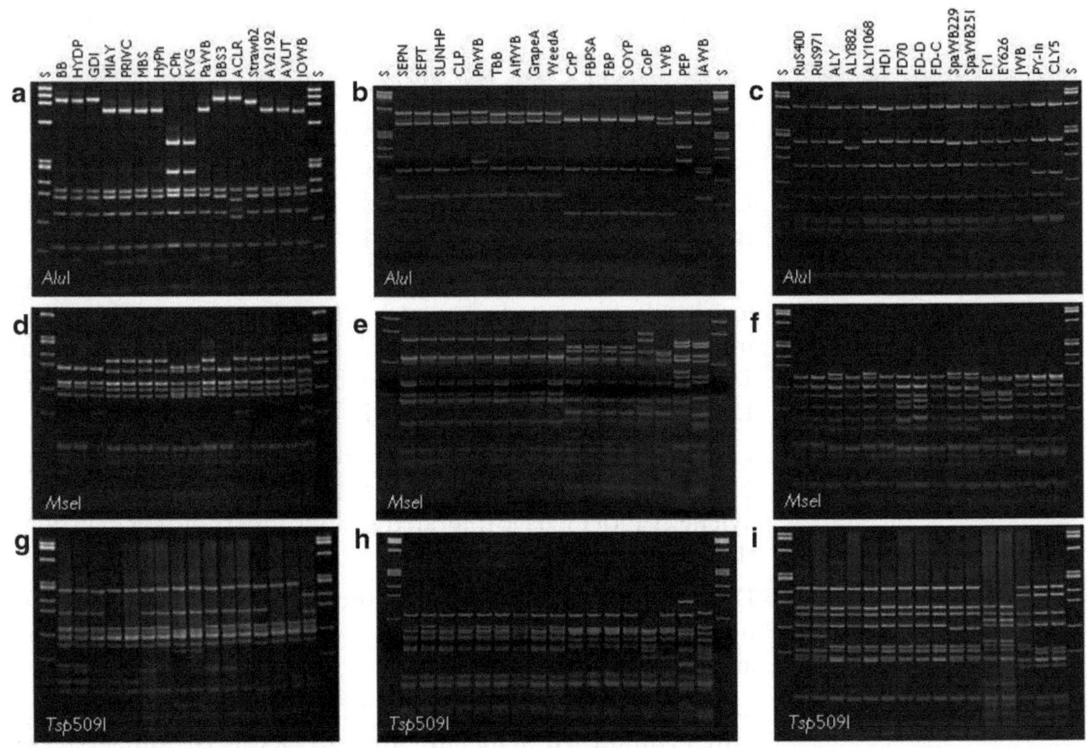

Fig. 1. (**a**, **d**, **g**) RFLP profiles of *rp* operon sequence (1.2 kb, containing *rpl22* and *rps3* genes), amplified by nested PCR with primer pair rpF1/R1 followed by primer pair rp(I)F1A/rp(I)R1A, from representative phytoplasma strains in the AY phytoplasma group (16SrI). PCR products were digested with *AluI*, *MseI*, or *Tsp*509I, respectively.(**b**, **e**, **h**) RFLP profiles of *rp* operon sequence (1.2 kb, containing *rpl22* and *rps3* genes), amplified by seminested PCR with primer pair rp(II)F1/rp(I)R1A followed by primer pair rp(II)F2/rp(I)R1A, from representative phytoplasma strains (*SEPN* SEPT sesame phyllody; *SUNHP* sunn hemp witches'-broom; *CLP Cleome* phyllody; *PnWB* peanut witches'-broom; *TBB* Australian tomato big bud; *AlfWB* alfalfa witches' broom; *GrapeA* Australian grapevine yellows; *WeedA* weed virescence; *CrP Crotalaria* phyllody; *FBPSA Crotalaria saltiana* phyllody; *FBP* faba bean phyllody; *SOYP* soybean phyllody; *CoP* cotton phyllody; *LWB* lime witches'-broom; *PEP Picris echioides* phyllody; *IAWB* Italian alfalfa witches'-broom) in the peanut witches'-broom phytoplasma group (16SrII). PCR products were digested with *Alu*I, *Mse*I, or *Tsp*509I, respectively. (**c**, **f**, **i**) RFLP profiles of *rp* operon sequence (1.2 kb, containing *rpl22* and *rps3* genes), amplified by nested PCR with primer pair rp(V)F1/R1 followed by primer pair rp(V)F1A/rp(V)R1A, from representative phytoplasma strains in the EY phytoplasma group (16SrV). PCR products were digested with *Alu*I, *Mse*I, or *Tsp*509I respectively. Digests in (**a**) were separated by electrophoresis through a 5% polyacrylamide gel; all other digests were separated through 12% polyacrylamide gels. Lanes S, ΦX174 DNA-*Hae*III digest; fragment sizes (bp) from top to bottom: 1353, 1078, 872, 603, 310, 281, 271, 234, 194, 118, 72.

4. Notes

1. Combine equal volumes of 100 mM stock solutions of dATP, dCTP, dGTP, and dTTP in a 1.5 mL microfuge tube, mix thoroughly, and then dilute the mixture 1:10 with nuclease-free water. Store at −20°C.
2. Dilute 200 μM primer stock solutions 1:10 with nuclease-free water.
3. A complete set of micropipettes is normally dedicated for PCR use only. The most commonly used micropipettes are the Gilson P10, P20, P100, P200, and P1000 models (or equivalents).
4. In the past few years, there has been increasing interest in the use of alternative DNA intercalating dyes such as GelRed™ (Biotium) or SYBR®Safe DNA Gel Stain (Invitrogen) as replacements for staining with the hazardous ethidium bromide.
5. Normally only nested PCR products are analyzed by agarose gel electrophoresis.
6. To reduce exposure to acrylamide and bisacrylamide, it is preferable to purchase a ready-made solution of acrylamide: bisacrylamide (29:1) available from different suppliers, instead of acrylamide and bisacrylamide powders. Always wear gloves to handle acrylamide:bisacrylamide solution, which is toxic.
7. Small aliquots (1 mL) of 10% APS can be freshly prepared and stored at 4°C for up to 1 month.
8. The semiuniversal forward primers rpL2F3 and rpF1C are degenerate primers and are used in combination with the reverse primer rp(I)R1A. Primers rpF1C/rp(I)R1A can also be used in seminested PCR after amplification with rpL2F3/rp(I)R1A for a more sensitive semiuniversal PCR. Primer pairs rp(I)F1A/rp(I)R1A (16SrI group-specific), rp(I-A)F1/rp(I-A)R1 (16SrI-A subgroup-specific), rp(I-B)F1/rp(I-B)R2 (16SrI-B subgroup-specific) can be used in nested PCR following direct PCR with rpF1/rpR1 primer pair. The group-specific primer pair rp(II)F1/rp(II)R1 can be used in nested PCR following direct PCR with rpF1C/rp(I)R1A, whereas group-specific primer pairs rp(III)F1/rp(III)R1, rp(VI)F2/rp(VI)R2, rp(VIII)F2/rp(VIII)R2, rp(IX)F2/rp(IX)R2 can be used in nested PCR following direct PCR with either rpL2F3/rp(I)R1A or rpF1C/rp(I)R1A. For amplification of partial *rp* operon from phytoplasma strains of 16SrII group seminested PCR with primer pairs rp(II)F/rp(I)R1A followed by rp(II)F2/rp(I)R1A can also be used. Partial *rp* operon from phytoplasma strains of 16SrV group can be amplified in direct PCR with primer pair rp(V)F1/rpR1 followed by seminested PCR

with rp(V)F2/rpR1 or nested PCR with rp(V)F1A/rp(V)R1A. The latter primer pair amplifies a longer and therefore more informative *rp* operon fragment. For amplification of partial *rp* operon from apple proliferation phytoplasma strains of 16SrX-A subgroup primer pair rpAP15f/rpAP15r can be used in direct PCR or in nested PCR following direct PCR with rpAP15f2/rp(I)R1A. Finally, partial *rp* operon of stolbur phytoplasmas (subgroup 16SrXII-A) can be amplified by seminested PCR with primers rpStolF/rpStolR followed by rpStolF2/rpStolR.

9. Hot-start *Taq* polymerase (such as AmpliTaq Gold DNA polymerase) is especially recommended in PCR reactions with degenerate semiuniversal primers; standard *Taq* polymerases from different manufacturers can be also used in PCR reactions with group- or subgroup-specific primers.
10. Positive controls (phytoplasma reference strains) are required to monitor the efficiency of the PCR, whereas negative controls are required to detect contamination with the DNAs that contain the target sequence.
11. Add 20 mL of 50× TAE to 980 mL of distilled water in a 1 L graduated cylinder, seal the cylinder with Parafilm, and mix the solution inverting the cylinder.
12. In a flask or a glass bottle add 1 g of standard agarose to 100 mL of 1× TAE buffer.
13. Heat the slurry for the minimum time required allowing agarose to dissolve; undissolved agarose appears as translucent chips floating in the solution. Carefully swirl the flask or bottle from time to time to make sure that any grains of undissolved agarose do not stick to the walls but enter the solution.
14. The thickness of the gel should be between 3 and 5 mm. No air bubbles should be present under or between the teeth of the comb.
15. 1 μL of 6× gel loading buffer can be deposited on a piece of Parafilm forming a drop and then 5 μL of the sample can be added and mixed on the Parafilm before loading the sample.
16. Distance (cm) is measured between the positive and negative electrodes.
17. The presence of GelRed™ or SYBR®Safe DNA Gel stain in the gel allows you to examine the gel by UV illumination at any stage during electrophoresis. The gel tray may be removed and placed directly on the transilluminator. You can also soak the gel after electrophoresis in GelRed™ or SYBR®Safe DNA Gel stain solution diluted to 1× strength in electrophoresis buffer if you did not add the stain to the agarose.
18. The aim is to form a watertight seal between the plates and the spacers so that the gel solution does not leak out.

19. The spacer thickness varies from 0.5 to 2 mm; in general thinner gels are preferred, as they produce the sharpest and flattest bands of DNA.
20. For RFLP analyses of *rp* gene-based PCR products a 5% polyacrylamide gel is normally used with all restriction enzymes. However, for frequent cutter enzymes such as *Mse*I and *Tsp*509I, the use of a 12% polyacrylamide gel is recommended for improved resolution.
21. Draw the solution with a 50-mL syringe, introducing the nozzle of the syringe between the two glass plates or with the beaker you used to prepare the solution, introducing the pouring lip between the two glass plates.
22. When polymerization is complete, a Schlieren pattern will be visible just beneath the teeth of the comb.
23. The notched plate should face inward towards the buffer reservoir.
24. It is important to use the same batch of electrophoresis buffer in both the reservoirs and in the gel, because small differences can greatly distort the DNA migration. To prepare 1× TBE solution, add 100 mL of 10× TBE to 900 mL of distilled water in a 1 L graduated cylinder, seal the cylinder with Parafilm, and mix the solution inverting the cylinder.
25. Use a syringe or a Pasteur pipette with thin tip to rinse out the wells with electrophoresis buffer. It is important to wash the wells out thoroughly as soon as the comb is removed, otherwise small amounts of acrylamide solution trapped by the comb will polymerize in the wells, producing irregularly shaped surfaces that give distorted bands of DNA.
26. The same tip can be used to load many samples, on condition that it is thoroughly washed between each loading.
27. Use the glass plate as a support.

References

1. Zhao Y et al (2010) Recent advances in 16S rRNA gene-based phytoplasma differentiation, classification and taxonomy. In: Weintraub PG, Jones P (eds) Phytoplasmas: genomes, plant hosts and vectors. CAB International, Wallingford, pp 64–92
2. Lee I-M et al (2011) '*Candidatus* Phytoplasma costaricanum' a new phytoplasma associated with a newly emerging disease in soybean in Costa Rica. Int J Syst Evol Microbiol. doi:10.1099/ijs.0.029041-0
3. Lee I-M, Zhao Y, Davis RE (2010) Prospects of multiple gene-based systems for differentiation and classification of phytoplasmas. In: Weintraub PG, Jones P (eds) Phytoplasmas: genomes, plant hosts and vectors. CAB International, Wallingford, pp 51–63
4. Lee I-M et al (1998) Revised classification scheme of phytoplasmas based on RFLP analyses of 16S rRNA and ribosomal protein gene sequences. Int J Syst Bacteriol 48:1153–1169
5. Lee I-M et al (2004) '*Candidatus* Phytoplasma asteris', a novel phytoplasma taxon associated with aster yellows and related diseases. Int J Syst Evol Microbiol 54:1037–1048
6. Lee I-M et al (2004) Classification of phytoplasma strains in the elm yellows group (16SrV) and proposition of '*Candidatus* Phytoplasma

ulmi' for the phytoplasma associated with elm yellows. Int J Syst Evol Microbiol 54:337–347

7. Martini M et al (2002) Genetic variability among flavescence dorée phytoplasmas from different origins in Italy and France. Mol Cell Probes 16:197–208
8. Martini M et al (2007) Ribosomal protein gene-based phylogeny for finer differentiation and classification of phytoplasmas. Int J Syst Evol Microbiol 57:2037–2051
9. Lim P-O, Sears BB (1992) Evolutionary relationships of a plant-pathogenic mycoplasmalike organism and *Acholeplasma laidlawii* deduced from two ribosomal protein gene sequences. J Bacteriol 174:2606–2611
10. Lee I-M et al (2003) Ecological implications from a molecular analysis of phytoplasmas involved in an aster yellows epidemic in various crops in Texas. Phytopathology 93:1368–1377
11. Martini M (2004) Ribosomal protein gene-based phylogeny: a basis for phytoplasma classification. PhD dissertation, University of Udine, Udine, Italy, p 106
12. Martini M et al (2008) Molecular differentiation of '*Candidatus* Phytoplasma mali' and its spreading in Friuli Venezia Giulia region (north-east Italy). Acta Hortic 781:395–402
13. Gundersen DE et al (1996) Genomic diversity and differentiation among phytoplasma strains in 16S rRNA group I (aster yellows and related phytoplasmas) and III (X-disease and related phytoplasmas). Int J Syst Bacteriol 46:64–75
14. Griffiths HM et al (1999) The phytoplasma associated with ash yellows and lilac witches'-broom: '*Candidatus* Phytoplasma fraxini'. Int J Syst Bacteriol 49:1605–1614
15. Durante G et al (2008) Bois noir in Lombardy (northern Italy): identification of molecular markers for diagnosis and characterization of 16SrXII-A phytoplasmas. Proceedings of the: 4th National Workshop on Phytoplasma Diseases. Rome, Italy 28–30 May 2008, p 300–303
16. Streten C, Gibb KS (2005) Genetic variation in *Candidatus* Phytoplasma australiense. Plant Pathol 54:8–14

Chapter 16

Tuf and *secY* PCR Amplification and Genotyping of Phytoplasmas

Xavier Foissac, Jean-Luc Danet, Sylvie Malembic-Maher, Pascal Salar, Dana Šafářová, Pavla Válová, and Milan Navrátil

Abstract

Tuf and *secY* genotyping techniques have been developed to distinguish phytoplasma strains. *Tuf* polymerase chain reaction sequence analyses are available for phytoplasma taxonomic groups 16SrI, 16SrV, 16SrXII-A, and XII-B. In addition to their use to confirm the taxonomic status of phytoplasma strains, they allow the spread of phytoplasma strains in host plants and insect vectors to be traced. *SecY* is more variable than *tuf* and is therefore more discriminatory than *tuf*, but *secY* and *tuf* phylogenies show congruence.

Key words: Epidemiology, Genetic diversity, Polymerase chain reaction, Phylogenetic analysis, secY, tuf

1. Introduction

Phytoplasma classification is based on the genetic variability of the 16S rRNA gene, with a species demarcation threshold of 97.5% similarity, as well as by distinctive biological, phytopathological, and genomic properties (1, 2). In many cases, the 16S rDNA sequence does not allow fine enough differentiation to undertake specific epidemiological studies based on strain discrimination. In recent years, interest has focused on conserved and less-conserved nonribosomal genes, such as *secA*, *tuf*, *map*, *secY*, *pnp* and ribosomal protein genes (3–5), or variable genes encoding surface proteins such as *vmp1* (6–9), *imp* (10–14), *amp* (15), *stamp* (16), and *hfl*B (17, 18). Genetic and phylogenetic analyses of these genes have led to precise phytoplasma strain classification and differentiation, with discoveries in population genetics such as the existence of interspecies recombination in group 16SrX (11).

Matt Dickinson and Jennifer Hodgetts (eds.), *Phytoplasma: Methods and Protocols*, Methods in Molecular Biology, vol. 938, DOI 10.1007/978-1-62703-089-2_16,

The *tuf* gene encodes the translation elongation factor EF-Tu. EF-Tu, when activated, provides energy to the protein translation process. *Tuf* has been used successfully as a marker for ecological, epidemiological, and taxonomic studies. There is a specific association of phytoplasmal *tuf* genotypes with specific host plants and insect vector ecotypes (4).

The nonribosomal gene *secY* encodes a translocase that is part of the translocation system for secreting proteins across the cytoplasmic membrane (19). *SecY* shows a similar degree of variability to that of ribosomal protein (*rp*) genes and is used as a molecular marker to differentiate phytoplasma strains in taxonomic groups 16SrI, 16SrV, 16SrX, and 16SrXII-A (7, 11, 20–29). The significance of *secY* gene sequences and their ability to group phytoplasmas into the same subgroups as the more commonly used 16S rRNA, *tuf*, *secA*, and *rp* genes was recently demonstrated for the taxonomic group 16SrV (30).

The polymerase chain reaction (PCR) is a method for *in vitro* amplification of specific regions of a target DNA using oligonucleotide primers, dNTPs, and a heat-stable DNA polymerase. PCR involves the following basic steps: (1) denaturation at about 95°C, during which double helix DNA template separates into single strands; (2) annealing between 45 and 72°C depending on the melting temperature (T_m) of the primers, which allows the hybridization of the primers onto complementary DNA sites flanking the sequence to be amplified; (3) elongation at 66–72°C, during which DNA polymerase synthesizes a complementary strand from the 3′ end of the primers. Steps (1)–(3) are repeated 25–40 times (PCR cycles) in order to exponentially amplify the sequenced flanked by the two primers. As a result, a PCR product (amplicon) of a defined size is obtained. To enhance the sensitivity of phytoplasma detection (particularly in woody hosts), a second round of PCR cycles (nested PCR) can be performed with internal primers.

The restriction fragment length polymorphism (RFLP) analysis of PCR-derived amplicons has become increasingly important in phytoplasma identification as well as in typing for epidemiological studies. RFLP typing can be performed by cutting DNA using restriction endonucleases (wet RFLP) and/or by using sequence information combined with computer analysis (*in silico* RFLP). The genes and/or genome fragments amplified by PCR can also be used for subsequent sequencing and phylogenetic analysis.

2. Materials

2.1. PCR Reagents

1. Ultrapure deionized water (ddH_2O, 18 MΩ conductivity at 25°C).

2. Thermostabile *Taq* DNA polymerase with appropriate DNA polymerase buffer and $MgCl_2$.
3. 5 mM dNTP stock solution.
4. *tuf* primers for groups 16SrI ('*Candidatus* Phytoplasma asteris') and 16rXII-A (stolbur phytoplasma) (Table 1): fTufl, rTufl, fTufAy, rTufAy.
5. *tuf* primers for group 16SrXII-B ('*Ca.* Phytoplasma australiense') and to distinguish tuf 1 and tuf 2 subtypes: PaTC1F 1, PaTC1R 1, PaTC1F, PaTC1R, PaTC2F, PaTC2R.
6. *tuf* primers for group 16SrV: FDTUF-F1, FDTUF-R1, FDTUF-F2, FDTUF-R2.
7. *secY* primers for group 16SrI ('*Ca.* Phytoplasma asteris'): AYsecYF1, AYsecYR1.
8. *secY* primers for group 16SrII ('*Ca.* Phytoplasma aurantifolia'): SecYF1(II), SecYR1(II), SecYF2(II).
9. *secY* primers for group 16SrIII (Western X phytoplasma group): SecYF1(III), SecYR1(III), SecYF2(III).
10. *secY* primers for group 16SrVI ('*Ca.* Phytoplasma trifolii'): SecYF1(VI), SecYR1(VI), SecYF2(VI).
11. *secY* primers for group 16SrV: either FD9f and FD9r, or FD9r and FD9f2L; FD9r2L, FD9f3L, sequencing primer FD9ri.
12. *secY* primers for group 16SrX ('*Ca.* Phytoplasma mali', '*Ca.* Phytoplasma pyri', and '*Ca.* Phytoplasma prunorum'): SecYMalF1, SecYMalR1, SecYMalF2, SecYMalR2.
13. *secY* primers for group 16SrXII-A (Stolbur 'Bois noir' phytoplasma): PosecF1, PosecR1, PosecF3, or PosecN2 (as alternative to PosecF3 if DNA smearing occurs on gels due to the high GC content of the PosecF3 3′ end), PosecR3.
14. Template DNA: total DNA extracted from phytoplasma-enriched fraction (33) or the CTAB extraction (34).

2.2. Agarose Gel Electrophoresis

1. 0.5 M EDTA, pH 8.0: 186.1 g disodium ethylenediaminetetraacetic acid dihydrate (EDTA), 800 mL of deionized water. Adjust pH to 8.0 using NaOH to dissolve it. Adjust to a final volume of 1 L with deionized water.
2. 50× TAE buffer (Tris-acetate-EDTA): 242 g Tris base, 500 mL deionized water. Add 57.2 mL acetic acid and 100 mL 0.5 M EDTA pH 8.0, and mix carefully. Adjust to a final volume of 1 L. (The pH of this buffer should be 8–8.5, it is not necessary to adjust it).
3. 1× TAE buffer: 20 mL of 50× TAE buffer, 980 mL of deionized water.
4. 1% agarose gel: 1 g of agarose (in a 250 mL flask or bottle), 100 mL of 1× TAE buffer. Melt in a microwave oven for

Table 1
Primer sequences for *tuf* and *secY* genes used in this chapter

Gene	Group	Primer	Sequence (5′-3′)	Reference
tuf	16SrI and 16rXII-A	fTufl	CACATTGACCACGGTAAAAC	(5)
		rTufl	CCACCTTCACGAATAGAGAAC	(5)
		fTufAy	GCTAAAAGTAGAGCTTATGA	(5)
		rTufAy	CGTTGTCACCTGGCATTACC	(5)
tuf	16SrXII-B, distinguish tuf 1 and tuf 2 subtypes	PaTC1F	CATTCTAGTTGTTTCTGGT	(31)
		PaTC1R	GTCTTCTCGGTTAATC	(31)
		PaTC2F	TATTTTAGTTGTTTCTGGA	(31)
		PaTC2R	GTCTTCGCGGTTAATA	(31)
tuf	16SrV	FDTUF-F1	ATTGGTCATGTAGACCATGG	(30)
		FDTUF-R1	GTTCTTCCGCCTTCACGTAC	(30)
		FDTUF-F2	ATTGGTCATGTAGACCATGG	(30)
		FDTUF-R2	CTTGTTCCTTCTTCGATCGC	(30)
secY	16SrI	AYsecYF1	CAGCCATTTTAGCAGTTGGTGG	(29, 32)
		AYsecYR1	CAGAAGCTTGAGTGCCTTTACC	(29, 32)
secY	16SrII	SecYF1(II)	CGCGTATAGGTTTTGAAGGTG	(28)
		SecYR1(II)	CCTGCCATTTTCATTATAGCG	(28)
		SecYF2(II)	TGAAGGTGGTCAAACTCCT	(28)
secY	16SrIII	SecYF1(III)	CTAGACCAGGTTTTGAAGG	(28)
		SecYR1(III)	GACCTGCTTTTCTCATTATAGC	(28)
		SecYF2(III)	TGAAGGYGGACAAATCCCT	(28)
secY	16SrVI	SecYF1(VI)	CTAGATTAGGATTYGAGGG	(28)
		SecYR1(VI)	GACCRCCAAAACCTTGATAATC	(28)
		SecYF2(VI)	ATTYGAGGGYGGYCAAACAC	(28)
secY	16SrV	FD9f	GAATTAGAACTGTTTGAAGAC	(27)
		FD9r	TTTGCTTTCACATCTTGTATCG	(27)
		FD9r	TTTGCTTTCATATCTTGTRTCG	(21)
		FD9f2L	GTTTTAGCTAAAGGTGATTTAAC	(21)
		FD9r2L	TAAAAGACTAGTCCCRCCAAAAG	(21)
		FD9f3L	AATAAGGTAGTTTTATATGACAAG	(21)
secY	16SrX	SecYMalF1	TTAGGACGTAGTATACAAATCCCNTT	(11)
		SecYMalR1	ACAATAATTAAAAATCCTGTNCC	(11)
		SecYMalF2	AAGAATGGCGTGAACARGGNGA	(11)
		SecYMalR2	GCATCTTGTTTAGATAAATGTTC	(11)
secY	16SrXII-A	PosecF1	TCTGCTTTGCCTTTGCCTTT	(7)
		PosecR1	ATTAGTAAACTAGTTCCTCC	(7)
		PosecF3	GGATTGATAGATGCTGCCCC	(7)
		PosecN2	CCATCAAAACTTTTTGGTTTAGGC	
		PosecR3	GCCCCTATAACGGTGATTTTGA	(7)

about 2–3 min until agarose is totally dissolved but without overheating the gel mixture (see Note 1).

5. Ethidium bromide (3, 8-diamino-5-ethyl-6-phenylphenanthridinium bromide, EtBr): 2 mg/mL (see Notes 2 and 3).
6. Any suitable standard DNA loading dye.
7. DNA markers of known sizes as standards for electrophoretic migration.

2.3. Equipment

1. Micropipettes.
2. Thermo-cycler.
3. Electrophoresis unit.
4. Power supply.
5. UV transilluminator.
6. Documentation system.
7. Vortexer.
8. Microfuge.
9. Balance.
10. Microwave oven.

3. Methods

Keep all components and samples on ice. Thaw, vortex, and centrifuge all of them before use.

3.1. PCR Amplification Components

1. Prepare the 25 μL PCR assay by mixing (pipetting) the following compounds into a 0.2 mL PCR tube: 2.5 μL of 10× PCR buffer without $MgCl_2$, 2 μL of 25 mM $MgCl_2$, 17.85 μL of nuclease-free ddH_2O, 1 μL of 5 mM dNTPs, 0.25 μL of 100 μM primer 1, 0.25 μL of 100 μM primer 2, 0.2 μL *Taq* DNA polymerase (5 U/μL). A master mix can be prepared according to the number of assays to be processed, including water control and healthy plant negative controls and an appropriate phytoplasma-infected plant DNA positive control. Mix by vortexing and centrifuging at low speed with a bench microfuge and then dispense as aliquots of 24 μL into 0.2 mL PCR tubes.
2. Pipette 1 μL (i.e., 20–50 ng) of DNA template into the reaction mix.
3. Vortex gently to mix sample with the PCR reaction mix and centrifuge briefly in microfuge.

4. Program the PCR cycler and process the first (direct) PCR reaction according to the recommendations in Subheading 3.2.
5. For the second (nested) PCR, 1 μL of first round PCR product can be added to the nested PCR reaction with the same composition of components as in step 1 before. Sometimes, it is preferable to dilute the DNA: dilute the PCR product 40-fold in nuclease-free ddH_2O and use it as template in the following nested PCR reaction (see Note 4).
6. Repeat steps 3 and 4.
7. Keep the PCR products at 4°C for electrophoresis or sequencing. Long-term storage can be achieved by keeping the PCR reactions at –20°C.

3.2. PCR Cycling Conditions

3.2.1. Tuf Group 16SrI and 16SrXII-A (Fig. 1)

1. Use primers fTufl/rTufl for direct PCR with the conditions:

   Initial denaturation: 94°C for 90 s.

   Amplification: 35 cycles (94°C for 30 s, 45°C for 30 s, 72°C for 60 s).

   Final extension: 72°C for 10 min.

   Product size: 1,038 bp.
2. Use primers rTufAy/rTufAy for nested PCR with the conditions:

   Initial denaturation: 94°C for 4 min.

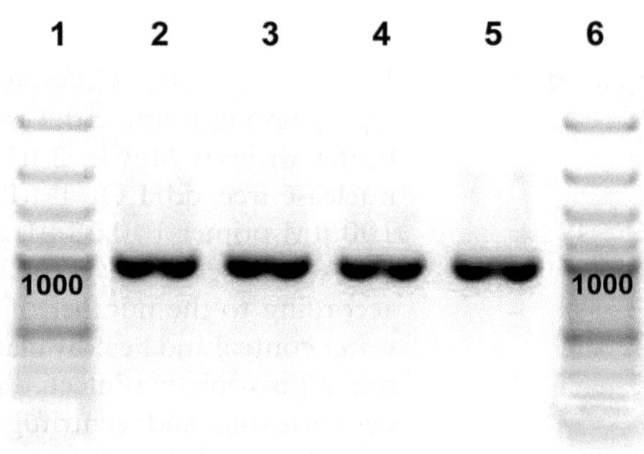

Fig. 1. *Tuf* gene PCR products amplified by the fTufAy/rTufAy primer pair. Lane 1 and 6: GeneRuler™ 100 bp Plus DNA Ladder (Fermentas); lanes 2–5: stolbur-infected plants of bindweed (2), nettle (3), grapevine (4), and tomato (5).

Amplification: 35 cycles (94°C for 30 s, 53°C for 30 s, 72°C for 60 s).

Final extension: 72°C for 5 min.

Product size: 940 bp.

3.2.2. Tuf Group 16SrXII-B

1. Use primers PaTC1F/PaTC1R (tuf 1 subtype) or PaTC2F/PaTC2R (tuf 2 subtype) with the conditions:

   Preamplification: 1 cycle (92°C for 2 min, 51°C for 1 min, 72°C for 2 min).

   Amplification: 35 cycles (92°C for 45 s, 51°C for 1 min, 72°C for 2 min).

   Product size: 553 bp.

3.2.3. Tuf Group 16SrV

1. Use primers FDTUF-F1/FDTUF-R1 for direct PCR with the conditions:

   Initial denaturation: 92°C for 60 s.

   Amplification: 35 cycles (92°C for 30 s, 56°C for 30 s, 66°C for 60 s).

   Final extension: 66°C for 5 min.

   Product size: 1,079 bp.

2. Use primers FDTUF-F2/FDTUF-R2 for nested PCR with the same conditions as the direct PCR.

   Product size: 998 bp.

3.2.4. SecY Group 16SrI (Ca. Phytoplasma asteris)

1. Use primers AYsecYF1/AYsecYR1 with the conditions:

   Initial denaturation: 94°C for 10 min.

   Amplification: 38 cycles (94°C for 1 min, 55°C for 2 min, 72°C for 3 min).

   Final extension: 72°C for 7 min.

   Product size: ~1.4 kb.

3.2.5. SecY Group 16SrII (Ca. Phytoplasma aurantifolia)

1. Use SecYF1(II)/SecYR1(II) primers for direct PCR with the conditions:

   Initial denaturation: 94°C for 1 min.

   Amplification: 35 cycles (94°C for 30 s, 50°C for 1 min, 68°C for 5 min).

   Final extension: 72°C for 10 min.

   Product size: ~1.7 kb.

2. Use SecYF2(II)/SecYR1(II) primers for nested PCR with the same conditions as the direct PCR:

   Product size: ~1.7 kb.

3.2.6. SecY Group 16SrIII (Western X Phytoplasma Group)

1. Use SecYF1(III)/SecYR1(III)primers for direct PCR with the conditions: Initial denaturation: 94°C for 1 min:

   Amplification: 35 cycles (94°C for 30 s, 50°C for 1 min, 68°C for 5 min).

   Final extension: 72°C for 10 min.

   Product size: ~1.7 kb.

2. Use SecYF2(III)/SecYR1(III) primers for nested PCR with the conditions:

   Initial denaturation: 94°C for 1 min.

   Amplification: 35 cycles (94°C for 30 s, 55°C for 1 min, 68°C for 5 min).

   Final extension: 72°C for 10 min.

   Product size: ~1.7 kb.

3.2.7. SecY Group 16SrVI (Ca. Phytoplasma trifolii)

1. Use SecYF1(VI)/SecYR1(VI) primers for direct PCR with the conditions:

   Initial denaturation: 94°C for 1 min.

   Amplification: 35 cycles (94°C for 30 s, 50°C for 1 min, 68°C for 5 min).

   Final extension: 72°C for 10 min.

   Product size: ~1.8 kb.

2. Use SecYF2(VI)/SecYR1(VI) primers for nested PCR with the conditions:

   Initial denaturation: 94°C for 1 min.

   Amplification: 35 cycles (94°C for 30 s, 55°C for 1 min, 68°C for 5 min).

   Final extension: 72°C for 10 min.

   Product size: ~1.7 kb.

3.2.8. SecY Group 16SrV

1. For direct PCR, either use FD9f/FD9r primers with the conditions:

   Initial denaturation: 92°C for 90 s.

   Amplification: 40 cycles (92°C for 30 s, 54°C for 30 s, 72°C for 80 s).

   Product size: 1.3 kb; or use FD9r/FD9f2L primers with the conditions:

   Initial denaturation: 92°C for 1 min.

   Amplification: 40 cycles (92°C for 1 min, 55°C for 1 min, 66°C for 30 s).

   Final extension: 66°C for 5 min.

   Product size: 1,343 bp.

2. Use FD9r2L/FD9f3L primers for nested PCR with the conditions:

   Initial denaturation: 92°C for 1 min.

   Amplification: 40 cycles (92°C for 1 min, 55°C for 1 min, 66°C for 30 s).

   Final extension: 66°C for 5 min.

   Product size: 1,174 bp.

3.2.9. SecY Group 16SrX (Ca. Phytoplasma mali, Ca. Phytoplasma pyri, Ca. Phytoplasma prunorum)

1. Use SecYMalF1/SecYMalR1 primers for direct PCR with the conditions:

   Initial denaturation: 95°C for 3 min.

   Amplification: 20 cycles (94°C for 30 s, 50°C for 30 s, 66°C for 45 s).

   Final extension: 66°C for 7 min.

   Product size: 664 bp.
2. Use SecYMalF2/SecYMalR2 primers for nested PCR with conditions as for the direct PCR but use 35 cycles of amplification:

   Product size: 691 bp.

3.2.10. Group 16SrXII-A (Stolbur 'Bois noir' Phytoplasma) (Fig. 2)

1. Use PosecF1/PosecR1 primers for direct PCR: with the conditions:

   Initial denaturation: 94°C for 3 min.

   Amplification: 35 cycles (94°C for 30 s, 62°C or 54°C for 30 s, 72°C for 60 s).

   Final extension: 72°C for 10 min.

   Product size: 1,052 bp.
2. Use PosecF3 or PosecN2 with PosecR3 for nested PCR with the conditions as for direct PCR

   Product size: 998 bp.

3.3. Agarose Gel Electrophoresis

1. Pour about 50–100 mL of melted 1% agarose (in 1× TAE buffer) into a sealed plastic tray in order to obtain approximately 5 mm depth of liquid gel. Immediately insert plastic comb to produce the wells for depositing the PCR samples. Remove any bubbles which may form by pricking them with a pipette tip.
2. While the gel is cooling to room temperature, assemble the electrophoresis apparatus.
3. Allow the gel to solidify at room temperature for 30–60 min.
4. Insert the tray with solidified gel into electrophoresis apparatus.
5. Fill the buffer reservoir with the 1× TAE buffer until it covers the gel; it should be 0.2–0.5 cm depth over the gel.

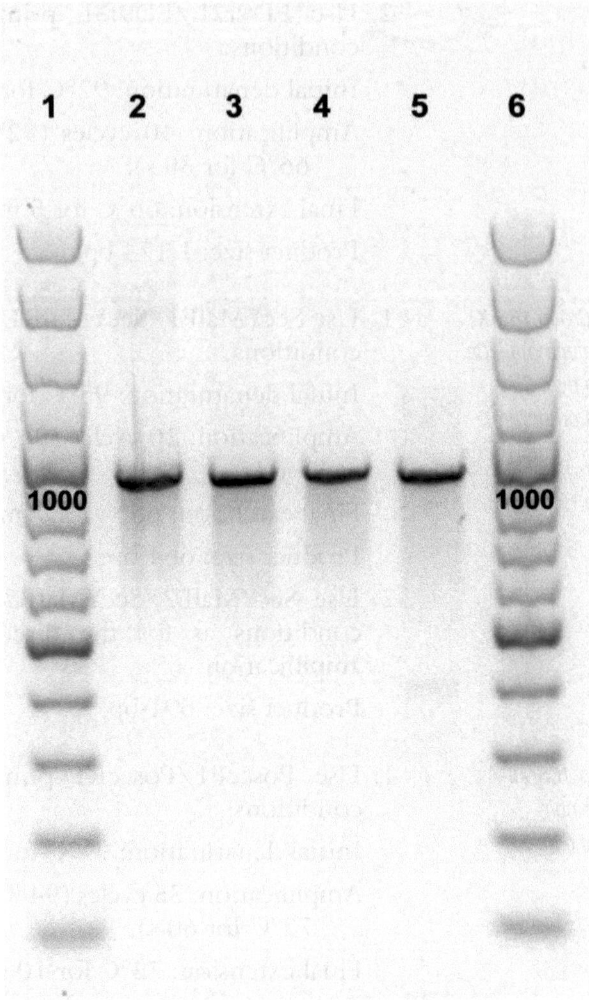

Fig. 2. *SecY* gene PCR products amplified by the PosecN2/PosecR3 primer pair. Lane 1 and 6: GeneRuler™ 100 bp Plus DNA Ladder (Fermentas); lanes 2–5: stolbur-infected plants bindweed (2), nettle (3), grapevine (4), and tomato (5).

6. Gently remove the comb.
7. Transfer 5–10 μL of PCR product into a new tube and add 0.5–1 μL of 10× loading dye previously warmed at 40°C, mix by vortexing or pipetting. If needed, collect droplets by brief centrifuging samples in a microfuge.

8. Load 2–10 μL of DNA ladder into the first and last wells of the gel and subsequently load 5.5–11 μL of sample into the other wells.
9. Assemble electrophoresis apparatus and connect electrodes to the power supply.
10. Run the gel at 80 V (about 8 V/cm of gel length). Stop electrophoresis when the dye front (bromophenol blue band) has traveled through two thirds of the gel.
11. Switch off power supply, unplug the electrophoresis apparatus, and transfer the gel for 15 min into a staining buffer containing 2 μg/mL ethidium bromide in nuclease-free ddH_2O (see Note 5).
12. Destain in 200 mL ddH_2O for 15 min.
13. Visualize the PCR products in the gel using a UV transilluminator at wavelength 302 nm (see Note 6) and take a photo for records (e.g., with a gel documentation system).
14. Stained gels can be dried at room temperature after use and must be disposed of by methods that protect the environment.

3.4. RFLP Analysis

1. RFLP analysis of PCR products is performed by cutting with restriction endonucleases according to the manufacturer's recommendations.
2. Restriction profiles can be visualized by gel electrophoresis (4, 28) (Fig. 3).

3.5. Sequencing of Amplicons and Further Analysis

1. Sequencing reactions of PCR products can be performed by methods adapted from the classical Sanger method followed by analysis on capillary sequencing instruments. All amplicons are directly sequenced on both strands to achieve 2× coverage.
2. Raw sequence chromatograms are assembled and edited using a variety of software packages, e.g., preGAP4 and GAP4 software (Staden Package, http://staden.sourceforge.net) or the PHRED-PHRAP-CONSED bioinformatic package (http://www.phrap.org).
3. Sequences can be RFLP-typed *in silico* using pDraw, MapDraw (Lasergene), or other software packages (Fig. 4).
4. Multiple alignments are performed using ClustalW software and the phylogenetic analyses using the method of maximum parsimony can be carried out with MEGA software (http://megasoftware.net) (Fig. 5). Validity of tree branches is evaluated by bootstrapping.

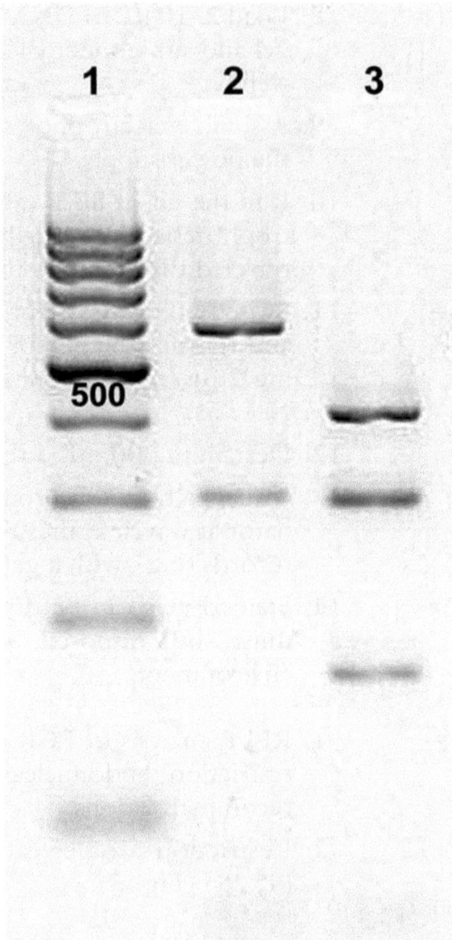

Fig. 3. *Hpa*II RFLP profiles of *tuf* gene amplified by the fTufAy/rTufAy primer pair. Lane 1: GeneRuler™ 100 bp DNA Ladder (Fermentas); lane 2: stolbur-infected bindweed plant, lane 3: stolbur-infected grapevine.

4. Notes

1. A larger volume can be prepared and stored at room temperature or 4°C until needed.
2. Ethidium bromide is a strong mutagen. Always use gloves when handling ethidium bromide-containing solutions. Solutions, gels, and all contaminated material, gloves, paper towels, tubes etc. should be disposed as a hazardous material.

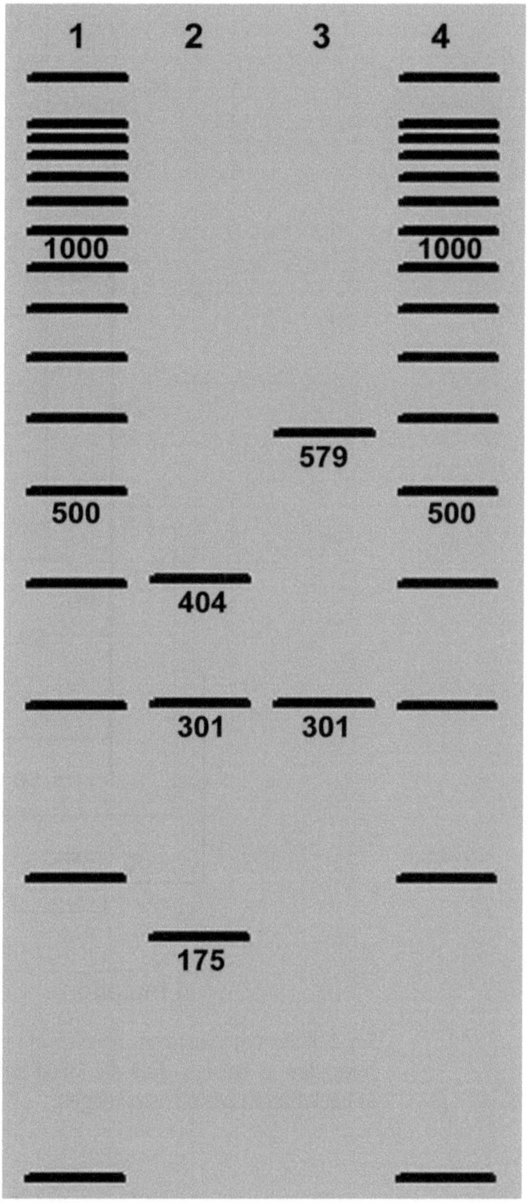

Fig. 4. Computer-simulated *Hpa*II RFLP profile of *tuf* gene; lanes 1 and 4 size standard, lane 2; '*Ca.* Phytoplasma solani' isolate R49/15 (GenBank no. FJ394551), lane 3: '*Ca.* P. solani' isolate R47/5 (GenBank no. FJ394552).

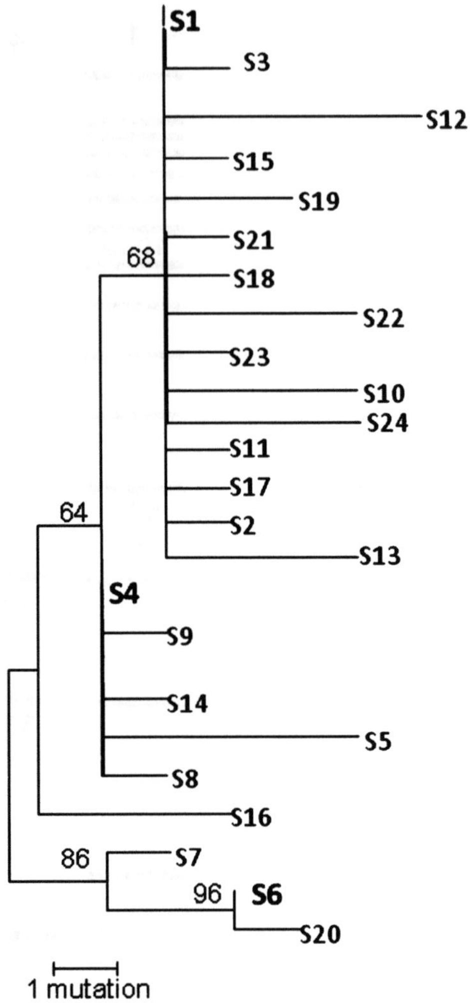

Fig. 5. Maximum parsimony analysis of reference *secY* sequences (S1–S24) from stolbur phytoplasma isolates using the MEGA software. Numbers on main branches correspond to bootstrap values as percentages.

If possible, purchase a liquid stock solution to avoid having to handle the powder.

3. It is recommended to use alternative nonmutagenic nucleic acid stains, such as GelRed™ Nucleic Acid Stain (Biotium Ltd.), GoodView™ Nucleic acid stain (SBS), Ultra Power™ Nucleic Acid Stain (BioTeke Corporation).
4. This is used because of the high efficiency of the first PCR, the dilution step between PCR and nested PCR is, however, a source of aerosol DNA contamination producing false positives. A recommended alternative is to reduce the number of cycles of the first PCR.

5. The ethidium bromide stain solution can be used 2 months if protected from light. It must be disposed of by methods that protect the environment.
6. UV light is carcinogenic. Always use a protective faceshield, glasses, and gloves to avoid the direct exposure of eyes and skin to UV light.

References

1. The IRPCM Phytoplasma/Spiroplasma Working Team—Phytoplasma taxonomy group (2004) '*Candidatus* Phytoplasma', a taxon for the wall-less, non-helical prokaryotes that colonize plant phloem and insects. Int J Syst Evol Microbiol 54:1243–1255
2. Firrao G, Gibb K, Streten C (2005) Short taxonomic guide to the genus '*Candidatus* Phytoplasma'. J Plant Pathol 87:249–263
3. Hodgetts J et al (2008) Phytoplasma phylogenetics based on analysis of *secA* and 23S rRNA gene sequences for improved resolution of candidate species of 'Candidatus Phytoplasma'. Int J Syst Evol Microbiol 58:1826–1837
4. Langer M, Maixner M (2004) Molecular characterisation of grapevine yellows associated phytoplasmas of the stolbur-group based on RFLP-analysis of non-ribosomal DNA. Vitis 43:191–199
5. Schneider B, Gibb KS, Seemüller E (1997) Sequence and RFLP analysis of the elongation factor Tu gene used in differentiation and classification of phytoplasmas. Microbiology 143:3381–3389
6. Cimerman A et al (2009) Striking diversity of *vmp1*, a variable gene encoding a putative membrane protein of the stolbur phytoplasma. Appl Environ Microbiol 75:2951–2957
7. Fialová R et al (2009) Genetic variability of stolbur phytoplasma in annual crop and wild plant species in south Moravia (Czech Republic). J Plant Pathol 91:411–416
8. Murolo S et al (2010) Genetic variability of the stolbur phytoplasma *vmp1* gene in grapevines, bindweeds and vegetables. J Appl Microbiol 109:2049–2059
9. Pacifico D et al (2009) Characterization of Bois noir isolates by restriction fragment length polymorphism of a Stolbur-specific putative membrane protein gene. Phytopathology 99: 711–715
10. Berg M et al (1999) Isolation of the gene encoding an immunodominant membrane protein of the apple proliferation phytoplasma, and expression and characterization of the gene product. Microbiology 145:1937–1943
11. Danet JL et al (2011) Multilocus sequence analysis reveals the genetic diversity of European fruit tree phytoplasmas and the existence of inter-species recombination. Microbiology 157:438–450
12. Danet JL et al (2007) *Imp* and *secY*, two new markers for MLST (multilocus sequence typing) in the 16SrX phytoplasma taxonomic group. Bull Insectol 60:339–340
13. Kakizawa S et al (2009) Cloning of immunodominant membrane protein genes of phytoplasmas and their *in planta* expression. FEMS Microbiol Lett 293:92–101
14. Kakizawa S, Oshima K, Namba S (2006) Diversity and functional importance of phytoplasma membrane proteins. Trends Microbiol 14:254–256
15. Kakizawa S et al (2006) Positive selection acting on a surface membrane protein of the plant-pathogenic phytoplasmas. J Bacteriol 188: 3424–3428
16. Fabre A, Danet JL, Foissac X (2011) The stolbur phytoplasma antigenic membrane protein gene *stamp* is submitted to diversifying positive selection. Gene 472:37–41
17. Schneider B, Seemüller E (2009) Strain differentiation of *Candidatus* Phytoplasma mali by SSCP- and sequence analyses of the *hflB* gene. J Plant Pathol 91:103–112
18. Seemüller E, Schneider B (2007) Differences in virulence and genomic features of strains of 'Candidatus Phytoplasma mali', the apple proliferation agent. Phytopathology 97:964–970
19. Economou A (1999) Following the leader: bacterial protein export through the Sec pathway. Trends Microbiol 7:315–320
20. Angelini E et al (2003) Phylogenetic relationships among Flavescence doreé strains and related phytoplasmas determined by heteroduplex mobility assay and sequence of ribosomal and nonribosomal DNA. Plant Pathol 52: 663–672
21. Arnaud G et al (2007) Multilocus sequence typing confirms the close genetic inter-relatedness of three distinct flavescence dorée phytoplasma strain clusters and group 16SrV

phytoplasmas infecting grapevine and alder in Europe. Appl Environ Microbiol 73: 4001–4010

22. Clair D et al (2003) A multiplex nested-PCR assay for sensitive and simultaneous detection and direct identification of phytoplasma in the Elm yellows group and Stolbur group and its use in survey of grapevine yellows in France. Vitis 42:151–157
23. Daire X et al (1992) Cloned DNA probes for detection of grapevine Flavescence doreé mycoplasma-like organism (MLO). Ann Appl Biol 121:95–103
24. Daire X et al (1993) Diversity among mycoplasma-like organisms inducing grapevine yellows in France. Vitis 32:159–163
25. Daire X et al (1997) Survey for grapevine yellows phytoplasmas in diverse European countries and Israel. Vitis 36:53–54
26. Daire X et al (1993) Occurrence of diverse MLOs in tissues of grapevine affected by grapevine yellows in different countries. Vitis 32:247–248
27. Daire X et al (1997) Detection and differentiation of grapevine yellows phytoplasmas belonging to the elm yellows group and to the stolbur subgroup by PCR amplification of non-ribosomal DNA. Eur J Plant Pathol 103:507–514
28. Lee IM et al (2010) Phylogenetic analysis and delineation of phytoplasmas based on *secY* gene sequences. Int J Syst Evol Microbiol 60: 2887–2897
29. Lee IM, Zhao Y, Bottner KD (2006) *SecY* gene sequence analysis for finer differentiation of diverse strains in the aster yellows phytoplasma group. Mol Cell Probes 20:87–91
30. Malembic-Maher S et al (2011) Genetic diversity of European phytoplasmas of the 16SrV taxonomic group and proposal of '*Candidatus* Phytoplasma rubi'. Int J Syst Evol Microbiol 61:2129–2134
31. Andersen MT et al (2006) Phylogenetic analysis of "*Candidatus* Phytoplasma australiense" reveals distinct populations in New Zealand. Phytopathology 96:838–845
32. Lee IM et al (2004) '*Candidatus* Phytoplasma asteris', a novel phytoplasma taxon associated with aster yellows and related diseases. Int J Syst Evol Microbiol 54:1037–1048
33. Ahrens U, Seemüller E (1992) Detection of DNA of plant pathogenic mycoplasmalike organisms by a polymerase chain reaction that amplifies a sequence of the 16S rRNA gene. Phytopathology 82:828–832
34. Maixner M, Ahrens U, Seemuller E (1995) Detection of the German grapevine yellows (Vergilbungskrankheit) MLO in grapevine, alternative hosts and a vector by a specific PCR procedure. Eur J Plant Pathol 101: 241–250

Chapter 17

PCR Analysis of Phytoplasmas Based on the *secA* Gene

Matt Dickinson and Jennifer Hodgetts

Abstract

Conventionally, diagnostics and phylogenetics of phytoplasmas have been primarily based on the 16S rRNA gene, for which "universal" primers are available that amplify from most phytoplasma 16Sr groups. However, there has been a drive in recent years to develop "universal" primers for other genes that can be used to complement the use of the 16S rRNA gene. This chapter details the use of primers based on the phytoplasma *secA* gene and describes how these primers can be used in both a single or nested PCR approach for amplification. It also notes the use of appropriate controls that should be undertaken and provides a source for phytoplasma *secA* sequences that are available in databases that can be used for phylogenetic analyses.

Key words: Diagnostics, Polymerase chain reaction, *secA* gene, *cox* gene, Phylogenetic analysis

1. Introduction

Phytoplasma diagnostics and phylogenetics have historically been based on the 16S rRNA gene because of the availability of "universal" primers for this region (1) (*see also* Duduk et al. (2), Chapter 14). However, it has also been noted that the standard "universal" primers can amplify products from closely related bacteria such as *Bacillus* spp., leading to occasional false positives and misdiagnosis (3). For phylogenetic analysis of organisms, it has now been recognized that phylogenetics based on a single highly conserved gene such as the 16S rRNA has limitations, particularly when it comes to defining subgroups and strain differences within subgroups, and the approach that is now being used for most other organisms is to rebuild trees by combining sequence data from a range of different genes. Although it has been possible to find

Matt Dickinson and Jennifer Hodgetts (eds.), *Phytoplasma: Methods and Protocols*, Methods in Molecular Biology, vol. 938, DOI 10.1007/978-1-62703-089-2_17,

primers for a number of phytoplasma genes that can be used to amplify from subgroups within a 16Sr group and also to sometimes amplify from closely related groups, finding a gene that can be used for designing universal primers that amplify from all phytoplasma groups has proven difficult because of the sequence divergence between strains. Primers based on the ribosomal protein (*rp*) genes have been developed by Martini et al. (4) and Martini and Lee (5), Chapter 15, and these have been validated for most 16Sr groups, although they do not appear to amplify from members of the 16SrII and 16SrXI groups. Here we provide details of primers based on the *secA* gene which, like the *rp* gene primers, can amplify from all major 16Sr groups including 16SrII but, like the *rp* gene primers, they are also unreliable on 16SrXI phytoplasmas.

The *secA* gene, which encodes SecA, the ATP-dependent force generator in the bacterial precursor protein translocation cascade system, was originally sequenced from the onion yellows phytoplasma as part of the genome sequencing project (6). In subsequent work, we aligned the onion yellows phytoplasma *secA* gene sequence with that of aster yellows witches' broom phytoplasma and the equivalent gene from coconut lethal yellowing phytoplasma, which is in the 16SrIV group (7). Based on these alignments, three degenerate primers were designed that can be used in a seminested PCR assay: SecAFor1, SecAFor2, and SecARev3. When primer pair SecAFor1/SecARev3 were used in PCR, products of about 840 bp were generated from samples where phytoplasma was present in high titer. Reamplification of diluted first-round products with SecAFor2/SecARev3 resulted in a product of 480 bp, including for samples where the phytoplasma was not detected in the first round. However, as reported in (7), these nested primers can also result in the presence of other PCR products of different sizes, which can lead to confusion in the diagnosis. We have therefore subsequently redesigned and developed new primers (SecAFor5 and SecARev2) to produce a fully nested approach in which the nested PCR results in a clear and unambiguous product of about 600 bp (8). Furthermore, we have noted in unpublished observations that while the first-round primers can sometimes produce products from other endosymbiotic bacteria such as *Wolbachia* spp., particularly from insect samples, when these samples are subjected to nested PCR there is no further amplification, and to date we have never found products from any other organisms apart from phytoplasmas using the fully nested approach.

To complement the use of the *secA* gene primers, we also report here on a supplementary assay based on the plant cytochrome oxidase (*cox*) gene (9) that we recommend for use on all samples from plants that appear to be negative following PCR with the *secA* primers. These primers have been shown to amplify a product of approximately 200 bp from all plant species that we have tested them on. They can therefore be used to confirm that DNA samples support PCR and do not contain PCR inhibitors, and therefore to rule out false negatives. Finally, we note and describe a method for rapid DNA cleanup that can be used to remove PCR inhibitors in samples that do not support PCR, based on the method of Cullen and Hirsch (10).

2. Materials

All stock solutions should be prepared with sterile deionized water and in an area that is distinct from where plant and phytoplasma DNA is being handled. Gilson pipettes (or equivalent) specifically dedicated to non-DNA work should be used for all primer and PCR reaction preparations, with a separate set used for pipetting of DNA and gel electrophoresis.

2.1. PCR

1. PCR tubes (200 or 500 μL).
2. Eppendorf tubes (1.5 mL).
3. Primers (10 μM) (Table 1).
4. PCR reagents. We generally use either IllustraPuReTaq Ready-To-Go™ PCR Beads (GE Healthcare, Amersham, UK) or MangoMix™ (Bioline, London UK) (see Note 1).
5. Vortex mixer.
6. PCR thermocycler.
7. Pipettors (20, 200, and 1,000 μL).
8. Filter tips (20, 200, and 1,000 μL).
9. Sterile distilled water (ddH_2O).

2.2. Gel Electrophoresis

1. Agarose.
2. Horizontal electrophoresis apparatus.
3. TBE buffer (10× stock): weigh 108 g Trizma base and 55 g boric acid. Add 40 mL 0.5 M EDTA (ethylenediaminetetra-acetic acid) (see Note 2) and make up to 1 L with distilled water. Store at room temperature. For 1× buffer, dilute 10 mL 10× buffer with 90 mL distilled water.
4. GelRed™ (Biotium, Cambridge, UK) nucleic acid stain. We use the 10,000× stock solution (see Note 3).
5. DNA ladder (1 kb): In this protocol we use Promega 1 kb ladder, but others may be used.
6. UV transilluminator and gel capture photography system.

Table 1
Primers sequences for the *secA* gene used in this protocol

Name	Sequence (5′-3′)
SecAFor1	GARATGAAAACTGGRGAAGG
SecARev3	GTTTTRGCAGTTCCTGTCATNCC
[a]SecAFor5-u	ASTCGTGAAGCTGAAGG
[a]SecAFor5-1	AGCTAAAAGAGAATTTGAAGG
[a]SecAFor5-LY	CTGATAGAGAAGCTAATGG
SecARev2	CCNTCRCTAAATTGNCGTCC
CoxF3	TATGGGAGCCGTTTTTGC
CoxB3	ACTGCTAAGRGCATTCC

[a]The SecAFor5 primer is a mixture of these 3 primers, each at 10 μM

2.3. DNA Cleanup

1. Micro Bio-Spin chromatography column (3 cm) (Bio-Rad, Hemel Hempstead, UK).
2. Poly(vinylpolypyrrolidone) (PVPP) powder.
3. Microfuge.
4. Microcentrifuge tubes (1.5 mL).
5. Sterile distilled water (ddH_2O).

3. Methods

Carry out all procedures at room temperature unless otherwise specified.

3.1. Preparation of secA PCR Reaction Mix and Performing PCR

1. Label PCR tubes.
2. Make PCR mastermixes for $n+1$ samples. The volumes will depend on how many samples are to be tested including a positive and water control (n). For example, if there are 6 samples to be tested, then $n=8$ once the positive and water controls are included. The mastermix would therefore be made for $n+1=9$ reactions. If there are 20 samples to be tested, $n=22$, and a mastermix would be made for 23 reactions, etc.
3. The composition of the mastermix will depend on whether PuReTaq Ready-To-Go™ PCR Beads or MangoMix™ is being used. For PuReTaq Ready-To-Go™ PCR Beads it will consist of $(n+1)\times(22$ μL ddH_2O + 1 μL SecAFor1 primer + 1 μL SecARev3 primer). For MangoMix™, the mastermix will

consist of $(n+1) \times (9.5\ \mu L\ ddH_2O + 12.5\ \mu L$ MangoMix™ + 1 μL SecAFor1 primer + 1 μL SecARev3 primer).

4. Wearing gloves, and in an area that is away from where plant and phytoplasma DNA are being handled, and using Gilson pipettes (or equivalent) specifically dedicated to non-DNA work with filter tips, pipette the required volume of mastermix reagent into a 1.5 mL Eppendorf tube. For example, if 6 samples are to be tested plus a positive and negative control ($n=8$ and $n+1=9$), pipette 198 μL ddH_2O, 9 μL SecAFor1 primer, and 9 μL SecARev3 primer (for PuReTaq Ready-To-Go™ PCR Beads) or pipette 85.5 μL ddH_2O, 112.5 μL MangoMix™, 9 μL SecAFor1 primer, and 9 μL SecARev3 primer (for MangoMix™). Vortex to mix and then aliquot 24 μL onto the PuReTaq Ready-To-Go™ PCR Beads in labeled tubes (if using beads) or directly into labeled PCR tubes (if using MangoMix).
5. The tubes are now ready for the addition of sample DNA, which may have been prepared by a range of strategies (see protocols for preparation of DNA). Close the lids on the PCR tubes and take them from the phytoplasma DNA-free area to an area in which phytoplasma DNA can be handled. Using a different Gilson pipette to the one used for making the mastermixes, with filter tips, add 1 μL DNA for each sample into the appropriate PCR tube. For the positive control, add 1 μL DNA from a sample that has previously been proven to work in the secA PCR protocol; for the negative control, add 1 μL ddH_2O.
6. Close the lids and place in the thermocycler. PCR is conducted as follows: 1 cycle at 94°C for 2 min; 30 cycles of 94°C for 30 s, 53°C for 1 min, and 72°C for 1 min; and 1 cycle at 72°C for 10 min.
7. Following the first round of PCR, the products may be stored long term in the freezer, visualized by agarose gel electrophoresis (see Subheading 3.2), or used for nested PCR before gel electrophoresis. We generally undertake the nested PCR before any gel electrophoresis.
8. For nested PCR, make up mastermixes as before in the phytoplasma DNA-free area, but using primers SecAFor5 and SecARev2 instead of SecAFor1 and SecARev3. Aliquot 24 μL aliquots into labeled PCR tubes as before, close the lids, and transfer to the area in which phytoplasma DNA is permitted. Also, in the phytoplasma DNA-free area, aliquot 39 μL aliquots of ddH_2O into $n \times 0.5$ mL microfuge tubes, close the lids, and transfer these tubes to the area in which phytoplasma DNA is permitted.
9. In the area in which phytoplasma DNA can be used, dilute the first-round PCR products 40-fold by adding 1 μL into the 39 μL aliquots of ddH_2O in the 0.5 mL microfuge tubes.

Gently vortex and then add 1 μL from these dilutions into the nested PCR reagents.

10. Close the lids and place in the thermocycler. PCR is conducted using the same conditions as for the first round: 1 cycle at 94°C for 2 min; 30 cycles of 94°C for 30 s, 53°C for 1 min, and 72°C for 1 min; and 1 cycle at 72°C for 10 min.
11. Following the PCR, samples may be stored in the freezer or used for agarose gel electrophoresis.

3.2. Agarose Gel Electrophoresis

1. To prepare 100 mL of a 1 % agarose gel, weigh out 1 g agarose into a 500 mL Duran bottle (or equivalent) and add 100 mL 1× TBE. Dissolve the agarose carefully in a microwave, then cool to about 50°C. Add 10 μL 10,000× GelRed™ solution, mix and pour into a gel tray. Insert comb, remove air bubbles, and allow to solidify.
2. Once cool, remove the comb and place the gel in an electrophoresis chamber. Submerge the gel in 1× TBE buffer.
3. Load 5 μL PCR products into the wells. If the PCR has been performed using MangoMix, no gel loading dye will need to be added to samples. However, if the samples have been prepared using PCR beads, loading dye will need to be added (see Note 4). Always add the DNA ladder in the initial and/or final wells (see Note 5).
4. Connect the electrodes to the power source (5 V/cm) and run the gel until the loading dye is near the bottom of the gel. Remove the gel from the tank and visualize over a UV transilluminator (see Note 6). Photograph and print/store in computer/lab book.
5. SecA PCR products for first-round PCR should be approximately 840 bp, while those from the nested PCR should be approximately 600 bp. If these products are seen in the positive control samples and nothing is present in the water control sample, it indicates that the PCR amplification has been successful. If nothing is visible in the positive control samples, the PCR will need to be repeated. If products are seen in the water control, the experiment will also need repeating but equipment may need cleaning and/or solutions replacing first.
6. For the test samples, if PCR products of the correct size are seen in first-round PCR and nested PCR, it indicates that the phytoplasma is present at high titer. If PCR products of the correct size are seen in nested PCR only, it indicates that the phytoplasma is present at lower titer. If PCR products of the correct size are seen after the first round but not after nested PCR, this suggests the product is not from phytoplasma

but from different bacteria such as a *Wolbachia* spp. If no PCR products are seen in a test sample, this indicates that either there is no phytoplasma in the sample or that the DNA does not support PCR (possibly due to the presence of PCR inhibitors). We therefore advocate that all DNA samples which are negative following nested PCR should be retested with the *cox* primers if derived from plant DNA samples to confirm that they support PCR (see Note 7).

3.3. Cox Primer PCR Validation

1. Label PCR tubes and make PCR mastermixes for $n+1$ samples as in Subheading 3.1, but using primers CoxF3 and CoxB3 instead of SecAFor1 and SecARev3. Aliquot the mastermix into the PCR tubes, close lids, and transfer to the DNA-permissive area.
2. Add 1 μL DNA sample to PCR mix as previously as well as preparing appropriate positive and water controls.
3. Close the lids and place in the thermocycler. PCR is conducted as follows: 1 cycle at 94°C for 2 min; 30 cycles of 94°C for 30 s, 53°C for 1 min, and 72°C for 1 min; and 1 cycle at 72°C for 10 min.
4. Visualize PCR products by agarose gel electrophoresis as described in Subheading 3.2. PCR products should be approximately 200 bp. Assuming the positive and negative controls have worked, a PCR product of the correct size in a test sample indicates that the DNA supports PCR and does not contain inhibitors. Therefore, if a sample has been shown to be positive with the *cox* primers but negative with the *secA* primers, it would indicate that it contains no phytoplasma and is a true negative (see Note 8). However, if the sample is negative with both *cox* and *secA* primers, this indicates the presence of inhibitors of PCR in the DNA sample and that it needs to be further cleaned (see Subheading 3.4).

3.4. DNA Cleanup Procedure

1. Take a Micro Bio-Spin Chromatography column (3 cm) and snap off the thin flat end. Place the column into a 1.5 mL Eppendorf tube with the lid removed.
2. Fill the bottom thin section of the spin columns 2/3 of the way up with PVPP powder.
3. Add 400 μL of ddH_2O and centrifuge at 1000×g in a benchtop microfuge for 3 min.
4. Discard the flow-through.
5. Add 200 μL of ddH_2O to the column and centrifuge at 1000×g for 3 min.

6. Transfer the column into a clean (labeled) 1.5 mL Eppendorf tube with the lid removed.
7. Add the DNA extract in a 200 μL volume (see Note 9) to the column and centrifuge at 1000 × g for 3 min.
8. Transfer this cleaned-up DNA to a new (labeled) Eppendorf tube. This DNA can now be used directly in PCR with the *cox* and/or *secA* primers as described in Subheadings 3.1 and 3.3 or it can be ethanol precipitated to concentrate the DNA. Normally one round of cleanup is sufficient to remove PCR inhibitors from a DNA sample. However, if the sample still does not support PCR after cleanup we recommend that fresh DNA samples are prepared from the original plant material.

3.5. Phylogenetic Analysis of secA Sequences

1. Following successful PCR of samples using the *secA* primers to confirm the presence of phytoplasmas, further analysis of the sample can be undertaken by cloning and sequencing, followed by phylogenetic analysis and comparison with known phytoplasma *secA* gene sequences.
2. Methods for cloning and sequencing of PCR products have been described in detail by others so will not be described here, and kits are available for such procedures from many different suppliers. We recommend that PCR products are cleaned using PCR cleanup kits, then cloned into appropriate cloning vectors and sequenced in both directions using standard procedures.
3. Following sequencing and editing of sequences to remove plasmid vector sequence, the nucleic acid sequence should be put through DNA translation software (see Note 10) to confirm that it has an open reading frame running through the length of the sequence. This is a useful validation of the DNA sequence to confirm that errors have not been introduced during the sequencing.
4. The sequence can then be compared using the BLAST program to NCBI databases. However, there are only a relatively small number of phytoplasma *secA* gene sequences available in NCBI databases. We therefore recommend that a better procedure for comparison of sequences and confirmation of the 16Sr group is to make a FASTA file of the *secA* sequences that have been published in Hodgetts et al. (7) and then use the MEGA4 software (11) (see Note 11) to align your test sample(s) sequence with the other phytoplasma sequences and construct phylogenetic trees.

4. Notes

1. Ready-prepared PCR mixes such as PCR beads or MangoMix are recommended because fewer stock solutions and less pipetting are required, reducing the risk of contamination. PCR beads are particularly recommended in situations where reagents need to be sent to remote locations and/or electricity supplies and storage in freezers is unreliable, because the beads can be stored long term at room temperature. However, other sources of *Taq* polymerase and PCR mixes can be used.
2. To prepare 0.5 M EDTA, add 186.1 g EDTA ($Na_2EDTA.2H_2O$) to 800 mL of ddH_2O. Using a pH meter, add about 20 g of NaOH pellets while stirring to adjust the pH to 8.0. Add the last few grams slowly to avoid overshooting the pH and make sure the pellets are completely dissolved before adding the next one. Note that the EDTA will not completely dissolve until the pH is around 8.0. Once prepared, the EDTA solution should be autoclaved and stored at room temperature.
3. In this protocol, we recommend the use of GelRed™ rather than ethidium bromide for staining of DNA in gels. GelRed can be visualized using transilluminators at the same wavelength as ethidium bromide, but is considered a much safer product such that no special safety procedures are required for handling or disposal of gels, solutions, etc. Ethidium bromide can be used instead of GelRed, but ensure that appropriate local safety regulations are followed for the correct use and disposal of ethidium bromide contaminated materials.
4. MangoMix and many equivalent PCR reagents already contain loading dyes so no additional dyes have to be added to such products before gel electrophoresis. However, if PCR beads are used, a loading dye mixture such as bromophenol blue/ xylene cyanol or Orange G should be used (12).
5. DNA ladders such as 1 kb ladders should be used according to manufacturer's instructions.
6. Transilluminators are used to visualize GelRed/ethidium bromide-stained DNA in agarose gels. Many laboratories now use gel documentation systems in which the transilluminators are enclosed with the camera system such that images are then captured onto computer screens, so the operator is not exposed to the UV light source. However, if the system does expose the operator directly to the UV light source, ensure that appropriate safety visors and goggles are worn for protection.
7. The *cox* primers have been developed to amplify from most plant DNA samples and will produce a product of approximately 200 bp from such samples. Therefore, they should be

used to verify DNA extracts that have been made from plants. For DNA extracts that have been made from insects, other primers that are universal for insect DNA would need to be used, but we have not developed a protocol for such assays.

8. Validation work at the University of Nottingham has shown that the *secA* primers work effectively for all phytoplasma phylogenetic groups tested (16SrI, II, III, IV, V, VI, VII, IX, X, XI, XII, XIII, XIV, XXII as defined by Wei et al. (13) apart from the 16SrXI group). Therefore, a negative result with these primers does not rule out the possibility that the sample contains a 16SrXI phytoplasma.
9. This DNA cleanup procedure is intended for 200 μL volume DNA samples with a concentration of approximately 0.1–10 mg/mL.
10. The ExPASy online DNA translation tool at http://web.expasy.org/translate/ is a convenient means of translating sequences.
11. The MEGA4 software can be downloaded free from http://www.megasoftware.net/.

Acknowledgement

We would like to thank Defra, BBSRC, DfiD, and the Leverhulme Society for funding various aspects of this work, and Dr Nigel Harrison, University of Florida, Fort Lauderdale, USA for providing us with the *secA* sequence of the 16SrIV lethal yellows phytoplasma that enabled us to design the initial primers. We would also like to thank all individuals and organizations that have provided us with phytoplasma DNA/infected plant samples to enable us to validate the assays.

References

1. Smart CD et al (1996) Phytoplasma-specific PCR primers based on sequences of the 16S–23S rRNA spacer region. Appl Environ Microbiol 62:2988–2993
2. Duduk B, Paltrinieri S, Lee IM, Bertaccini A (2012) Nested PCR and RFLP analysis based on the 16S rRNA gene. In: Dickinson M, Hodgetts J (eds) Methods in molecular biology. Springer, New York
3. Harrison NA, Womack M, Carpio ML (2002) Detection and characterization of a lethal yellowing (16SrIV) group phytoplasma in Canary Island date palms affected by lethal decline in Texas. Plant Dis 86:676–681
4. Martini M et al (2007) Ribosomal protein gene-based phylogeny for finer differentiation and classification of phytoplasmas. Int J Syst Evol Microbiol 57:2037–2051
5. Martini M, Lee IM (2012) PCR and RFLP analyses based on the ribosomal protein operon. In: Dickinson M, Hodgetts J (eds) Methods in molecular biology. Springer, New York
6. Kakizawa S et al (2001) Cloning and expression analysis of *Phytoplasma* protein transloca-

tion genes. Mol Plant Microbe Interact 14:1043–1050

7. Hodgetts J et al (2008) Phytoplasma phylogenetics based on analysis of *secA* and 23S rRNA gene sequences for improved resolution of candidate species of '*Candidatus* Phytoplasma'. Int J Syst Evol Microbiol 58:1826–1837
8. Bekele B et al (2011) Use of a real-time LAMP isothermal assay for detecting 16SrII and XII phytoplasmas in fruit and weeds of the Ethiopian Rift Valley. Plant Pathol 60:345–355
9. Tomlinson JA et al (2005) On-site DNA extraction and real-time PCR for detection of *Phytophthora ramorum* in the field. Appl Environ Microbiol 71:6702–6710
10. Cullen DW, Hirsch PR (1998) Simple and rapid method for direct extraction of microbial DNA from soil for PCR. Soil Biol Biochem 30:983–993
11. Tamura K et al (2007) MEGA4: molecular evolutionary genetics analysis (MEGA) software version 4.0. Mol Biol Evol 24:1596–1599
12. Sambrook J, Russell DW (2001) Molecular cloning: a laboratory manual, 3rd edn. Cold Spring Harbour Press, New York
13. Wei W et al (2007) Computer-simulated RFLP analysis of 16S rRNA genes: identification of ten new phytoplasma groups. Int J Syst Evol Microbiol 57:1855–1867

Chapter 18

Single-Strand Conformation Polymorphism Analysis for Differentiating Phytoplasma Strains

Martina Šeruga Musić and Dijana Škorić

Abstract

Single-strand conformation polymorphism (SSCP) analysis is a sensitive and rapid technique for detecting DNA polymorphisms and mutations in PCR-amplified fragments. Due to its technical simplicity, it is widely used as a screening tool in various investigations, ranging from clinical diagnosis of human hereditary diseases to the characterization of microbial communities. This method can also be used successfully on phytoplasmas as a tool for the detection of molecular variability in conserved housekeeping genes such as 16S rRNA and *tuf*, as well as in more variable genes, revealing the presence of polymorphisms undetected by routine RFLP analyses. The reliability of SSCP has been confirmed by multiple alignments and phylogenetic analyses of representative sequences showing different SSCP profiles. However, it is not broadly applied in phytoplasma research yet. The technique provides an inexpensive, convenient, and sensitive method for determining sequence variation and to differentiate phytoplasma strains, and is particularly suitable for epidemiological studies or as a fast screening, typing tool when dealing with a large number of field samples.

Key words: Fingerprinting, Genetic polymorphism, Mutations, PCR, Sequence, SSCP, Typing, Variability

1. Introduction

Single-strand conformation polymorphism (SSCP) analysis is one of the technically simplest methods for rapid detection of nucleotide changes in any given DNA sequence. The technique is based on the differing electrophoretic mobilities of single-stranded DNA (ssDNA) fragments that have different primary structures (1). Fragments of DNA, usually amplified by PCR, are chemically and/or heat-denaturated and then subjected to electrophoresis through a non-denaturing polyacrylamide gel (Fig. 1). Single-stranded DNAs of differing primary structures fold into different conformations as a result of self-complementarity and intramolecular

Matt Dickinson and Jennifer Hodgetts (eds.), *Phytoplasma: Methods and Protocols*, Methods in Molecular Biology, vol. 938, DOI 10.1007/978-1-62703-089-2_18,

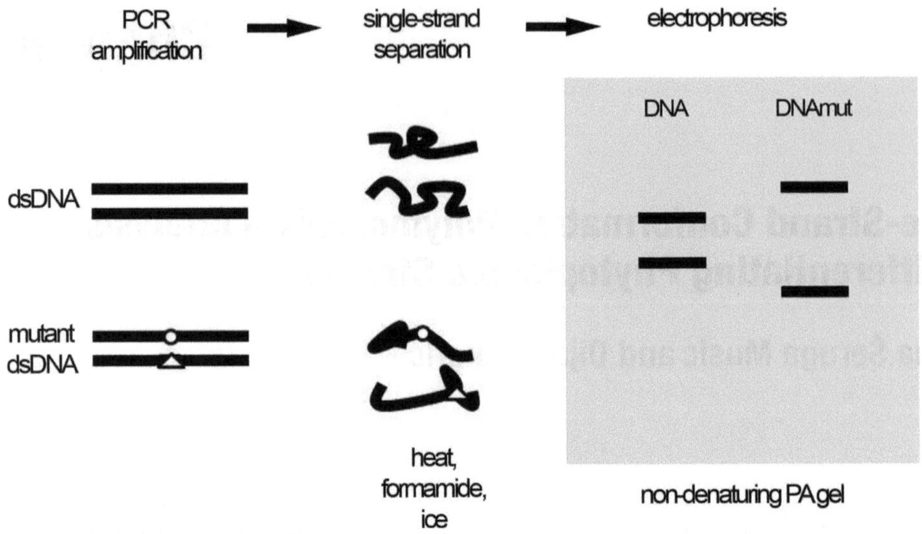

Fig. 1. Scheme of underpinning principles of single-strand conformation polymorphism (SSCP) analysis.

interactions. Under the appropriate electrophoretic conditions, conformationally distinct ssDNAs migrate at different speeds, and the presence of mutations is detected as a band shift or as a change in the number of bands. Most single base changes in up to 200-base fragments may be detected as mobility shifts (2), although there are mutations that can be "invisible" to SSCP analysis and remain undetected (3). Apart from the primary structure of ssDNA itself, there are other factors that could affect the electrophoretic mobility, such as: the size of an analyzed DNA fragment, the gel and buffer compositions, the temperature at which the electrophoresis is performed, etc. Thus, in order to obtain satisfactory reproducibility, the conditions for SSCP analysis should be established empirically for the each set of fragments (3, 4). Nevertheless, the accuracy of this method in detecting nucleotide substitutions was estimated as being between 80 and 95% (5, 6). Since this method was first published, improvements and modifications have been made in order to increase its sensitivity, speed, and reproducibility. In the original protocol, radioactive isotope labeling of DNA followed by autoradiography was used (1). Visualization of DNA bands by ethidium bromide staining (7, 8) and the more sensitive silver staining (4, 9) were introduced soon afterwards. Recently, high-throughput automated capillary electrophoresis systems have also been used instead of conventional slab gel electrophoresis, enabling shorter analyses time, less sample and reagent consumption and higher resolution, sensitivity, and reproducibility (10). A combined SSCP and heteroduplex analysis (HA) using capillary electrophoresis system has also been described (11). Nevertheless, SSCP is still a widely used technique. It is often the method of choice in medical research and clinical diagnosis of human hereditary diseases and cancer (5, 12, 13), clinical and therapeutic studies of human viruses (14, 15), veterinary and animal research (16, 17) as well as the detection and genotyping of fungi (18, 19). As even non-cultivatable bacterial species from environmental samples can be successfully analyzed by SSCP, it is also often used in studies of bacterial communities (20). In plant pathology, SSCP is widely used for genotyping of strains and variants, as well as population structure studies of viruses (21, 22) and characterization of viroids (9). The applicability of this method as a fingerprinting tool in the analysis of phytoplasma genes was first assessed by Šeruga Musi et al. (23), where the SSCP analysis of 16S rRNA, *tuf*, and *dnaB* genes revealed the presence of molecular variability among isolates from the same ribosomal subgroup that was undetected by RFLP analysis. The phylogenetic analysis of phytoplasma sequences sharing similar SSCP profiles corroborated the results. Subsequently, Schneider and Seemüller (24) successfully applied the method to differentiate '*Candidatus* Phytoplasma mali' strains by SSCP analysis of the *hflB* gene, detecting more than 20 different profiles. However, SSCP is still not widely used in routine analysis of phytoplasma genes.

2. Materials

2.1. Separation of DNA Strands and Polyacrylamide Gel Electrophoresis

1. Denaturing buffer: 90% formamide, 25 mM EDTA, 0.05% bromophenol blue, 0.05% xylene cyanol. Prepare with deionized water. Store at 4°C, in a tube wrapped with aluminum foil (see Note 1).
2. TBE buffer (1×): 90 mM Tris, 90 mM boric acid, 1 mM EDTA, pH 8.3.
3. 30% Acrylamide/bis-acrylamide solution (40:1): 30 g of acrylamide monomer, 0.75 g of bis (*N*, *N*′-methylenebisacrylamide). Transfer to a 100 mL graduated cylinder containing about 40 mL of water and mix on a magnetic stirrer for about 30 min. Make up to 100 mL with water and filter through a 0.45 μm pore filter. Store at 4°C, in a bottle wrapped with aluminum foil.
4. Ammonium persulphate: 10% solution in water. Prepare fresh and store at 4°C for a short period of time.
5. *N*, *N*, *N*, *N*′-tetramethyl-ethylenediamine (TEMED). Store at 4°C.
6. Glycerol.

2.2. Visualization of DNA by Silver Staining

1. Fixing solution: 50% ethanol, 10% glacial acetic acid.
2. Staining solution: 12 mM silver nitrate. Store in a dark bottle at 4°C.

3. Developing solution: 0.75 M potassium hydroxide, 0.28% formaldehyde. Always prepare fresh (see Note 2).
4. Stop solution: 0.07 M sodium carbonate. Prepare fresh.

3. Methods

As PCR fragments of different lengths amplified with different primers will be analyzed, the conditions for the SSCP analysis should be optimized in order to obtain the best separation and resolution of particular ssDNA products. The following parameters can be varied: acrylamide concentration, presence of glycerol in a gel, and/or the duration of the electrophoresis. For fragments of ~500 bp, we have found 14% polyacrylamide (PA) gels with 5% glycerol run for 7–8 h to be optimum conditions; for longer fragments of ~800–1,250 bp, we use 8% PA gels with 2.5% glycerol run for 3–4 h (Fig. 2; (23)). Protocols other than the silver staining described can also be used successfully for visualization of the DNA (25).

3.1. Separation of DNA Strands and Polyacrylamide Gel Electrophoresis

1. To prepare a 14% PA gel containing 5% glycerol, mix 1 mL of 10× TBE buffer, 3.7 mL of water, 4.67 mL of acrylamide/bis-acrylamide solution, 0.5 mL of glycerol and stir gently. Add 160 μL of ammonium persulphate and 16 μL of TEMED, stir gently and cast in a 10 cm × 10.5 cm × 1 mm gel cassette. To prepare an 8% PA gel containing 2.5% glycerol, mix 1 mL of 10× TBE buffer, 5.95 mL of water, 2.67 mL of acrylamide/

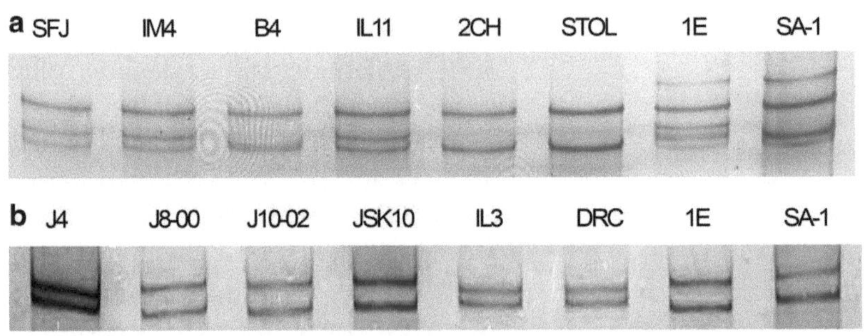

Fig. 2. SSCP patterns of phytoplasma DNA amplified from grapevine extracts. (**a**) 16S rRNA (510 bp) and (**b**) *tuf* gene (850 bp) amplicons were obtained by using primer pairs $16R_{738f}/16R_{1232r}$ and fTufu/rTufu, respectively. Analyses were performed in 14% (**a**) or 8% (**b**) polyacrylamide gels with silver staining after electrophoresis. Lanes contain: SFJ, cv. Debet, Sveti Filip i Jakov; IM4, cv. Chardonnay,I mbriovec; B4, cv. Chardonnay, Brodski Stupnik; 2CH, cv. Chardonnay, FYR of Macedonia; DRC, cv. Chardonnay, Drniš; 1E, cv. Chardonnay, Erdut; J4, J8-00, J10-02, cv. Chardonnay, Jazbina; JSK10, cv. Chardonnay, Jaska; IL3, cv. Chardonnay, Ilok; IL11, cv. Riesling, Ilok; SA-1, STOL, phytoplasma reference strains of the 16SrXII-A ribosomal subgroup (stolbur) (23).

bis-acrylamide solution, 0.25 mL of glycerol and stir gently. Add 120 μL of ammonium persulphate and 12 μL of TEMED, stir gently, and cast in a 10 cm × 10.5 cm × 1 mm gel cassette (see Note 3).

2. Mix 1–4 μL of PCR product depending on the intensity of the band (see Note 4) with 7 volumes of denaturing buffer and incubate for 10 min at 99°C. Samples should be put on ice immediately and loaded on a polyacrylamide gel. Electrophoresis is performed in 1× TBE buffer at 4°C and at a constant voltage of 200 V (see Note 5).

3.2. Visualization of DNA by Silver Staining

1. Following electrophoresis, transfer the gel into fixing solution and incubate on an orbital shaker for at least 1 h (see Note 6).
2. Rinse the gel three times with redistilled water.
3. Incubate the gel in silver nitrate staining solution for 1 h.
4. Rinse the gel three times with double distilled water.
5. Add developing solution and agitate gently.
6. When a satisfactory intensity of bands is achieved, discard the developing solution and add stop solution.
7. Gels should be documented immediately (see Note 7).

4. Notes

1. EDTA has poor solubility below pH 8.0. Stock solution (e.g., 0.5 M pH 8.0) will need its pH adjusting using NaOH.
2. Pellets of KOH should be dissolved first in a small volume of water.
3. Gels can be prepared 1 day in advance and stored at 4°C.
4. Use 4 μL of PCR product only if the band is very faint; normally, 1–2 μL of PCR product achieves the best results.
5. Usually, the electrophoresis should be terminated immediately after the xylene cyanol dye migrates out of the gel.
6. Incubation in a fixing solution can be extended overnight. Avoid contact with the gel, not only for health and safety reasons but also for the clearest visualization. Touching a gel with a gloved hand may leave dark stains on the gel. To avoid touching the gel when discarding solutions during the silver staining procedure, use suction devices.
7. Gels can be put between two sheets of transparent film and kept for a short period of time or dried for long-term storage.

References

1. Orita M et al (1989) Detection of polymorphisms of human DNA by gel electrophoresis as single-strand conformation polymorphisms. Proc Natl Acad Sci U S A 86:2766–2770
2. Orita M et al (1989) Rapid and sensitive detection of point mutations and DNA polymorphisms using the polymerase chain reaction. Genomics 5:874–879
3. Telenti A (1994) Detection of mutations by PCR-SSCP (single-strand conformation polymorphism). In: Dale JW, Sanders PG (eds) Methods in gene technology, vol 2. JAI Press Ltd, Middlesex, UK, pp 241–252
4. Rubio L et al (1996) Differentiation of citrus tristeza closterovirus (CTV) isolates by single-strand conformation polymorphism analysis of the coat protein gene. Ann Appl Biol 129: 479–489
5. Ravnik-Glavač M, Glavač D, Dean M (1994) Sensitivity of single-strand conformation polymorphism and heteroduplex method for mutation detection in the cystic fibrosis gene. Hum Mol Genet 3:801–807
6. Hall JS et al (2001) Structure and temporal dynamics of populations within wheat streak mosaic virus isolates. J Virol 75:10231–10243
7. Yap EP, McGee JO (1992) Nonisotopic SSCP detection in PCR products by ethidium bromide staining. Trends Genet 8:49
8. Hongyo T et al (1993) 'Cold SSCP': a simple, rapid and non-radioactive method for optimized single-strand conformation polymorphism analyses. Nucleic Acids Res 21: 3637–3642
9. Palacio A, Duran-Vila N (1999) Single-strand conformation polymorphism (SSCP) analysis as a tool for viroid characterisation. J Virol Methods 77:27–36
10. Ren J-C (2000) High-throughput single-strand conformation polymorphism analysis by capillary electrophoresis. J Chromatogr B 741:115–128
11. Kozlowski P, Krzyzosiak WJ (2001) Combined SSCP/duplex analysis by capillary electrophoresis for more efficient mutation detection. Nucleic Acids Res 29:e71
12. Papp T et al (2007) Mutational analysis of *Chk1*, *Chk2*, *Apaf1* and *Rb1* in human malignant melanoma cell lines. Oncol Rep 17:135–140
13. Silveira APF et al (2011) Gene mutations and polymorphisms of *TP53* and *FHIT* in chronic esophagitis and esophagealcarcinoma. Anticancer Res 31:1685–1690
14. Mackiewicz V et al (2005) Rapid investigation of hepatitis A virus outbreak by single strand conformation polymorphism analysis. J Med Virol 76:271–278
15. Di Liberto G et al (2006) Clinical and therapeutic implications of hepatitis C virus compartmentalization. Gastroenterology 131:76–84
16. Simsek S et al (2011) Molecular discrimination of sheep and cattle isolates of *Echinococcus granulosus* by SSCP and conventional PCR in Turkey. Vet Parasitol 178:367–369
17. Schaschl H, Wegner KM (2006) Polymorphism and signature of selection in the MHC class I genes of the three-spined stickleback *Gasterosteus aculeatus*. J Fish Biol 69(suppl B):177–188
18. Dassanayake RS, Samaranayake LP (2003) Amplification-based nucleic acid scanning techniques to assess genetic polymorphism in *Candida*. Crit Rev Microbiol 29:1–24
19. Nahiyan ASM et al (2011) PCR-SSCP analysis of *Fusarium* diversity in asparagus decline in Japan. Eur J Plant Pathol 130:197–203
20. Oto M, Suda W, Shinoyama H (2006) 16S rRNA gene-based analysis of microbial community by whole-genome amplification and minigel-single-strand conformation polymorphism technique. J Biosci Bioeng 102: 482–484
21. Turturo C et al (2005) Genetic variability and population structure of *Grapevine leafroll-associated virus 3* isolates. J Gen Virol 86: 217–224
22. Cerni S et al (2008) Stem pitting and seedling yellows symptoms of *Citrus tristeza virus* infection may be determined by minor sequence variants. Virus Genes 36:241–249
23. Šeruga Musić M, Krajačić M, Škorić D (2008) The use of SSCP analysis in the assessment of phytoplasma gene variability. J Microbiol Methods 73:69–72
24. Schneider B, Seemüller E (2009) Strain differentiation of *Candidatus* Phytoplasma mali by SSCP- and sequence analyses of the *hfl*B gene. J Plant Pathol 91:103–112
25. Igloi GL (1983) A silver stain for the detection of nanogram amounts of tRNA following two-dimensional electrophoresis. Anal Biochem 134:184–188

Chapter 19

Microarrays for Universal Detection and Identification of Phytoplasmas

Mogens Nicolaisen, Henriette Nyskjold, and Assunta Bertaccini

Abstract

Detection and identification of phytoplasmas is a laborious process often involving nested PCR followed by restriction enzyme analysis and fine-resolution gel electrophoresis. To improve throughput, other methods are needed. Microarray technology offers a generic assay that can potentially detect and differentiate all types of phytoplasmas in one assay. The present protocol describes a microarray-based method for identification of phytoplasmas to 16Sr group level.

Key words: Phytoplasma, Microarray, DNA chip, Detection, Diagnosis, 16S rRNA

1. Introduction

Phytoplasma identification on 16Sr group level is currently performed by nested PCR, most often using the universal primers P1/P7 (1, 2) and R16F2n/R16R2 (3), followed by restriction fragment length polymorphism (RFLP) analysis and high resolution gel electrophoresis (4). This method is laborious, requires skilled technicians, and a number of reference strains.

DNA microarrays are powerful tools for identification and differentiation of microorganisms, including plant pathogens, as assays for many organisms can be performed in parallel. Examples include microarrays for the detection of nematodes (5), fungi (6, 7), perenosporomycetes (8), bacteria (9, 10), phytoplasmas (11), and viruses (12). Most DNA arrays for pathogen detection are based on oligonucleotide probes spotted on glass slides, allowing one sample to be analyzed on each slide. However, formats for higher throughput have been developed, such as multiple arrays on each slide (this protocol and (11)) and arrays in microtitre plates or

Matt Dickinson and Jennifer Hodgetts (eds.), *Phytoplasma: Methods and Protocols*, Methods in Molecular Biology, vol. 938, DOI 10.1007/978-1-62703-089-2_19, © Springer Science+Business Media, LLC 2013

in microcentrifuge tubes, where up to 96 samples can be analyzed (Alere Technologies GmbH). These approaches are, however, currently only feasible for low-density arrays.

The keystone in microarray development is probe design. Microarray probe design starts with identification of the most suitable target region for probe design. As PCR is most often used for amplification of target DNA, the DNA region of interest should be easily amplified from all strains of interest and the region should contain sufficient variation for discrimination at the desired level. After selection of a useable region, relevant sequences should be aligned and potential probes should be selected that discriminate strains. Probes of different lengths can be used on microarrays depending on the application. Short probes (15–25 nt) exhibit reduced sensitivity but show better discrimination of minor sequence differences, whereas longer probes (>50 nt) show increased sensitivity but do not discriminate between target sequences with only few sequence differences. Ideally, all probes on the array should have the same melting temperature (T_m). This can be attempted by allowing (minor) differences in probe length and by choosing sequences within the region of interest that meet melting temperature criteria.

The main technology for microarray manufacture is robotic spotting of pre-synthesized oligonucleotide probes on glass slides with special surfaces to allow irreversible binding of probes to the surface. For a review, see ref. (13).

Generally a pre-microarray amplification step, such as PCR, is required to enhance sensitivity, especially for low titre pathogens such as phytoplasmas. To allow detection of the hybridization event, the PCR product must be labeled. Several approaches are available to label PCR products with Cy3 or other dyes. Cy3-labeled primers may be used; however, several studies have found a "position effect" so that only probes hybridizing close to the Cy3 molecule will result in a readable signal, putting some constraints on probe design (14). PCR products may be labeled during their amplification by including Cy3-coupled nucleotides; however, it has been noted that this may decrease PCR efficiency. Alternatively, products may be biotin-labeled during PCR, followed by incubation with Cy3-streptavidin after hybridization. Post-PCR labeling procedures include random-primed labeling of the PCR product with Cy3-coupled nucleotides (11). Franke-Whittle et al. (15) compared different labeling methods and found that, in their hands, labeling with a Cy3-coupled forward primer gave the best results. A comparison of different post-hybridization secondary detections steps for signal amplification has been performed (16).

The present protocol is based on the method developed by Nicolaisen and Bertaccini (11), where it was shown that short oligonucleotide probes can identify most phytoplasma 16Sr groups, although a few showed too little sequence variation to be identified at 16Sr group level. Only a limited number of probes were designed in this work; however, more probes could be designed for each 16Sr group to improve reliability. Other probes designed for specific purposes such as identification of closely related strains in specific 16Sr groups or '*Candidatus* Phytoplasma' species could be designed from other gene regions as long as it is possible to design flanking primers for PCR amplification.

2. Materials

1. Microarray spotter (e.g., Genetix QArray Mini).
2. Split pin (e.g., Telechem XSMP2.5) (see Note 1).
3. Microarray slides (Nexterion Slide E, Schott UK Ltd, Stafford, UK) (see Note 2).
4. 384-well plates (Genetix #X7020).
5. Slideholders.
6. SpeedVac concentrator.
7. Specific oligonucleotide probes with oligo d(T) linker and an NH_2-group attached at the 5′ end (Table 1) (see Note 3).
8. Non-specific oligonucleotide probes as negative controls (see Note 4).
9. SSC (20× stock solution): dissolve 175.3 g of NaCl and 88.2 g of sodium citrate in 800 mL of water. Adjust pH to 7.0 with 10 M NaOH. Adjust volume to 1 L and sterilize by autoclaving.
10. Spotting buffer: 1.5 M betaine, 3× SSC.
11. Wash 1: 0.1% Triton X-100.
12. Wash 2: 1 mM HCl.
13. Wash 3: 100 mM KCl.
14. Blocking buffer: 50 mM ethanolamine, 0.1% SDS, 0.1 M Tris, pH 9.
15. Desiccation chamber.
16. Hybridization oven.
17. Hybridization frames (FAST FRAME for 4 slides, Sigma-Aldrich #Z721212).
18. Healthy and infected plant controls, preferably of same species and tissue as samples.
19. PCR reagents (see Note 5).
20. Primers: Pp array fwd 5′-AGT GGC GAA CGG GTG AGT AAC-3′, Pp array rev 5′-CGT TTA CGG CGT GGA CTA CCA G-3′ (for amplification of the 5′ end of the 16S rRNA gene).

Table 1
Probes used on the microarray used in this protocol

Probe	Target 16Sr group	Sequence 5′–3′
PpI-465	I	TATTAGGGAAGAATAAATGATGGAAAAATC
PpII-471	II	CGAACCATTTGTTTGCCGGTA
PpII-629	II	CGTTGTCCGGCTATTGAAACTGC
PpIII-478	III	GTGGAAAAACTCCCTTGACGGTACTTAAT
PpIV-630	IV	CTTAACGTTGTCCTGCTAGAGAAACTGTT
PpV-221	V	AGACCTTCTTCGGAGGGTATGCTTAA
PpVI-276	VI	TTAGTTGGTAGAGTAAAAGCCTACCAAGAC
PpVII-621	VII	TATAGAAACTACCTTGACTAGAGTTAGATAGAG
PpVIII-634	VIII	ACGCTTAACGTTGTTTTGTTATAGAAACTG
PpIX-609	IX	AACGCTGTAsCGCTATAGAAACTGTCTG
PpX-183	X	GGATAGGAAGTTTTAAGGCATCTTGAAAC
PpX-224	X	AGGGTATGCTAAGAGATGGGCTTGC
PpXII-465	XII	GGGAAGAAAAGATGGTGGAAAAACC
PpXII-480	XII	GTGGAAAAACCATTATGACGGTACCT
PpXIV-585	XIV	GAAACTATCAGACTAGAGTGAGATAGAGGCAAG
Pp-148	All	TTTCGGCAATGGAGGAAACTCTGAC
Pp-502	All	AGGCGGCTyrCTGGGTCTTTACT
Tub1	Neg	GGTGGCAGCTCAACAACAATGC
Tub2	Neg	GCGATAGGTTCACCTTCAGACCG

In a few probes, degenerate nucleotides are used (s = G or C; y = C or T; r = A or G)

21. BioPrime® Array CGH Genomic Labeling Module (Invitrogen, #18095-012).
22. Cy3-dCTP (Amersham Biosciences, #PA53021).
23. MinElute Reaction Cleanup Kit (QIAGEN, #28204).
24. Hybridization buffer: 20% formamide, 5× SSC, 0.1% SDS, 0.1 μg/μL sonicated salmon sperm DNA.
25. Saturated NaCl solution.
26. Wash 4: 2× SSC, 0.1% SDS.
27. Wash 5: 1× SSC.
28. Wash 6: 0.1× SSC.
29. Compressed air (oil-free).
30. Microarray scanner (e.g., GenePix 4100A Microarray Scanner from Molecular Devices).

3. Methods

3.1. Oligonucleotide Probes

Oligonucleotide probes in this protocol were designed from the 16S rRNA region (Table 1) (see Note 6). Add random probes of similar T_m as negative controls.

3.2. Preparation of Microarray Slides

The array described in this protocol allows for 16 samples to be hybridized on each slide (Fig. 1). Using the Genetix arraying robot, it is most convenient to use only one pin when printing identical arrays on each slide (see Note 7).

1. Load 2 μL of each oligonucleotide probe (100 μM) into a Genetix 384-well plate.
2. Dry the plate to completion in a Speed Vac concentrator and dissolve oligonucleotides in 20 μL of spotting buffer. Cover the plate with adhesive film and store at −20°C until use. Just before use, centrifuge the plate at low speed (1,000 × *g*) for 2 min (see Note 8).
3. Design a printing program that prints 16 identical arrays that fit on the FAST frame. Each probe should be printed in duplicate (at least). The pin should be thoroughly washed between each oligonucleotide application, e.g., water for 8 s, followed by ethanol for 2 s, and drying for 5 s. Test the designed program by using a dummy plate and standard microscope slides using spotting buffer (see Note 9).
4. Adjust room (or arraying robot) humidity to 50% (see Note 10).

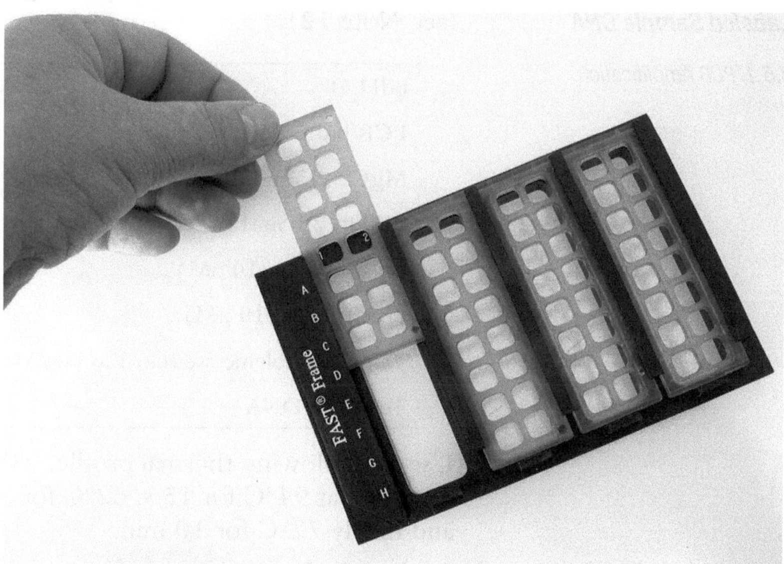

Fig. 1. Four slides attached to frames in the FAST frame.

5. Prepare the spotting robot by mounting the plate with the oligonucleotide probes, an appropriate number of slides including dummy slides for blotting, and a split pin (e.g., XSMP2.5).
6. Start the printing process.
7. Dismantle slides and confirm that printing was satisfactory using a stereomicroscope (see Note 11).
8. Store the slides in a desiccator.

3.3. Immobilization of Probes and Blocking of Slides

1. Immediately before hybridization, immobilize probes by incubation at 90% humidity (a box with wet tissue) for 30 min.
2. Wash slide(s) at room temperature. Do not dry slides between steps:

   0.1% Triton X-100 for 1 × 5 min.

   1 mM HCl for 2 × 2 min.

   100 mM KCl for 1 × 10 min.

   dH_2O for 1 × 1 min.
3. Block slides in blocking buffer at 50 °C for 15 min, and wash in dH_2O at room temperature for 1 min.
4. Air dry slides.

3.4. DNA Extraction

Extraction of phytoplasma DNA is beyond the scope of this chapter. There are several methods for phytoplasma DNA extraction; we use the method of Prince et al. (17). Healthy and negative controls should be included.

3.5. Preparation of Labeled Sample DNA

3.5.1. PCR Amplification

1. Amplify sample DNA (~20 ng) using the following components (see Note 12):

ddH_2O	10.4 μL
PCR buffer (5×, Promega)	5.0 μL
$MgCl_2$ (25 mM)	1.5 μL
dNTP (2.5 mM each)	2.0 μL
Pp array fwd (10 μM)	2.5 μL
Pp array rev (10 μM)	2.5 μL
Taq DNA polymerase (Go Taq Flexi) (5 U/μL)	0.125 μL
Template DNA	1.0 μL

2. Use the following thermal profile: 94°C for 3 min, followed by 35 cycles at 94°C for 15 s, 62°C for 30 s, and 72°C for 1 min, and finally 72°C for 10 min.
3. Analyze 5 μL of the reaction on a agarose gel. An amplification product of ~0.7 kb should be visible after staining.

3.5.2. Labeling of PCR Product

1. Label 1 μL of PCR product using BioPrime® Array CGH Genomic Labeling Module with Cy3-dCTP (see Note 13). Mix the following components:

2.5× random primers	5 μL
dH_2O	3.25 μL
PCR product	1 μL

2. Incubate at 95°C for 5 min then place on ice.
3. Add the following as a mastermix:

10× dCTP nucleotide mix	1.25 μL
1 mM Cy3-dCTP	0.25 μL
1 mM dCTP	0.5 μL
Exo-Klenow fragment	0.25 μL

4. Mix and incubate at 37°C for 2 h.
5. Add 1.25 μL stop buffer.
6. Add 7.25 μL H_2O.
7. Purify labeling reactions using MinElute Reaction Cleanup Kit (QIAGEN) according to the manufacturer's instructions. Store purified products at −20°C until use. Protect from light (see Note 14).

3.6. Hybridization

1. Mount slides with a FAST Frame (Schleicher & Schuell, Dassel, Germany) array grid with 16 wells. Mark slides for orientation.
2. Denature labeled PCR product at 95°C for 5 min and place immediately on ice.
3. Add 80 μL hybridization buffer to each hybridization well.
4. Add labeled PCR product to each well.
5. Cover FAST frame with a microtitre plate cover or similar.
6. Place in a moist chamber over a saturated NaCl solution at 48°C overnight in the dark.

3.7. Wash

1. Remove the hybridization mix from each well with a pipette without touching the slide.
2. Place FAST frame with slides in a container with 500 mL 2× SSC, 0.1% SDS for 2 min, rocking the container.
3. Dismantle slides from FAST frame and place in slide holder. Wash in 2× SSC, 0.1% SDS for 5 min at 42°C.
4. Wash 2×2 min in 1× SSC at room temperature.
5. Wash 2×2 min in 0.1× SSC at room temperature.

6. Remove excess liquid and dry using oil-free compressed air.
7. Store slides in the dark and under desiccation (see Note 15).

3.8. Scanning

Scanning parameters will depend on the scanner used. We use the GenePix Personal 4100A scanner. Scan slide using default settings at 532 nm. Define spot area and background area according to the software for that particular scanner type and according to the spotting pattern. After aligning individual features and spots, analyze and export results as a text file (see Note 16).

4. Notes

1. To allow for higher density arrays, small tip diameter pins are recommended. A split pin allows for multiple prints from one liquid acquisition whereas the use of solid pins will prolong the printing procedure significantly. The pin should be cleaned regularly. Pins can be cleaned in a sonication bath with distilled water. Never touch the pin tips.
2. We found that Nexterion Slide E, which is an epoxysilane based slide, worked well in our experiments. Detailed information can be found at http://www.schott.com/nexterion/english/products/coated_slides/epoxysilane/epoxysilane.html.
3. An oligo d(T) tail at the 5′ end will improve hybridization efficiency, particularly for short probes, as the probe sequence will be more accessible to the target DNA. In this protocol we used an oligo $d(T)_{10}$ tail.
4. Any probes that are not expected to hybridize to sample DNA (e.g., plant or bacterial DNA) can be used as negative controls. The T_m of these probes should be similar to the T_m of the specific probes. We used two probes from a fungal β-tubulin gene on this microarray.
5. In this protocol, Promega GoTaq works well for PCR amplification; however other PCR enzymes may be used.
6. The probes used in this protocol were designed on basis of alignments of the phytoplasma 16S rRNA using the CLC DNA Workbench. Probes for phytoplasma group discrimination were identified manually with preferably at least four mismatches to non-target groups. Furthermore, two universal probes were designed to hybridize to 16S rRNA gene sequences from all known phytoplasmas. The length of oligonucleotides was adjusted to give a T_m of 59 ±2°C, calculated using the nearest-neighbor method. The ability to form dimers and hairpin structures was checked using the CLC DNA Workbench, and probes with a low tendency to form such structures were chosen.

7. In most protocols, only one array is printed on each slide; however, this reduces throughput significantly. By printing more arrays on each slide and using frames to separate arrays, throughput can be significantly increased. The position of each subarray must be adjusted carefully to fit on the FAST Frame. In the Genetix spotter we use the following settings for a 16 × 16 spot array: Spotting pitch 330 μm; No Fields X 8; No Fields Y 2; Offset Field 1X 4510; Offset Field 1Y 5310; Offset Field X 3720; Offset Field Y 3720; Size X 75500; Size Y 24200. Test the calculated parameters using spotting buffer and standard microscope slides in a test run.
8. Printing plates with dissolved oligonucleotides can be used for printing for a few times. However, each printing procedure will reduce the volume of probe solutions, mainly as a result of evaporation. Check that volumes are sufficient before each printing process.
9. We use 330 μm as spot distance, the stamp time is 0 ms, and the inking time is 1 s. Twenty stamps are printed on a dummy slide before printing slides to avoid excess liquid deposition, and thereafter up to 200 stamps can be made.
10. Humidity should be constant within and between printing sessions. Although some printers can control humidity, we prefer a humidity-controlled room, as this creates a more uniform environment. The humidity should be adjusted well before spotting begins to ensure a stable humidity. The microarrayer should be placed in a clean room. If this is not possible, try to place the arrayer in a room with limited traffic.
11. Due to salts in the printing buffer, individual spots are visible before washing slides. An examination should confirm that all probes are printed and that spots are well separated.
12. Precautions should be taken to avoid contamination of samples and reagents at all stages while setting up PCR. Use only dedicated pipettes for preparing master mixes and another set for adding template DNA. Use only filter tips. Wear gloves and change these frequently. Use only clean, sterile plasticware. Master mix preparation and the addition of template DNA should be performed in separate places.
13. To save valuable reagents, volumes of all reagents have been reduced to one quarter.
14. Half of the reaction can be saved as backup.
15. The signal intensity of Cy3 will decrease over time. However, slides still show acceptable signals after some weeks, if properly stored.
16. A range of microarray scanners and software is available for recording and analysis of microarray hybridizations. Usually only one sample is hybridized to each (sub)array and thus single-color scanning is used. Resolution is set at the highest

level. To analyze signal intensities after scanning, a grid of circles is usually superimposed on the image so that the individual circles each define a microarray spot: this may require adjustment of individual circles. The median signal intensity is quantified (median is preferred over mean intensity because it excludes extremes) and the background is subtracted (in GenePix Pro this is done automatically). For background subtraction in GenePix Pro, a default local subtraction method is used: the median intensity of an area centered on the feature that is three times the diameter of the feature (but excluding neighboring features) is subtracted from the median intensity of the feature. A probe is considered positive if its background-subtracted signal is above an arbitrarily defined or empirically determined threshold. We set the threshold for defining positives as 3 × (median spot intensity + the standard deviation of the signals from negative control probes).

References

1. Deng S, Hiruki C (1991) Amplification of 16S rRNA genes from culturable and nonculturable mollicutes. J Microbiol Meth 14:53–61
2. Smart CD et al (1996) Phytoplasma-specific PCR primers based on sequences of the 16S–23S rRNA spacer region. Appl Environ Microbiol 62:2988–2993
3. Gundersen DE, Lee I-M (1996) Ultrasensitive detection of phytoplasmas by nested-PCR assays using two universal primer pairs. Phytopathol Mediterr 35:144–151
4. Lee I-M et al (1998) Revised classification scheme of phytoplasmas based an RFLP analyses of 16S rRNA and ribosomal protein gene sequences. Int J Syst Bacteriol 48:1153–1169
5. François C et al (2006) Towards specific diagnosis of plant-parasitic nematodes using DNA oligonucleotide microarray technology: a case study with the quarantine species *Meloidogyne chitwoodi*. Mol Cell Probes 20:64–69
6. Nicolaisen M et al (2005) An oligonucleotide microarray for the identification and differentiation of trichothecene producing and non-producing *Fusarium* species occurring on cereal grain. J Microbiol Meth 62:57–69
7. Lievens B et al (2006) Detecting single nucleotide polymorphisms using DNA arrays for plant pathogen diagnosis. FEMS Microbiol Lett 255:129–139
8. Tambong JT et al (2006) Oligonucleotide array for identification and detection of *Pythium* species. Appl Environ Microbiol 72:2691–2706
9. Pelludat C, Duffy B, Frey JE (2009) Design and development of a DNA microarray for rapid identification of multiple European quarantine phytopathogenic bacteria. Eur J Plant Pathol 125:413–423
10. Fessehaie A, De Boer SH, Lévesque CA (2003) An oligonucleotide array for the identification and differentiation of bacteria pathogenic on potato. Phytopathology 93:262–269
11. Nicolaisen M, Bertaccini A (2007) An oligonucleotide microarray-based assay for identification of phytoplasma 16S ribosomal groups. Plant Pathol 56:332–336
12. Boonham N, Tomlinson J, Mumford R (2007) Microarrays for rapid identification of plant viruses. Annu Rev Phytopathol 45:307–328
13. Dufva M (2005) Fabrication of high quality microarrays. Biomol Eng 22:173–184
14. Zhang L, Hurek T, Reinhold-Hurek B (2005) Position of the fluorescent label is a crucial factor determining signal intensity in microarray hybridizations. Nucleic Acids Res 33:e166
15. Franke-Whittle IH et al (2006) Comparison of different labeling methods for the production of labeled target DNA for microarray hybridization. J Microbiol Meth 65:117–126
16. Vora GJ et al (2008) Comparison of detection and signal amplification methods for DNA microarrays. Mol Cell Probes 22:294–300
17. Prince JP et al (1993) Molecular detection of diverse mycoplasmalike organisms (MLOs) associated with grapevine yellows and their classification with aster yellows, X-disease, and elm yellows MLOs. Phytopathology 83:1130–1137

Chapter 20

T-RFLP for Detection and Identification of Phytoplasmas in Plants

Jennifer Hodgetts and Matt Dickinson

Abstract

Conventionally, detection of phytoplasmas has been performed by PCR of the 16S rRNA gene, followed by either RFLP or DNA sequencing to determine the phytoplasma 16Sr group. This chapter demonstrates the technique of terminal restriction fragment length polymorphism (T-RFLP), a fingerprinting technique which combines both detection and identification in a single method, with the added benefit of inbuilt controls which removes the risk of false negative results and in addition highlights potential false positive results.

Key words: T-RFLP, Diagnostics, 23S rRNA gene, 16Sr group identification

1. Introduction

Phytoplasma diagnostic testing has historically been based on the 16S rRNA gene because of the availability of "universal" primers for this region (1) (see also Duduk et al. (2), Chapter 14). However, it has also been noted that the standard "universal" primers can amplify products from closely related bacteria such as *Bacillus* spp., leading to occasional false positives and misdiagnosis (3). Due to this other genes have been studied as potential diagnostic targets such as the *rp* operon, *tuf*, *secY*, and *secA* genes (see Chapters 15, 16, and 17). Nevertheless whilst these can be less prone to false positive results, most of these targets do not give universal detection of all phytoplasmas. Outside of the realm of plant pathology the 23S rRNA is a target which is commonly used for broad range detection of a large array of bacterial species (4). However this target has not been commonly utilized for phytoplasma detection aside from some conventional PCR primers which amplify the 16S rRNA also extending into the start of the 23S rRNA.

Matt Dickinson and Jennifer Hodgetts (eds.), *Phytoplasma: Methods and Protocols*, Methods in Molecular Biology, vol. 938, DOI 10.1007/978-1-62703-089-2_20,

Terminal restriction fragment length polymorphism analysis (T-RFLP) is a technique first developed in 1997 for the profiling of microbial communities based upon the location of restriction enzyme sites within a stretch of DNA (5). It combines PCR amplification with one fluorescently labeled primer, restriction enzyme digestion with one or more enzymes, and separation of the fragments on an automated DNA sequencer through high-resolution electrophoresis to allow the detection of the fluorescent dye-labeled products, called terminal restriction fragments (TRFs) (Fig. 1). In our lab we have modified the technique for the phytoplasma 23S rRNA so that it allows combined detection of phytoplasmas and identification of the 16Sr group in one process. Whilst in standard T-RFLP each TRF would represent a different bacterial species in a microbial population, in this method each TRF indicates a single phytoplasma 16Sr group (6). Once the TRFs have been determined by interrogation of the output electropherogram (or T-RFLP profiles) these are compared to a table of expected TRFs which allows the 16Sr group to be identified. T-RFLP differs from typical RFLP analysis (see Chapters 14 and 15) in that only the terminal fragment is detected due to the presence of the fluorescent dye. The other (nonterminal) fragments have no dye and are therefore not detected, compared to standard RFLP where the pattern of all of the fragments is visualized.

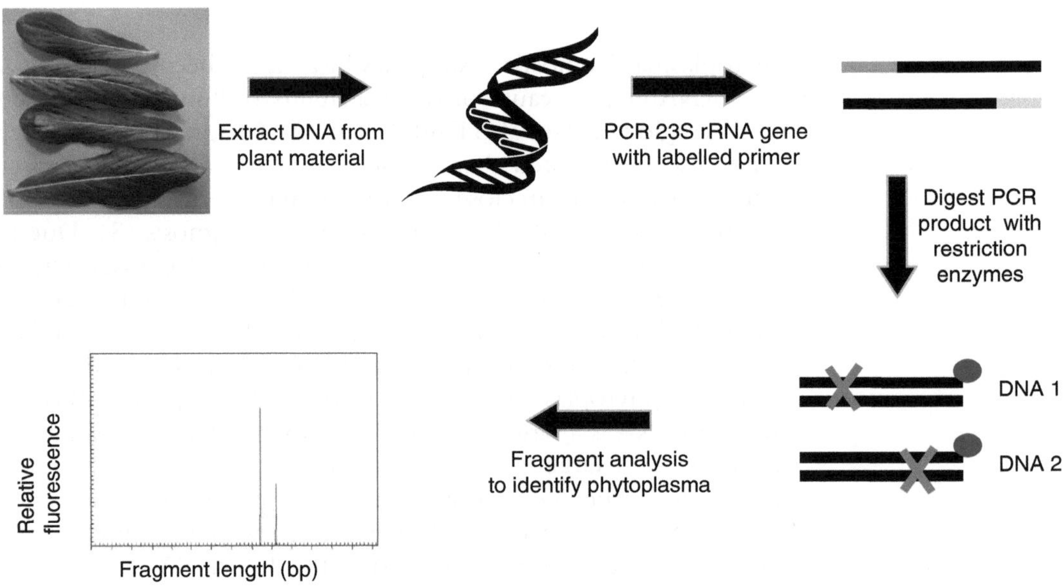

Fig. 1. Schematic illustration of the various steps in conducting terminal restriction fragment length polymorphism (T-RFLP) on plant samples.

Only recently with the increased uptake of real-time PCR has it become common place for control assays to be routinely run which demonstrate that DNA extraction has been successful and that the DNA preparation supports PCR amplification. Whilst this is a vital step, by both conventional and real-time PCR this typically requires two individual assays to be performed. A primary advantage of T-RFLP is that in a single reaction one set of primers amplifies both the plant and phytoplasma DNA, and TRFs are generated which represent both. This means that easy and rapid confirmation of the absence of false negatives is achieved confirming the quality of the DNA and the absence of inhibitors. The lack of both plant and phytoplasma TRFs indicates either inhibitors or nonamplifiable DNA, and rules out false negative samples. Just as important as detection of false negatives is the problem of false positives. In T-RFLP this is eliminated as each 16Sr group produces TRFs which are distinctive and characteristic with the combination of two restriction enzymes. If any other TRFs were present then this would indicate either a new phytoplasma group or a possible nontarget species, thereby illuminating a potential false positive sample. To establish the correct TRFs for a given group, the 23S rRNA sequence needs to be determined, and then in silico digestions with the given enzymes can be performed to predict the TRF sizes. This allows the technique to continue evolving as new species of phytoplasmas are detected.

The assay described targets the 23S rRNA, a multicopy gene. This target affords primers which are universal for all known phytoplasmas yet also maintains a reasonable level of sensitivity. The sensitivity of T-RFLP is equivalent to conventional PCR and midway between this and real-time PCR.

The combination of all these factors (universal detection, determination of false negatives and false positives, sensitivity) make T-RFLP a tool that is particularly suited for situations such as higher throughput screening, for example of an outbreak site, where large numbers of samples require screening and confirmation of the 16Sr group of the phytoplasma. In some situations this may be more appropriate than real-time PCR (the standard high-throughput technique) as it identifies the phytoplasma in one step. Compared to real-time PCR either a universal assay is used which then requires downstream identification to the group level, or group specific primers are used which fail to detect infections with other phytoplasmas (or dual infections). T-RFLP also has the ability to detect mixed infections that involve phytoplasmas from different phylogenetic groups since they can be detected in the same assay. However T-RFLP is less suited for identification of a new phytoplasma group, where techniques which confer the highest levels of sensitivity would be more appropriate.

2. Materials

All stock solutions should be prepared with sterile deionized water and in an area that is distinct from where plant and phytoplasma DNA is being handled. Gilson pipettes (or equivalent) specifically dedicated to non-DNA work should be used for all primer, PCR reaction and restriction enzyme digestion reagent preparations, with a separate set used for pipetting of DNA and PCR products.

2.1. PCR

1. PCR tubes (200 or 500 μL).
2. Eppendorf tubes (1.5 mL).
3. Primers (10 μM) (Table 1). The reverse primer (23Srev) should be labeled with WellRED dye-D4 at the 5′end (available from Sigma-Proligo or Integrated DNA Technologies) (see Note 1).
4. PCR reagents (without loading dye). We generally use Fermentas PCR Master Mix 2X (see Note 2).
5. Vortex mixer.
6. PCR thermocycler.
7. Pipettors (2, 20, 200, and 1,000 μL).
8. Filter tips (10, 20, 200, and 1,000 μL).
9. Sterile distilled water (ddH_2O).
10. DNA extract of sample to be tested (see Note 3).

2.2. Restriction Enzyme Digestion

1. Eppendorf tubes (1.5 and 0.6 mL).
2. Restriction enzyme *Mse*I (T'TAA).
3. Restriction enzyme *Bsh*12361 (CG'CG).
4. 10× restriction enzyme buffers supplied with *Mse*I and *Bsh*12361 restriction enzymes.
5. Vortex mixer.
6. Heat block or water bath (with thermometer).
7. Pipettors (2, 20, and 200 μL).

Table 1
Primers used to perform terminal restriction fragment length polymorphism (T-RFLP) in this protocol

Primer name	Sequence (5′–3′)	Label
MJD5	GGCACTAAGAGCCGATGAAGG	None
23Srev	TTCGCCTTTCCCTCACGGTACT	D4 Beckman dye (Sigma-Proligo)

8. Filter tips (10, 20, and 200 μL).
9. Sterile distilled water (ddH_2O).
10. PCR products from Subheading 2.1.

2.3. T-RFLP Analysis

1. Eppendorf tubes (0.6 and 1.5 mL).
2. Mineral oil.
3. Sterile distilled water (ddH_2O).
4. Chilled rack (4°C) for DNA sequencer plate (96-well).
5. Vortex mixer.
6. GenomeLab DNA Size Standard Kit—600 (Beckman Coulter).
7. GenomeLab Sample Loading Solution (Beckman Coulter).
8. Pipettors (2, 20, 200, and 1,000 μL).
9. Filter tips (10, 20, 200, and 1,000 μL).
10. 96-well Microtiter plate suitable for DNA sequencer.
11. Restriction digest products from Subheading 2.2.
12. Fragment analysis service on a CEQ 8000 DNA analysis system.

3. Methods

Carry out all procedures at room temperature unless otherwise specified.

3.1. Preparation of PCR Reaction Mix and Performing PCR

1. Label PCR tubes.
2. Make the PCR mastermix for $n+1$ samples. The volumes will depend on how many samples are to be tested including a positive control (sample known to be positive by PCR and restriction enzyme digestion) and negative control (water/no template) (n). For example, if there are 6 samples to be tested, then $n=8$ once the positive and negative controls are included. The mastermix would therefore be made for $n+1=9$ reactions. If there are 20 samples to be tested, $n=22$, and a mastermix would be made for 23 reactions etc.
3. The composition of the mastermix will depend on the PCR reagents being used. For Fermentas PCR Master Mix 2X each reaction will be composed of: 12.5 μL PCR Master Mix 2X+9.5 μL ddH_2O+1 μL MJD5 primer+1 μL 23Srev primer.
4. Wearing gloves, in an area that is away from where plant and phytoplasma DNA are being handled, and using Gilson pipettes (or equivalent) specifically dedicated to non-DNA work with

filter tips, pipette the required volume of mastermix reagent into a 1.5 mL Eppendorf tube (*see* **Note 4**). For example, if 19 samples are to be tested plus a positive and negative control ($n=21$ and $n+1=22$), pipette 209 μL ddH_2O, 275 μL PCR Master Mix 2X, 22 μL MJD5 primer, and 22 μL 23Srev primer. Vortex to mix and then aliquot 24 μL into labeled PCR tubes.

5. The tubes are now ready for the addition of sample DNA. Close the lids on the PCR tubes and take them from the phytoplasma DNA-free area to an area in which phytoplasma DNA can be handled. Using a different Gilson pipette to the one used for making the mastermixes, with filter tips, add 1 μL DNA for each sample into the appropriate PCR tube. For the positive control, add 1 μL DNA from a sample that has previously been proven to work in the T-RFLP protocol; for the negative control, add 1 μL ddH_2O.
6. Close the lids and place in the thermocycler (if the thermocycler does not have a heated lid then add a layer of mineral oil on top of the mastermix in each tube). PCR is conducted as follows: 1 cycle at 94°C for 2 min; 35 cycles of 94°C for 30 s, 56°C for 60 s and 72°C for 90 s; and 1 cycle at 72°C for 15 min.
7. Once the PCR has completed the samples can be stored at −20°C prior to restriction enzyme digestion (see Note 5).

3.2. Restriction Enzyme Digestion

1. Label 0.6 mL Eppendorf tubes.
2. Prepare a mastermix for the restriction enzyme digestion for $n+2$ samples (see Note 6). For example if a total of 8 samples were processed by PCR (including the controls) then $n+2=10$ reactions. The PCR positive control is included to act as a positive control for the digestion. The "+2" allows preparation of a negative control where the restriction enzyme is omitted from the mastermix of an aliquot of the PCR positive control, and to allow for pipetting errors.
3. Following the manufacturer's protocol, mix the appropriate volume of buffer, water and use 1 U of restriction enzyme (see Note 7) in a 1.5 mL Eppendorf tube. Votex well and aliquot the appropriate volume into the labeled 0.6 mL Eppendorf tubes.
4. Add 10 μL of the appropriate PCR product to the mastermix (see Note 8).
5. Incubate samples at 37°C for 2 h using either a water bath or heat block (see Note 9).
6. Once the digestion has completed the samples can be stored at −20°C prior to T-RFLP analysis (see Note 10).

3.3. T-RFLP Reaction Setup

1. Label 0.6 mL Eppendorf tubes.
2. Dilute 1 μL each of the digest products in 5 μL ddH_2O in 0.6 mL Eppendorf tubes and vortex well.
3. Place the 96-well microtiter plate into a chilled block.
4. In each well of the 96-well microtiter plate which will be used add 38.5 μL of GenomeLab Sample Loading Solution and 0.5 μL of GenomeLab DNA Size Standard Kit—600.
5. Add 1 μL of each of the diluted digest (from step 2) to an individual well.
6. Overlay each well with a drop of mineral oil.

3.4. Running and Analyzing T-RFLP

1. Run samples on a CEQ 8000 DNA analysis system for fragment analysis following the manufacturer's protocols.
2. Analyze electropherograms using the CEQ systems software considering TRFs between 60 and 640 bp. TRFs which differ by more than 1.25 bp should be considered as distinct.
3. A TRF of 461 bp indicates that plant DNA is present and that the procedure has been successfully completed (Fig. 2a). The absence of this would indicate the DNA extraction failed, the DNA contained inhibitors or the PCR failed, and in this instance the sample should be repeated or the DNA re-extracted and/or cleaned-up.
4. Any additional TRFs indicate the presence of a phytoplasma (Fig. 2b-g). Compare the TRF size generated with both *Mse*I and *Bsh*12361 with Table 2 to determine the phytoplasma 16Sr group (see Note 11).

4. Notes

1. Primers can be purchased either lyophilised or at a range of concentrations (generally 50–200 μM) from various manufacturers. If lyophilised primers should be reconstituted following the manufacturer's recommendations in either molecular grade water or TE buffer. Typically a stock solution of 100 μM is maintained along with small aliquots (i.e., 100 μL) of a working solution of 10 μM (dilute 100 μM stock 1:10 in ddH_2O) at −20°C in the dark. It is wise to use aliquots of the working solution (rather than diluting all of the stock) to avoid multiple freeze-thaw cycles which can damage the primers, and also to help prevent contamination of the stocks. Fluorescently labeled primers should always be kept in the dark to prevent degradation of the fluorophore.

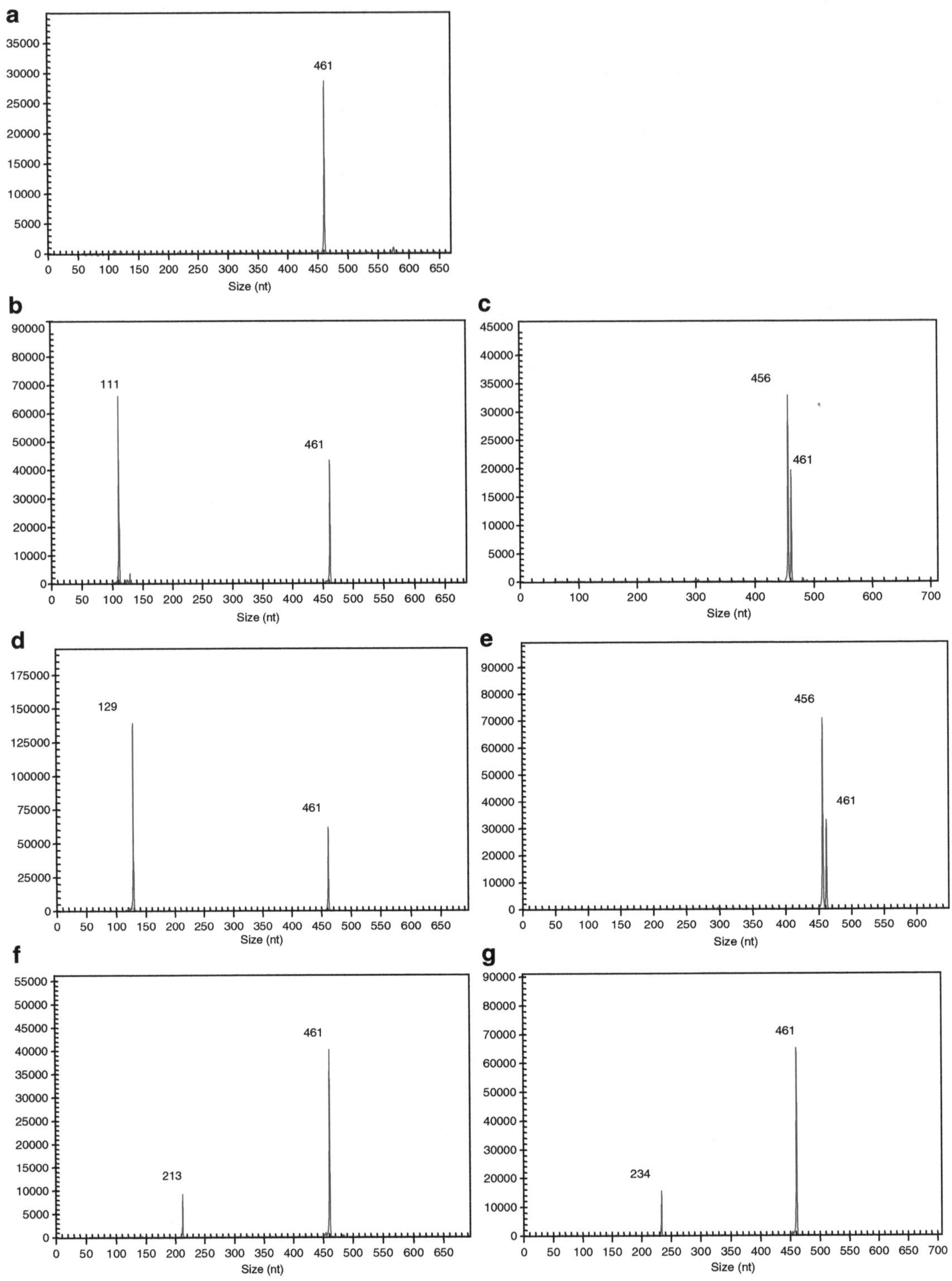

Fig. 2. T-RFLP profiles for *Catharanthus roseus* control (healthy) and infected plants with a range of different phytoplasma amplified with MJD5 and labeled 23Srev following digestion with either *Mse*I or *Bsh*12361. (**a**) *C. roseus* digested with *Bsh*12361; (**b**) FBP digested with *Mse*I; (**c**) FBP digested with *Bsh*12361; (**d**) TBB digested with *Mse*I; (**e**) TBB digested with *Bsh*12361. (**f**) FD-C digested with *Mse*I; (**g**) FD-C digested with *Bsh*12361. (see Table 2 for phytoplasma acronyms) The values on the *x* axis are the sizes in base pairs (as are the numbers on the peaks). The *y* axis gives the signal strength. Figure reproduced from (6) with permission from Wiley-Blackwell.

Table 2
Terminal restriction fragment (TRF) sizes for phytoplasmas following amplification with MJD5 and 23Srev and digestion with either *Mse*I or *Bsh*12361

Sample	Description	16Sr group	*Candidatus* species	TRF (bp)	
				*Mse*I	*Bsh*12361
CHRYM	Chrysanthemum yellows	IA	P. asteris	129	305
DIV	Diplotaxis virescence	IB	P. asteris	129	305
DAY	Dwarf aster yellows	IB	P. asteris	129	305
CACT	Aster yellows cactus	IB	P. asteris	129	305
EAY	European aster yellows	IB	P. asteris	129	305
KVE	Clover phyllody	IC	P. asteris	114	305
AYA	Apricot chlorotic leaf roll	IF	P. asteris	129	305
AY2192	Aster yellows	IL	P. asteris	129	305
AVUT	Atypical aster yellows	IM	P. asteris	129	305
CLP	Cleome phyllody	IIA	P. aurantifolia	129	456
FBP	Faba bean phyllody	IIC	P. aurantifolia	111	456
FBPSA	Crotalaria saltiana phyllody	IIC	P. aurantifolia	111	456
SOYP	Soyabean phyllody	IIC	P. aurantifolia	111	456
TBB	Australian tomato big bud	IID	P. australasiae	129	456
SPLL	Sweet potato little leaf	IID	P. australasiae	129	456
PEP	Picris echioides phyllody	IIE	P. aurantifolia	129	456
CoP	Cotton phyllody	IIF	P. aurantifolia	111	456
CX	Peach X disease	IIIA	P. pruni	129	450
GVX	Green valley X	IIIA	P. pruni	129	450
PYLV	Peach western X	IIIA	P. pruni	129	450
API	Euscelidius variegatus	IIIB	P. pruni	129	450
JRI	Poinsettia branching factor	IIIH	P. pruni	129	450
ULW	Elm witches broom	VA	P. ulmi	213	234
FD-C	Flavescence doree	VC	P. vitis	213	234
FD-D	Flavescence doree	VC	P. vitis	213	234
RuS	Rubus stunt	VE	P. ulmi	213	455
PWB	Potato witches broom	VIA	P. trifolii	425	237
BLL	Brinjal little leaf	VIA	P. trifolii	425	237

(continued)

Table 2 (continued)

Sample	Description	16Sr group	*Candidatus* species	TRF (bp) *MseI*	TRF (bp) *Bsh*1236I
CPS	Catharanthus phyllody	VIC	P. trifolii	425	237
ASHY-1	Ash yellows	VIIA	P. fraxini	328	234
LfWB	Loofah witches broom	VIIIA	P. luffae	207[a]	227[a]
AP-15	Apple proliferation	XA	P. mali	129	230
GSFY-1	German stone fruit yellows	XB	P. prunorum	129	230
PD	Pear decline	XC	P. pyri	129	230
BVK	Flower stunting	XIC	P. oryzae	129	456
NGS	Napier grass stunt	XI	P. oryzae	129	456
STOL	Stolbur of pepper	XIIA	P. solani	412	441
LDG	Ghanaian Cape St Paul wilt	XXII	P. cocosnigeriae	420	450

Table reproduced from (6) with permission from Wiley-Blackwell
[a]Indicates where the sizes have been estimated from the 23S accession

2. Ready-prepared PCR mixes are recommended because fewer stock solutions and less pipetting steps are required reducing the risk of contamination, although other sources of *Taq* polymerase and PCR mixes can be used. However it is important that any PCR reagents used for T-RFLP do not contain any loading dye as these can interfere with the fragment separation of the DNA sequencer.
3. DNA can be prepared by any suitable method for the matrix being tested.
4. Ensure all reagents are thoroughly defrosted prior to use and vortex well. Once defrosted it is wise to store reagents on ice.
5. Until a user is confident in the use of the PCR assay it would be prudent for the PCR products to be visualized (5 μL) by agarose gel electrophoresis (following standard practices) to ensure the PCR was successful prior to T-RFLP analysis. The PCR amplicon should be approximately 470 bp in size.
6. All PCR products including the positive and negative controls should be processed for T-RFLP to ensure the process has adequate controls for trouble shooting (if required).
7. Manufacturer's restriction enzymes and buffers are available at various concentrations so the exact composition of the mastermix will vary. Ensure that the manufacturer's protocol is

followed. Restriction enzymes are very sensitive to freeze-thaw cycles and should always be stored on ice when not in a freezer. It is wise to aliquot both the buffer and enzyme into smaller volumes upon delivery from the manufacturer.

8. Separate aliquots of each PCR product should be digested independently with *Mse*I and *Bsh*1236I and not as a dual digest. It is often convenient to analyze samples first with *Mse*I and then with *Bsh*1236I as the first enzyme will elucidate some of the 16Sr groups alone, whilst for others the results from both enzymes will be required.
9. Ensure the heat block/water bath temperature is accurate using a thermometer and does not vary more than ±2°C.
10. Until a user is confident in the use of the restriction digestion protocol it would be prudent for the products to be visualized (5 μL) by agarose gel electrophoresis to ensure the digestion was successful prior to T-RFLP analysis. Ensure that the restriction digestion negative control has not been fragmented compared to a nondigested PCR product.
11. Any TRF of sizes not present in Table 2 could be due to either; a new phytoplasma group for which the TRF size needs establishing or a false positive sample where a different bacterium has been amplified. To determine which the undigested PCR product should be studied by performing DNA sequencing. If this is to be completed then it must be remembered that the primers co-amplify plant DNA and generate a mixed PCR amplicon of plant and phytoplasma DNA. Therefore to generate a sample suitable for DNA sequencing the PCR product must be cloned (this can be done via any number of the commonly available cloning kits for example the Promega pGEM®-T easy vector system). Clones should then be screened by restriction digestion as detailed above and the digestion products analyzed by gel electrophoresis to allow the mixed PCR products to be resolved. A range of the clones with differing RFLP patterns should then be sequenced and this should represent both the plant and bacterial DNA present. Analysis of the DNA sequence will determine if the sample contains a new phytoplasma species or a bacterial species which has cross-reacted with the PCR primers.

Acknowledgement

This work was completed as part of a Defra Plant Health Fellowship for JH. We would also like to thank all individuals and organizations that have provided us with phytoplasma DNA/infected plant samples to enable us to validate the assay.

References

1. Smart CD et al (1996) Phytoplasma-specific PCR primers based on sequences of the 16S–23S rRNA spacer region. Appl Environ Microbiol 62:2988–2993
2. Duduk B, Paltrinieri S, Lee IM, Bertaccini A (2012) Nested PCR and RFLP analysis based on the 16S rRNA gene. In: Dickinson M, Hodgetts J (eds) Methods in molecular biology. Springer, New York
3. Harrison NA, Womack M, Carpio ML (2002) Detection and characterization of a lethal yellowing (16SrIV) group phytoplasma in Canary Island date palms affected by lethal decline in Texas. Plant Dis 86:676–681
4. Anthony RM, Brown TJ, French GL (2000) Rapid diagnosis of bacteremia by universal amplification of 23S ribosomal DNA followed by hybridisation to an oligonucleotide array. J Clin Microbiol 38:781–788
5. Liu W-T et al (1997) Characterization of microbial diversity by determining terminal restriction fragment length polymorphisms of genes encoding 16S rRNA. Appl Environ Microbiol 63:4516–4522
6. Hodgetts J et al (2007) Use of terminal restriction fragment length polymorphism (T-RFLP) for identification of phytoplasmas in plants. Plant Pathol 56:357–365

Chapter 21

Real-Time PCR for Universal Phytoplasma Detection and Quantification

Nynne Meyn Christensen, Henriette Nyskjold, and Mogens Nicolaisen

Abstract

Currently, the most efficient detection and precise quantification of phytoplasmas is by real-time PCR. Compared to nested PCR, this method is less sensitive to contamination and is less work intensive. Therefore, a universal real-time PCR method will be valuable in screening programs and in other setups in which large numbers of samples are tested.

Key words: Phytoplasma, Quantification, Real-time PCR, Universal detection

1. Introduction

As in vitro culturing of phytoplasmas is currently not possible, PCR has become the preferred method of detection and diagnosis. Several PCR primer combinations have been used to amplify phytoplasma DNA, usually the 16S rRNA gene. Some of these primers are universal so that they amplify DNA from all phytoplasma groups (1, 2). Most conventional PCR assays involve a nested PCR procedure to enhance sensitivity; however, this dramatically increases the risk of contamination and thus false positives. Real-time PCR, on the other hand, has been shown to be equally, or even more sensitive, than nested PCR (3), and, at the same time, real-time PCR is much less sensitive to contamination as no post-PCR processing is required. Only a few assays have been published for universal real-time PCR detection and diagnosis of phytoplasmas (3, 4), whereas several assays are available for the detection of specific groups of phytoplasmas (see Chapters 22 and 23 in this book).

Real-time PCR detection is based upon the measurement of emitted fluorescence during the PCR amplification. Two main types of molecules are used: intercalating agents or fluorogenic probes.

Matt Dickinson and Jennifer Hodgetts (eds.), *Phytoplasma: Methods and Protocols*, Methods in Molecular Biology, vol. 938, DOI 10.1007/978-1-62703-089-2_21, © Springer Science+Business Media, LLC 2013

SYBR Green® is a commonly used intercalating agent; however, it binds nonspecifically to all amplicons and thereby potentially compromises the accuracy of quantification and may even lead to false positives. Fluorogenic probes, such as TaqMan® probes, ensure a higher degree of specificity as they require not only hybridization of primers but also the TaqMan® probe to the template. For a more detailed description of real-time PCR, the reader is referred to some of the excellent books written on the subject.

As phytoplasma titre can vary significantly during the season and in different plant tissues, quantification of phytoplasmas can be important. Also, quantification is important in screening plants for resistance against phytoplasmas. On the other hand, quantification is not always essential in indexing programs for phytoplasmas or in diagnostic assays. When quantification is needed, dilution standards of known concentration should be run in parallel.

This protocol is based on the real-time PCR TaqMan® assay developed by Christensen et al. (3). The protocol was used for quantification of phytoplasmas in different plant organs in *Euphorbia pulcherrima* but has since proved to be of use in many other plant species (see ref. (3)). The protocol could easily be adapted to the primers used in Hodgetts et al. (4), although it should be noted that these primers do not amplify the 16S region, and thus the PCR product used for the standard curve in this protocol cannot be used and another one would need to be developed.

2. Materials

1. Tissue homogenizer (Geno/Grinder 2000, OPS Diagnostics, Bridgewater, NJ, USA).
2. Grinding tubes.
3. Grinding steel beads (5 mm diameter).
4. Liquid N_2.
5. DNA extraction kit (DNeasy Plant Mini Kit, QIAGEN).
6. NanoDrop 2000 (Thermo Scientific).
7. TaqMan® Universal PCR Master Mix (Applied Biosystems).
8. Oligonucleotide primers (Table 1).
9. FAM™-TAMRA™-labeled oligonucleotide probes (Table 1) (see Note 1).
10. PCR product from phytoplasma 16S rDNA target sequence for use as a standard.
11. MinElute® PCR Purification Kit (QIAGEN).
12. Healthy plant controls of same species and tissue as samples.
13. MicroAmp® Optical 384-Well Reaction Plate (Applied Biosystems).

Table 1
Sequence of primers and probes designed for detection of phytoplasma and plant DNA

	Phytoplasma 16S rDNA (5′–3′)	Plant 18S rDNA (5′–3′)
Forward primer	CGTACGCAAGTATGAAACTTAAAGGA	GACTACGTCCCTGCCCTTTG
Probe	TGACGGGACTCCGCACAAGCG	ACACACCGCCCGTCGCTCC
Reverse primer	TCTTCGAATTAAACAACATGATCCA	AACACTTCACCGGACCATTCA

Probes are labeled with FAM™ (5′) and TAMRA™ (3′)

14. MicroAmp® 96- & 384-Well Optical Adhesive Film (Applied Biosystems).
15. Centrifuge for PCR plates.
16. Real-time PCR instrument (Applied Biosystems, 7900HT).

3. Methods

3.1. DNA Extraction

1. Load the plant samples (0.5–1 g) into the grinding tubes together with eight grinding beads (see Note 2).
2. Freeze the closed tubes in liquid N_2.
3. Load the frozen grinding tubes into the Geno/Grinder homogenizer and shake at 1,500 strokes/min for 1 min.
4. Repeat steps 2 and 3 until the plant material is a fine powder, at least three cycles of freezing/grinding. Take care not to thaw the samples.
5. Load up to 100 mg of the homogenized frozen plant sample into prefrozen Eppendorf tubes using a prefrozen spatula (see Note 3).
6. Add 400 μL of buffer AP1 and 4 μL of RNase A from the DNeasy kit to the homogenized plant material and continue extraction according to the manufacturer's instructions.
7. Elute the DNA in 100 μL of the elution buffer and store the DNA at −20°C until required.
8. Just before use, an aliquot of the DNA should be diluted 10× (in water) for the real-time PCR (see Note 4).

3.2. Preparation of Standards for Quantification

Repeated freeze–thaw cycles are a major reason for degradation of DNA templates. To ensure consistent quality of the standards, aliquots of each standard dilution series should be prepared. The aliquots should be stored in PCR strips for convenient pipetting and should only be used a few times (see Note 5).

3.2.1. Phytoplasma Assay

1. Perform a nested PCR using P1/P7 primers followed by 16SF2n/16SR2 primers on DNA from a phytoplasma-infected sample (see Note 6).
2. Separate PCR fragments on a 1% agarose gel. A single band of 1,246 bp should be visible. Purify the PCR product using the MinElute® PCR Purification Kit (QIAGEN) according to the manufacturer's protocol.
3. Quantify the extracted PCR product by standard methods such as gel electrophoresis or NanoDrop analysis according to the manufacturer's instructions.
4. Calculate the number of molecules (the molecular weight of the amplified fragment is approximately 387 kDa). Adjust the concentration of the purified PCR product to 10^9 molecules per μL.
5. Make a 10× serial dilution to obtain 100 μL of each 10^2–10^8 molecules per 2 μL of the solution.
6. Prepare aliquots of each dilution (for example 20 μL).

3.2.2. Plant Assay

1. Extract DNA from healthy plant (see Note 7).
2. Determine the concentration and quality by gel electrophoresis and NanoDrop analysis. Adjust to appropriate concentrations: the first point in the dilution series should be undiluted or similar, for example 10 ng/μL. Prepare the dilution series to obtain 100 μL of each concentration, containing 10,000, 1,000, 100, 10, 1, 0.1, 0.01 pg DNA.
3. Prepare aliquots of each dilution (for example 20 μL).

3.3. Real-Time PCRs

At all stages while PCR reactions are being set up, precautions should be taken to avoid contamination of samples and reagents (see Note 8).

1. Once thawed, mix all reagents well (by inverting the tubes a number of times or for small volumes by flicking the tube) and spin briefly (~5 s) in a centrifuge. Keep all thawed reagents and plant DNA samples on ice prior to use. TaqMan® probes should be protected from light and should not be thawed until immediately prior to use.
2. Prepare two PCR master mix tubes: one for the phytoplasma assay and one for the plant normalization assay. The final volume of each master mix depends on the number of reactions required for each real-time PCR. To allow for pipetting errors prepare 10% excess of master mix (see Table 2 for preparing the master mixes).
3. Set up the real-time PCR in a 384-well plate with 8 μL of master mix per well (see Note 9). The first columns of the plate may be used for the phytoplasma assay and the following

Table 2
Master mix for real-time PCR assays

Reaction component	Stock concentration	Volume (μL) in each well (final total volume 10 μL)	Final concentration
Master mix	2×	5	1×
Forward primer	10 μM	0.3	300 nM
Reverse primer	10 μM	0.3	300 nM
TaqMan® Probe	2.5 μM	0.4	100 nM
ddH_2O		2	

Volumes are listed for one well. Volumes should be multiplied according to the number of required wells. Add 10% extra wells to allow for some error in pipetting

columns may be used for the plant assay. Each plate should include the dilution series of phytoplasma and plant standards, non-template (NTC) and negative controls, positive controls and test samples, each in triplicate.

4. Add 2 μL of 10× diluted sample DNA to each well (see Note 10). When pipetting is finished, seal the plate with optical adhesive film.
5. Centrifuge the reaction plate at low speed for 2 min and then load the plate onto an Applied Biosystems 7900HT real-time PCR instrument.

3.4. Program and Run Applied Biosystems 7900HT Instrument

Briefly, launch the SDS software and open a new plate template window to denote well locations on the 384-well plate for the standard dilutions, controls and test samples. Select detectors (FAM/TAMRA). Save the template window with the recorded data as an SDS run file. Start the run (see Notes 11 and 12).

3.5. Data Collection and Analysis

1. After the run is completed, remove the plate from the instrument, select baseline, and place the threshold line at the exponential phase of amplification (see Note 13). Analyze and save the results of the run.
2. Examine the standard curve results in the amplification plot and the standard curve plot views in the SDS software. The analysis of the phytoplasma and the plant assays must be done separately, because the software can only accept one standard curve at a time. Record concentrations of standard curves for wells containing sample dilutions for the phytoplasma assay first. The slope of the standard curve should ideally be −3.3 and the R^2 value should be close to 1.00. Finally, check that sample values are within the range of the standard curve values.

3. Export the results from the SDS software as a text file.
4. Repeat steps 2 and 3 using the plant DNA dilution as standard.
5. Import the text files into Microsoft Excel software for subsequent calculations.
6. Combine the phytoplasma and the plant results for each sample in the Excel file created for calculations.

The SDS software automatically calculates quantities from the Ct values for each sample according to the standard curves. Required calculations include (1) the calculation of average values for technical replicates and for replicate samples tested, and (2) normalization of the phytoplasma quantity in each sample using the plant DNA quantity. This calculation will give the number of phytoplasma cells/plant DNA amount; however, as phytoplasmas carry two copies of 16S, results should be divided by two (see Note 14).

4. Notes

1. Other systems can be used for grinding samples. If no tissue homogenizers are available, a mortar and pestle can be used; however, care should be taken to avoid cross-contamination between samples.
2. TaqMan® probes should be protected from light.
3. Use frozen tools, otherwise plant samples will stick to tools.
4. To minimize inhibitor effects, it is advisable to dilute sample DNA 1:10.
5. Quantification is not always necessary. In such cases, the dilution series of phytoplasma and plant standards can be omitted.
6. Nested amplification using P1/P7 (2, 5) followed by R16F2n/R16F2 (1) is widely used for amplification of most of the phytoplasma 16S rDNA gene (~1,246 bp). PCR conditions for P1/P7 are 94°C for 5 min, then 35 cycles at 94°C for 90 s, 55°C for 2 min and 72°C for 3 min, followed by 1 cycle at 72°C for 10 min and finally hold at 4°C; reaction conditions for R16F2n/R16F2 are 94°C for 2 min, then 35 cycles at 94°C for 1 min, 50°C for 2 min and 72°C for 3 min, followed by 1 cycle at 72°C for 10 min and finally hold at 4°C. Primer sequences are P1: 5′-AAGAGTTTGATCCTGGCTCAGGATT-3′, P7: 5′-CGTCCTTCATCGGCTCTT-3′, R16F2n: 5′-GAAACGACTGCTAAGACTGG-3′, R16F2: 5′-TGACGGGCGGTGTGTACAAACCCCG-3′. See Chapter 14 of this volume for more details.

7. DNA extraction is influenced by the method used and the type of starting material. DNA from different plant species and even plant organs will be extracted more or less efficiently even when the same method is used, as the tissues can contain different degrees of PCR inhibitors (for further details, see ref. (6)). It is therefore of paramount importance that the same method and same tissue are used for producing the standard extracts as for the samples.
8. At all stages while PCR reactions are being set up, precautions should be taken to avoid contamination of samples and reagents. Use only dedicated pipettes for preparing master mixes and another set for adding template DNA. Use only tips with filters. Wear gloves and change them frequently. Use only clean sterile plasticware. Master mix preparation and addition of template DNA should be performed in separate places with dedicated laboratory equipment.
9. Due to the high number of pipetting steps, equipment designed for handling large numbers of samples (such as robots, multichannel and multistep pipettes) is recommended.
10. Dilutions should only be made for an aliquot of the sample DNA, whereas the stock is kept as concentrated DNA for long-term storage.
11. Software program settings in the SDS:

    Run in 9600 emulation mode.

    Set reaction volume to 10 μL.

    Use FAM/TAMRA as detector/quencher.
12. Cycling parameters in the SDS:

    2 min at 50°C (UNG activation step).

    10 min at 95°C (polymerase activation).

    40 cycles of 15 s at 95°C (DNA denaturation) then 1 min at 60°C (annealing and extension).
13. In some cases, especially when analyzing plant 18S amplifications, signals are observed in the first cycles of the run. In such cases, baseline should be set from cycle 1.
14. Example calculation for one sample:

Ct (triplicates) phytoplasma	**Copies (calculated by SDS software)**	**Average number of copies**	**Cells**	**Ct (triplicates) plant**	**Amount (ng plant DNA)**	**Average (ng plant DNA)**	**Normalized amount (phytoplasma cells/ng plant DNA)**
25.43	8.5×10^3	8.9×10^3	4.5×10^3	15.17	11.3	12.0	375
25.52	8.0×10^3			14.78	14.3		
25.17	10.2×10^3			15.28	10.5		

References

1. Gundersen DE, Lee I-M (1996) Ultrasensitive detection of phytoplasmas by nested-PCR assays using two universal primer pairs. Phytopathol Mediterr 35:144–151
2. Smart CD et al (1996) Phytoplasma-specific PCR primers based on sequence of the 16S-23S rRNA spacer region. Appl Environ Microbiol 62:2988–2993
3. Christensen NM et al (2004) Distribution of phytoplasmas in infected plants as revealed by real-time PCR and bioimaging. Mol Plant Microbe Interact 17:1175–1184
4. Hodgetts J et al (2009) Panel of 23S rRNA gene-based real-time PCR assays for improved universal and group-specific detection of phytoplasmas. Appl Environ Microbiol 75:2945–2950
5. Deng SJ, Hiruki C (1991) Amplification of 16S ribosomal-RNA genes from culturable and non-culturable mollicutes. J Microbiol Meth 14:53–61
6. Demeke T, Jenkins GR (2010) Influence of DNA extraction methods, PCR inhibitors and quantification method on real-time PCR assay of biotechnology-derived traits. Anal Bioanal Chem 396:1977–1990

Chapter 22

A Real-Time PCR Detection System for the Bois Noir and Flavescence Dorée Phytoplasmas and Quantification of the Target DNA

Nataša Mehle, Nina Prezelj, Matjaž Hren, Jana Boben, Kristina Gruden, Maja Ravnikar, and Marina Dermastia

Abstract

The real-time PCR detection system for grapevine yellows phytoplasmas described here is composed of two assays for group-specific detection of flavescence dorée (FD) and bois noir (BN) phytoplasmas and a universal phytoplasma assay. It uses hydrolysis minor groove binder probes (TaqMan-MGB). The addition of an assay for amplification of plant DNA co-extracted with phytoplasma DNA provides a further quality control for the DNA extraction and PCR amplification for each sample. The detection system described is reliable, specific, sensitive, and easily applicable to fast, high-throughput diagnosis of grapevine yellows phytoplasmas. In addition to the detection system, an approach for the quantification of phytoplasmas in the sample is described.

Key words: Bois noir, Flavescence dorée, Grapevine yellows, MGB probes, Quantification, Real-time PCR, TaqMan

1. Introduction

Grapevine yellows diseases are associated with different phytoplasmas that occur in grape-growing areas worldwide. The main causal agents of grapevine yellows in Europe are flavescence dorée (FD) and bois noir (BN) (1). FD has the greatest economic impact and is recognized by the European and Mediterranean Plant Protection Organization (EPPO) as a quarantine pest (2). On the other hand, BN is present in most wine-producing regions in Europe. FD and BN are transmitted by insect vectors of the order Hemiptera: the only known natural vector of FD is the leafhopper *Scaphoideus titanus* Ball (3), while the main vector of BN is the planthopper *Hyalesthes obsoletus* Signoret (4).

Matt Dickinson and Jennifer Hodgetts (eds.), *Phytoplasma: Methods and Protocols*, Methods in Molecular Biology, vol. 938, DOI 10.1007/978-1-62703-089-2_22,

As the detection of phytoplasmas is not possible through the use of traditional microbiological approaches, molecular methods are required for their reliable diagnostics instead. In recent years several protocols based on a real-time PCR (qPCR) have been described for phytoplasma detection. Universal phytoplasma qPCR-based detection systems have been developed (5, 6). An FD-specific qPCR assay was developed by Bianco et al. (7). Galetto et al. (8) developed qPCR assays for the specific detection of FD, BN, and apple proliferation (AP) phytoplasmas (SYBR® Green), and an assay for the detection of 16SrV, 16SrX, and 16SrXII phytoplasma groups (TaqMan®). Angelini et al. (9) published qPCR assays for identification of phytoplasmas associated with FD, BN, and aster yellows (AY). A triplex qPCR for the simultaneous detection of FD and BN has been designed (10). Additionally, FD- and BN-specific reverse transcription qPCR detection assays, using the same crude extract as used for grapevine virus detection, have been developed (11).

In this chapter we describe a qPCR-based detection system for FD and BN phytoplasmas that can be used for diagnostic purposes (12). It includes two amplicons for group-specific detection of FD and BN and a universal phytoplasma amplicon. The universal amplicon also amplifies phytoplasmas belonging to other groups, e.g., 16SrI, 16SrIII, 16SrX, and 16SrXII (12). As a positive control to check the quality of extracted DNA for each sample, the amplification of plant DNA co-extracted with phytoplasma DNA is a part of the procedure. Inhibition of PCR is checked additionally by testing two different dilutions of extracted DNA. The use of TaqMan® (hydrolysis) MGB probes (Minor Groove Binder Probes) (13) in the assay greatly diminishes the possibility of detecting nonspecific target DNA. In comparison to the nested PCR/RFLP assays, this qPCR assay is a faster, less tedious and more sensitive detection means for phytoplasmas. Due to the shorter and simpler laboratory procedures, the risk of sample cross-contamination is also reduced. With some additional steps, such as the use of reference DNA material (a sample containing the target DNA sequence or a sample with a known copy number of plasmids containing target DNA sequence) and the use of a standard calibration curve, the system may be used for quantitative analysis. In this chapter a step-by-step analysis of the system, together with the precautions that should be taken to avoid contamination of samples and reagents, is described in detail. The high-throughput version of the system has been successfully used in the Slovenian official surveys of grapevine yellows since 2007.

Table 1
Primers and MGB probes for flavescence dorée (FD)-specific (FDgen), bois noir (BN)-specific (BNgen), and universal phytoplasma (UniRNA) amplicons (12)

Name	Orientation	Sequence (5′–3′)	Working concentration (μM)
UniRNA	Forward	AAA TAT AGT GGA GGT TAT CAG GGA TAC AG	10
	Reverse	AAC CTA ACA TCT CAC GAC ACG AAC T	10
	Probe	*FAM*-ACG ACA ACC ATG CAC CA-*NFQ*[a]	2.5
FDgen	Forward	TTA TGC CTT ATG TTA CTG CTT CTA TTG TTA	10
	Reverse	TCT CCT TGT TCT TGC CAT TCT TT	10
	Probe	*FAM*-ACC TTT TGA CTC AAT TGA-*NFQ*	2.5
BNgen	Forward	AAG CAG GTT TAG CGA TGG TTG T	10
	Reverse	TGG TAC CGT TGC TTC ATC ATT T	10
	Probe	*FAM*-TTA ATA CCA CCT TCA GGA AA-*NFQ*	2.5

[a] *NFQ* nonfluorescent MGB quencher

2. Materials

2.1. Samples and Controls

1. Extracted DNAs of samples (see Note 1).
2. Tenfold dilutions of extracted DNAs: 5 μL of DNA, 45 μL of sterile nuclease free water (see Notes 1 and 2).
3. Negative control of isolation (NCI): extraction method control prepared from material known to be uninfected (see Note 3).
4. No template control (NTC): sterile deionized water (the same as used for qPCR master mix preparation) (see Note 4).
5. Positive controls (PC) (sample containing target sequence): DNA from a known BN- and FD-infected plant (see Notes 3–5).
6. Series of tenfold dilutions of the reference DNA material (only if quantitative analysis is needed) (see Note 6).

2.2. qPCR Reagents

1. 2× TaqMan Universal PCR Master Mix (Applied Biosystems) (see Notes 7 and 8).
2. Human 18S rRNA (20×) (Applied Biosystems): 18S rRNA primer–probe mix. The 18S rRNA probe has a VIC fluorescent label (see Note 9).
3. Molecular biology grade water (nucleic acid- and nuclease-free) or sterile-filtered double deionized water (see Note 10).
4. Primers and probes (Table 1) (see Notes 11 and 12).

2.3. Equipment

1. qPCR reaction plates: e.g., 384-well optical reaction plates (Applied Biosystems) (see Notes 7 and 13).
2. Optical adhesive covers (e.g., Applied Biosystems).
3. Centrifuge with a rotor for plates.
4. Microfuge with a rotor for tubes.
5. Vortexer.
6. Real-time thermal cycler: e.g., ABI PRISM® 7900 HT Sequence Detection System (Applied Biosystems) (see Note 7).
7. Software (e.g., SDS 2.3, Applied Biosystems) for fluorescence acquisition and calculation of threshold cycles (quantification cycles; *Cq*) (see Note 7).
8. Two UV chambers (see Notes 14 and 15).
9. Two sets of pipettes.
10. DNase-free tubes, racks for tubes.
11. Microfuge tube opener.
12. Aerosol-barrier pipette tips.

3. Methods

3.1. Setting Up Plates

1. Arrange the samples in the reaction plate according to the sample number and amplicon. For each primer set, include the NTC, NCI, and PC as a part of the sample group. See Fig. 1 for an example of a typical plate plan. If quantitative analysis is needed, replace PC with a series of reference DNA samples diluted tenfold for each primer set to generate standard calibration curves. The reference DNA sample should contain the same target DNA as the samples (e.g., FD) and for the amplification of endogenous control DNA (e.g., 18S rRNA) used for normalization. See Fig. 2 for an example.
2. Calculate the volume of components needed for the assay master mixes according to Tables 2 and 3. Include sufficient reagents for the number of samples (undiluted and tenfold dilutions), reference DNA sample dilutions (if quantitative analysis is needed), and controls to be tested in duplicate. Include 10% extra volume for loss during pipetting and possible errors.
3. Remove all reagents from the freezer and allow them to thaw. Mix all thawed reagents well with the vortexer and spin them briefly (~5 s) in the microfuge.
4. Prepare four test assay master mixes (18S rRNA, UniRNA, FDgen, and BNgen) in the individual tubes. For each mix, the reagents are added in order and quantity calculated in step 2.

	18S rRNA				UniRNA				FDgen				BNgen											
	1	2	3	4	5	6	7	8	9	10	11	12	13	14	15	16	17	18	19	20	21	22	23	24
A	S1		S1 d		S1		S1 d		S1		S1 d		S1		S1 d									
B	S2		S2 d		S2		S2 d		S2		S2 d		S2		S2 d									
C	S3		S3 d		S3		S3 d		S3		S3 d		S3		S3 d									
D	S4		S4 d		S4		S4 d		S4		S4 d		S4		S4 d									
E	S5		S5 d		S5		S5 d		S5		S5 d		S5		S5 d									
F	S6		S6 d		S6		S6 d		S6		S6 d		S6		S6 d									
G	S7		S7 d		S7		S7 d		S7		S7 d		S7		S7 d									
H	S8		S8 d		S8		S8 d		S8		S8 d		S8		S8 d									
I	S9		S9 d		S9		S9 d		S9		S9 d		S9		S9 d									
J	S10		S10 d		S10		S10d		S10		S10 d		S10		S10 d									
K	S11		S11 d		S11		S11 d		S11		S11 d		S11		S11 d									
L	S12		S12 d		S12		S12 d		S12		S12 d		S12		S12 d									
M	S13		S13 d		S13		S13 d		S13		S13 d		S13		S13 d									
N	S14		S14 d		S14		S14 d		S14		S14 d		S14		S14 d									
O	NCI		NCI d		NCI		NCI d		NCI		NCI d		NCI		NCI d									
P	N 1	N 2	P	P	N 1	N 2	P	P	N 1	N 2	P	P	N 1	N 2	P	P								

Fig. 1. Typical plate plan of qPCR for grapevine phytoplasmas. The plate arrangement represents an assay in which 14 grapevine samples will be assayed for the presence of FD and BN. Each DNA sample is tested with all three amplicons for phytoplasmas (UniRNA, FDgen, BNgen) and an amplicon for an endogenous control (18S rRNA) in separate qPCR reactions. All reactions are performed in two replicate wells in two dilutions (undiluted and tenfold dilution) on the same 384-well plate. *SX (X= 1–14)* extracted DNAs of samples; *SX d* dilutions of extracted DNAs; *NCI* negative control of isolation; *NCI d* dilutions of negative control of isolation; *N 1* no template control—put sterile deionized water in the plate at the start of pipetting (NTC1); *N 2* no template control—put sterile deionized water in the plate at the end of pipetting (NTC2); *P* positive DNA target control (PC).

After completion, mix all reagents well with the vortexer and spin them briefly (~5 s) in a microfuge (see Note 16).

5. Take a 384-well qPCR plate. Fill the appropriate wells with 8 μL of each master mix according to the plate plan (see Note 17).
6. Add 2 μL of undiluted sample, sample diluted tenfold, reference DNA sample dilutions (for quantitative analysis), or control DNA into each test well. Always start by adding water for the first NTC (NTC1), continue with DNA samples, reference DNA sample dilutions and controls. Finish with water for the last NTC (NTC2). Using this order of DNA addition facilitates the identification of any possible source of contamination (see Note 17).

	18S rRNA								FDgen															
	1	2	3	4	5	6	7	8	9	10	11	12	13	14	15	16	17	18	19	20	21	22	23	24
A	S1		S1 d		R 10d				S1		S1 d		R 10d											
B	S2		S2 d		R 10^2d				S2		S2 d		R 10^2d											
C	S3		S3 d		R 10^3d				S3		S3 d		R 10^3d											
D	S4		S4 d		R 10^4d				S4		S4 d		R 10^4d											
E	S5		S5 d		R 10^5d				S5		S5 d		R 10^5d											
F	S6		S6 d						S6		S6 d													
G	S7		S7 d						S7		S7 d													
H	S8		S8 d						S8		S8 d													
I	S9		S9 d						S9		S9 d													
J	S10		S10 d						S10		S10 d													
K	S11		S11 d						S11		S11 d													
L	S12		S12 d						S12		S12 d													
M	S13		S13 d						S13		S13 d													
N	S14		S14 d						S14		S14 d													
O	NCI		NCI d						NCI		NCI d													
P	N 1	N 2							N 1	N 2														

Fig. 2. Example of a plate plan for quantitative qPCR for phytoplasma. The plate arrangement represents an assay in which 14 grapevine samples will be assayed, quantified, and compared for the presence of FD phytoplasma. Each DNA sample is tested with a target amplicon for phytoplasmas (in this case FDgen) and an amplicon for an endogenous control (18S rRNA) in separate qPCR reactions. All reactions are performed in two replicate wells in two dilutions on the same 384-well plate. *SX (X = 1–14)* extracted DNAs of samples; *SX d* dilutions of extracted DNAs; *NCI* negative control of isolation; *NCI d* dilutions of negative control of isolation; *N 1* no template control—put sterile deionized water in the plate at the start of pipetting (NTC1); *N 2* no template control—put sterile deionized water in the plate at the end of pipetting (NTC2); *R Yd (Y = 10–10⁵)* Y-times diluted reference DNA sample. If more samples are analyzed and you need more than one plate, put the series of reference DNA sample dilutions for all the plates analyzed. The results of reference DNA sample on all the plates should be the same (not differ more than 0.5–1 *Cq*) in order to be able to compare the samples on different plates.

7. After adding DNA, the qPCR plate should be covered with an optical adhesive cover and sealed with a plastic applicator (see Note 18).
8. Centrifuge the plate for 1 min at 1,000 × *g* to ensure that both the reaction mix and the DNA collect at the bottom of the wells (see Note 19).

3.2. Perform qPCR run

1. Switch on the thermal cycler in advance, according to the manufacturer's instructions to warm it up and to stabilize temperature and light sources.

Table 2
Master mixes for flavescence dorée (FD)-specific (FDgen), bois noir (BN)-specific (BNgen), and universal phytoplasma (UniRNA) amplicons

Components	Final concentration	Volume per well (μL)
Sterile nuclease-free water		0.2
2× TaqMan Universal PCR Master Mix	1×	5.0
Forward primer (10 μM)	900 nM	0.9
Reverse primer (10 μM)	900 nM	0.9
Probe (2.5 μM)	250 nM	1.0
Total volume		8.0

Table 3
Master mix for 18S rRNA amplicon

Components	Final concentration	Volume per well (μL)
Sterile nuclease-free water		2.5
2× TaqMan Universal PCR Master Mix	1×	5.0
Human 18S rRNA (20×)	1×	0.5
Total volume		8.0

2. Transfer the qPCR plate to the real-time thermal cycler and run qPCR using the following universal cycling program (consult the instrument manual for specific instructions, see Note 20):

1 cycle	2 min at 50°C (UNG activation step) 10 min at 95°C (polymerase activation)
45 cycles	15 s at 95°C (DNA denaturation) 1 min at 60°C (annealing and extension)

3.3. Analyze Data

1. Analyze and export raw data (see instrument manual for detailed instructions about document setup, baseline, and threshold settings) (see Note 21).
2. Import data into Microsoft Excel or equivalent spreadsheet program with statistical features. For each of the two replicates of a sample, calculate the average *Cq* value. If large differences between the replicates are observed, the analysis should be

repeated (see Note 22). All samples with an average *Cq* value lower than the cut-off value should be considered positive for that particular assay (see Note 23). If quantitative analysis using a standard calibration curve is performed, proceed with the calculation of standard calibration curves for all the amplicons (target DNA and endogenous control DNA) (see Notes 6 and 24). Calculate the amplification efficiencies of standard calibration curves and test samples for all the amplicons. Discard all the test samples with insufficient amplification efficiencies (see Note 24). From the standard calibration curves determine the amounts of target DNA and endogenous control DNA in the samples. For each sample normalize the amount of target DNA with the amount of endogenous control DNA in that sample. Compare the normalized amounts of target DNA between the samples (see Note 25).

3. If target is detected in the extraction blank control (NCI), contamination of reagents or samples during DNA extraction should be suspected. The results of sample analysis are therefore not reliable; the analysis is not valid and should be repeated, including the extraction of DNA.
4. NTC analysis of any assays must not result in a positive qPCR signal; otherwise contamination during preparation of qPCR should be suspected. The results of the sample analyses are not reliable and the qPCR analysis must be repeated (see Note 26).
5. The presence of all target sequences in the positive controls must be confirmed. If the target is not detected in these controls, the analysis is not valid and errors in the PCR reaction should be considered.
6. The *Cq* of the isolated DNA measured by endogenous controls (e.g., 18S rRNA, *cox*) must be in the predicted range (see Note 27). If not, then the analysis is not valid and some errors in the extraction procedure or the presence of inhibitors should be considered.
7. The *Cq* values of UniRNA should be checked to confirm the presence of FD and BN phytoplasmas or to detect other unidentified phytoplasma species (Table 4) (see Note 28).

4. Notes

1. In general, the total amount of DNA added to a PCR reaction should be less than 100 ng. Mix all DNA samples and their dilutions well, and spin them briefly (about 5 s) in a microfuge before use. If DNA samples were stored frozen, they should be allowed to thaw before mixing.

Table 4
Interpretation of results of real-time PCR

Primers used for amplification[a]			Phytoplasma present[b]		
BNgen	FDgen	UniRNA	Bois noir	Flavescence doreé	Unidentified
+	−	±	✓	□	
−	+	+	□	✓	
+	+	+	✓	✓	
−	−	+	□	□	✓
−	−	−	□	□	□

[a]+*Cq* value lower than cut-off value; −*Cq* value higher than cut-off value
[b]✓ presence shown; □ presence not shown

2. All samples should be tested at two dilutions. In our case, when the KingFisher automated DNA extraction procedure described in Chapter 12 of this volume is used, undiluted and tenfold diluted DNA, and in some cases 100-fold diluted DNA, are suitable for testing. Testing of dilutions is a PCR inhibition control, checking for the presence of inhibitors in the sample that can influence the efficiency of the qPCR amplification. If there is no inhibition, the shapes of the amplification curves are the same for undiluted and tenfold diluted DNA and the difference in the *Cq* values for the tenfold diluted DNA is greater by an extra 3.3 than the undiluted one with 100% efficiency in amplification. In the case of strong inhibition (e.g., the *Cq* value for the tenfold dilution of 18S rRNA is lower than that of undiluted 18S rRNA control), the qPCR reaction may give false negative results. Analyses should be repeated with DNA diluted more. If strong inhibition is still detected, DNA should be extracted using a different method. However, in our routine work this was only the case when clematis samples were used. When testing undiluted and DNA diluted tenfold, extracted with the procedure described in Chapter 12, no *Cq* values for 18S rRNA and FD amplicons were obtained; only when using DNA diluted 100-fold (for some clematis samples, even higher dilutions of DNA should be tested) were *Cq* values obtained. If no inhibition is detected whilst using the protocol for many samples of the same sample types, the use of only one DNA dilution is possible, but in that case we suggest that each sample is tested by qPCR in triplicate instead of in duplicate. For quantitative analysis, always use two dilutions of a sample, making sure there is no inhibition and that the difference in *Cq* between tenfold dilutions is between 2.9 and 4.0.

3. The assay is validated only for the grapevine, insects (*S. titanus*, *H. obsoletus*, *Orientus ishidae*, *Oncopsis alni*), and clematis samples. When the assay is used for other sample materials, a healthy sample and a spiked healthy sample of the same material must be included in the analysis to check for possible cross-reactions of primers and probes with the new sample material and to prove that detection of the target from the new sample material is possible, respectively.
4. A non-template control (NTC) and a positive control (PC) are performed in every qPCR run. If quantitative analysis is needed, replace the PC with a series of tenfold diluted reference DNA samples. NTC contains all reagents used for qPCR, but sterile deionized water is added into the reaction instead of the sample DNA. NTC should be used for the same amplicon as the sample in duplicate, one at the start of pipetting (NTC1) and one at the end (NTC2) for each amplicon.
5. The positive control (PC) is a sample that contains target sequence DNA. To make assays more uniform, we suggest preparing several aliquots of appropriate controls and putting them in the freezer so that a fresh aliquot can be used for every analysis to prevent degradation of DNA by repeated freezing and thawing.
6. For quantitative analysis, reference DNA material is needed. This can be any sample containing the target DNA sequence or a plasmid sample with a known concentration (or copy number) of plasmids containing target DNA sequence. You can prepare a reference DNA sample by making a DNA pool of all the samples to be analyzed. At least four (ideally five) tenfold dilutions of a reference sample must be used to generate a standard calibration curve (see also Note 24).
7. Use of names of chemicals or equipment in this chapter implies no approval of them to the exclusion of others. This information is given for the convenience of users and does not constitute an endorsement of the products or services. Equivalent products may be used if they can be shown to lead to the same results.
8. Master mix from Applied Biosystems and Fermentas were compared. The results obtained were very similar.
9. In order to control the DNA extraction procedure, an endogenous control must be included in the assays. This is preferably a conserved gene, which provides confirmation that the extraction from different plant species has been successful, and it should be present in high copy numbers in the genome in order to increase the sensitivity of the assay. In our case a gene encoding 18S rRNA is used to detect grapevine DNA. Instead of 18S rRNA other endogenous controls may be used, e.g.,

plant cytochrome oxidase (*cox*) (14). 18S rRNA may be used for the normalization in quantitative comparative analysis to ensure that the measured differences between samples are not attributed to disproportionate amounts of the starting material; *cox*, being a mitochondrially encoded gene, is not appropriate for such quantification purposes.

10. Use of molecular biology grade (nucleic acid- and nuclease-free) or sterile-filtered double deionized water in all recipes and protocol steps is essential. Do not use DEPC-treated water because the slightly acidic pH may promote primer degradation. It is easy to contaminate a PCR assay by reusing an old bottle of water, therefore aliquot the water into tubes and then use one tube per assay.
11. The FDgen amplicon was designed within the *secY* gene, which codes for a translocation protein. This amplicon is situated between the primer pairs FD9f/r (15) and FD9f3b/r2 (16), used for detection of FD by nested PCR. The BNgen amplicon was designed within the *Stol11* genomic fragment and it is situated between the primer pairs STOL11f2/rl (15) and STOL11f3/r2 (16), which are used for detection of BN by nested conventional PCR. The universal amplicon UniRNA amplifies FD and BN phytoplasmas (along with other phytoplasma strains, e.g., 16SrI, 16SrIII, and 16SrX phytoplasma groups), but with a distinct preference for FD. It is included in the protocol as a tool to increase the sensitivity of FD detection and to detect other phytoplasma types (12).
12. Purchase lyophilized oligonucleotides from any commercial source, synthesized at a 25 nmol scale. For these, standard desalting is sufficient and no additional purification is required. Briefly centrifuge the tubes of dried oligonucleotide to get the contents to the bottom. Resuspend the oligonucleotides in molecular biology grade water to 100 μM. Dilute each stock of forward and reverse primers to 10 μM, prepare aliquots of these working solutions and store them at −20°C. When stored properly and subjected to minimal freeze–thaw cycles, a primer stock may last at least 2 years. Depending on the vendor, the probe may be supplied in 100 μM solutions. In this case, dilute a small amount of 100 μM stock to 2.5 μM, prepare aliquots of these working solutions and store them at −20°C. Avoid repetitive freeze–thaw cycles, and thaw on ice before usage to preserve the integrity of the probe. Protect the probe from excessive exposure to light (e.g., put the tubes with probes in dark plastic bags/containers) to prevent photobleaching of the fluorescent dyes and evaporation.
13. Alternatively you can use 96-well optical reaction plates. If 96-well plates are used, the total reaction volume should be 25 μL (20 μL of master mix and 5 μL of sample DNA).

14. Reaction mixes and the addition of DNA samples should be prepared in separate UV chambers using dedicated laboratory equipment (pipettes, tips, tubes, microfuge tube opener, racks for tubes, lab coat, gloves): first, prepare reaction mixes and load them onto the qPCR plate (no DNA other than primers and probes should be present here); second, add DNA samples in another UV chamber. Both steps should be done in a separate location from the places used for homogenization of the samples and DNA extraction.
15. In all stages of setting up PCR reactions, the following precautions should be taken in order to avoid contamination of samples and reagents: (1) use only filter tips (aerosol barrier tips) and use a fresh tip for each pipetting step; (2) close reagent/sample tubes after the desired volume has been aspirated to the tip; (3) always wear gloves and change them if you suspect contamination; (4) use only clean, sterile plasticware; (5) all tubes should be opened by a microfuge tube opener.
16. This step may be prepared in advance and the tubes may be put on ice or stored at 4°C for up to a day before preparing the reaction plate.
17. An automated liquid handling system (e.g., Multiprobe® II PLUS EX, PerkinElmer) may be used for pipetting a large number of DNA samples and master mixes onto the 384-well plates. If the plates are pipetted manually, start with pipetting the master mixes onto the bottom of the wells and continue by adding DNA samples and controls to the side walls of the wells. This facilitates visual checking during the addition of samples. Avoid touching of master mixes inside the tubes with the pipette tip. It is advisable to use multistep pipettors. In this case use one tip to fill up all wells with the same master mix and one tip for two replicates when adding the DNA of the sample. In the case of liquid handling machines, the pipetting order of master mixes and DNA may be reversed: DNA is applied first (onto the bottom of the wells), the plate is then centrifuged to ensure the DNA is collected at the bottom, and then master mixes are added without touching the well bottoms, followed by plate sealing and centrifugation.
18. Do not touch the optical adhesive cover with fingers.
19. Once prepared, reactions must be processed immediately. If this is not possible, they may be stored in the dark at 2–8°C for a maximum of a few hours.
20. Before starting the run: (1) adjust the reaction volume, if needed; (2) assign samples to locations on the plate; (3) assign detectors to your samples (choose the correct filters according to the probes used in amplicons: FAM for phytoplasma amplicons and VIC for the 18S rRNA amplicon). The temperature

ramp rate should be similar to Applied Biosystems 7900 thermal cyclers. The standard mode for temperature rates on the 7900HT is: 1.6°C/s up and 1.6°C/s down. Do not use faster temperature ramp rates.

21. The transformation of the fluorescence signal into *Cq* data (e.g., Ct-threshold cycle, Cp-crossing point) (17), as well as methods for baseline and threshold settings, varies between instrument models. The specific instrument manual must be consulted. When analyzing the raw data it is important to adjust the cycle threshold (*Cq*) of the amplification plot to within the geometric (exponential) phase of amplification, preferably at the beginning of the geometric phase. At the log view this is the linear part of the amplification plot. This is critical for correct quantitative analysis because the geometric phase represents the point of the reaction at which *Cq* is quantitatively related to the amount of initial PCR template. We have experimentally determined that automatic baseline and threshold at 0.065 are usually suitable when using Applied Biosystems 7900 thermal cyclers. Manually check the multicomponent plots for all samples (wells). In wells containing positive samples, the fluorescence of a fluorescent reporter (in this case FAM dye) should be increasing with cycles. A *Cq* value where no increase of FAM fluorescence is observed must not be considered as positive.
22. It is always important to run duplicate or, preferably, triplicate reactions of single dilutions when performing a qPCR assay. If the *Cq* values are very high, the use of replicates is especially important to make sure the reaction is working properly. If data from the replicate reactions vary significantly, the reactions should be repeated, unless high variability in the *Cq* values are attributed to the stochastic effect of target copy distribution in replicate reactions. This occurs when less than ten copies of the target are present in the reaction (18, 19).
23. The cut-off value is determined for each assay in the process of assay validation before the assay is put into routine use. It is the highest *Cq* value still considered to be a positive result. The following guidelines should be followed for the determination of the cut-off value during validation of the assay. Analyze serial dilutions of known positive and negative samples in different sample types. At least five dilution steps should be analyzed (usually tenfold dilution steps), each containing a sample in five replicate wells. Next, search for the last group of samples in which there is no amplification in some of the replicates. Take the range of the highest *Cq* values observed. Round this *Cq* value up to the next half value and then add 0.5 to this *Cq* value to take into account the difference in threshold chosen between runs. This *Cq* value is the cut-off value, and should be

determined for each assay at the time of the method validation. The cut-off value for the same assay may vary from laboratory to laboratory for various reasons (e.g., the qPCR machine and software used). Therefore each laboratory should revalidate the cut-off values for any new assay prior to its implementation.

24. Standard calibration curves are calculated for each amplicon from the *Cq* values of reference sample dilutions. In the case of using plasmid DNA (with a known copy number of target DNA/endogenous control DNA) as a reference sample, plot the *Cq* values on the *x*-axis and log the DNA copy number on the *y*-axis. When a sample with unknown exact copy number of target DNA/endogenous control DNA is used as a reference sample, assign a copy number of 10 to the most concentrated dilution (log 10 = 1) and then assign copy numbers to other dilutions according to the dilution factor. Calculate the slope (*k*) of the log-linear portion of the calibration curve and determine the amplification efficiency. The R^2 value of the linear regression should be more than 0.99. Efficiency, calculated as $10^{-1/\text{slope}}$, should be in the range 1.85–2.15.

25. From the calibration curves and *Cq* values of the sample, determine the amounts of target DNA and endogenous control DNA in the sample. Use the formula log copy number = slope (*k*)* *Cq* + *y*-axis crossing point (*n*). The resulting copy number is not an absolute value when using a reference sample with unknown DNA copy number: in this case, you get a value relative to the reference sample. Relative quantification can be done by comparing these values between your samples to get a ratio of target DNA in sample 1 compared to sample 2.

26. If amplification of target occurs in the NTC, it may be a result of the contamination of reagents, degradation of the primer mix, or the formation of nonspecific PCR products. If contamination of the reagents is suspected, use fresh aliquots of all reaction components. In the case of 18S rRNA, high *Cq* values (>36) for NTC are not unusual: it is a result of DNA traces in bacterial polymerase. Even in cases in which all amplification products, including those that are non-specific, contribute to the fluorescence signal, they may be considered insignificant if the *Cq* values of the NTC or NCI are higher than the cut-off value.

27. Quality and quantity of isolated DNA greatly influence the detection of grapevine yellows. Monitoring the *Cq* values of endogenous controls (e.g., *cox*, 18S rRNA) allows the estimation of both parameters, thereby increasing the confidence in results. The *Cq* values for 18S rRNA in the DNA extraction procedure used in our laboratory by the KingFisher method (see Chapter 12 of Mehle et al. (20)) vary from 17 to 25 (while using *cox*, *Cq* values are up to 29). Higher *Cq* values are an

indicator of lower amounts of plant DNA and presumably lower amounts of phytoplasma DNA in the sample, which therefore decreases the confidence of results. This is a signal that such samples should be re-extracted and tested again. Occasionally, higher *Cq* values are only an indicator of inhibitors present in the sample, therefore higher dilutions of DNA sample should be retested.

28. The universal amplicon UniRNA amplifies FD and BN phytoplasmas along with other phytoplasma strains belonging to different groups, e.g., 16SrI, 16SrIII, and 16SrX. However, a distinct preference for FD generally results in lower *Cq* values for FD. In cases when only FD is present in the sample, the *Cq* values for UniRNA are less than four cycles below the FDgen *Cq* values. If only BN is present in the sample, however, the *Cq* value for the UniRNA amplicon is more than four cycles higher than for the BNgen. PCR inhibitors in extracted DNA, coinfection of FD and BN, or the presence of other species of phytoplasma in the sample usually result in other differences in *Cq* values. If only the UniRNA amplicon produces a signal, the sample may be considered to be infected with an unidentified phytoplasma. If the *Cq* values for UniRNA are above 30, even in undiluted DNA, it must be considered that FD is present in a very low concentration because the FD-specific amplicon is approximately ten times less sensitive than the universal one (see Table 4).

References

1. Angelini E et al (2001) Flavescence dorée in France and Italy—occurrence of closely related phytoplasma isolates and their near relationships to Palatinate grapevine yellows and an alder yellows phytoplasma. Vitis 40:79–86
2. EPPO/CABI (2003) Data sheets on quarantine pests—grapevine flavescence dorée phytoplasma. In: Smith IM, McNamara DG, Scott PR, Holderness M (eds) Quarantine pests for Europe. CABI International, Wallingford, UK, pp 1013–1021
3. Schwester D, Carle P, Moutous G (1963) Transmission de la flavescence dorée de la vigne par Scaphoideus littoralis Ball. Ann Epiphyt 14:175–198
4. Maixner M, Ahrens U, Seemüller E (1995) Detection of the German grapevine yellows (Vergilbungskrankheit) MLO in grapevine, alternative hosts and a vector by specific PCR procedure. Eur J Plant Pathol 101:241–250
5. Christensen NM et al (2004) Distribution of phytoplasmas in infected plants as revealed by real-time PCR and bioimaging. Mol Plant Microbe Interact 17:1175–1184
6. Hodgetts J et al (2009) Panel of 23S rRNA gene-based real-time PCR assays for improved universal and group-specific detection of phytoplasmas. Appl Environ Microbiol 75:2945–2950
7. Bianco PA, Casati P, Marziliano N (2004) Detection of phytoplasmas associated with grapevine flavescence dorée disease using real-time PCR. J Plant Pathol 86:257–261
8. Galetto L, Bosco D, Marzachì C (2005) Universal and group-specific real-time PCR diagnosis of flavescence doreé (16Sr-V), bois noir (16Sr-XII) and apple proliferation (16Sr-X) phytoplasmas from field-collected plant hosts and insect vectors. Ann Appl Biol 147:191–201
9. Angelini E et al (2007) A new TaqMan method for identification of phytoplasmas associated with grapevine yellows by real-time PCR assay. J Microbiol Methods 68:613–622
10. Pelletier C et al (2009) Triplex real-time PCR assay for sensitive and simultaneous detection of grapevine phytoplasmas of the 16SrV and 16SrXII-A groups with an endogenous analytical control. Vitis 48:87–95

11. Margaria P, Turina M, Palmano S (2009) Detection of Flavescence dorée and Bois noir phytoplasmas, *Grapevine leafroll associated virus-1* and *-3* and *Grapevine virus A* from the same crude extract by reverse transcription-RealTime Taqman assays. Plant Pathol 58: 838–845
12. Hren M et al (2007) Real-time PCR detection systems for Flavescence dorée and Bois noir phytoplasmas in grapevine: comparison with conventional PCR detection and application in diagnostics. Plant Pathol 56:785–796
13. Afonina I et al (1997) Efficient priming of PCR with short oligonucleotides conjugated to a minor groove binder. Nucleic Acids Res 25:2657–2660
14. Weller SA et al (2000) Detection of *Ralstonia solanacearum* strains with a quantitative, multiplex, real-time, fluorogenic PCR (TaqMan) assay. Appl Environ Microbiol 66:2853–2858
15. Daire X et al (1997) Detection and differentiation of grapevine yellows phytoplasmas belonging to the elm yellows group and to the stolbur subgroup by PCR amplification of non-ribosomal DNA. Eur J Plant Pathol 103: 507–514
16. Clair D et al (2003) A multiplex nested-PCR assay for sensitive and simultaneous detection and direct identification of phytoplasma in the Elm yellows group and Stolbur group and its use in survey of grapevine yellows in France. Vitis 42:151–157
17. Bustin SA et al (2009) The MIQE guidelines: minimum information for publication of quantitative real-time PCR experiments. Clin Chem 55:611–622
18. Morrison TB, Weis JJ, Wittwer CT (1998) Quantification of low-copy transcripts by continuous SYBR Green I monitoring during amplification. Biotech 24:954–962
19. Ellison SLR et al (2006) Routes to improving the reliability of low level DNA analysis using real-time PCR. BMC Biotechnol 6:33
20. Mehle N, Nikolić P, Rupar M, Boben J, Ravnikar M, Dermastia M (2012) Automated DNA extraction for large numbers of plant samples. In: Dickinson M, Hodgetts J (eds) Methods in molecular biology. Springer, New York

Chapter 23

Real-Time PCR for Specific Detection of Three Phytoplasmas from the Apple Proliferation Group

Nataša Mehle, Petra Nikolić, Kristina Gruden, Maja Ravnikar, and Marina Dermastia

Abstract

In this chapter, we describe a real-time PCR detection system for fast, reliable, specific, and sensitive detection and discrimination of '*Candidatus* Phytoplasma mali', '*Ca.* P. prunorum', and '*Ca.* P. pyri' from the 16SrX (apple proliferation-AP) group. These phytoplasmas are causal agents of fruit tree diseases within the *Rosaceae* family, namely apple proliferation, European stone fruit yellows, and pear decline. The assays use (hydrolysis) TaqMan® minor groove binder probes. The panel of assays comprises the same set of primers and specific probes for species-specific amplification, and an additional set of primers and probe for 18S rRNA as an endogenous quality control of DNA extraction. The assays described can be used in routine phytoplasma surveys and in certification programmes.

Key words: Apple proliferation, European stone fruit yellows, Minor groove binding probes, Pear decline, Real-time PCR, TaqMan

1. Introduction

In Europe, fruit trees of the *Rosaceae* family are seriously affected by phytoplasmas of the apple proliferation group (16SrX group), which include '*Candidatus* Phytoplasma mali', '*Ca.* Phytoplasma pyri', and '*Ca.* Phytoplasma prunorum', the causal agents of apple proliferation (AP), pear decline (PD), and European stone fruit yellows (ESFY), respectively (1). '*Ca.* Phytoplasma mali' and '*Ca.* Phytoplasma pyri' are quarantine organisms in Europe, in the EPPO A2 list (2).

Several molecular tools based on PCR/nested PCR/RFLP have been developed for the specific detection of all three species. New protocols based on real-time PCR (qPCR) were recently developed to detect the agents of AP, ESFY, and PD diseases. Some of the assays are based on SYBR® Green chemistry and allow either a detection of phytoplasmas in the 16SrX group (3), or specifically

Matt Dickinson and Jennifer Hodgetts (eds.), *Phytoplasma: Methods and Protocols*, Methods in Molecular Biology, vol. 938, DOI 10.1007/978-1-62703-089-2_23,

'*Ca.* Phytoplasma mali' (4, 5) and '*Ca.* Phytoplasma prunorum' (6). For detection of '*Ca.* Phytoplasma prunorum', TaqMan® chemistry was applied (7). Babini et al. (8) have reported on the specific detection of '*Ca.* Phytoplasma pyri' by multiplex assay. A similar multiplex assay has also been developed for the identification of '*Ca.* Phytoplasma mali' (9), which uses a TaqMan® minor groove binding (MGB) probe, and so allows the use of shorter oligodeoxynucleotides (10) and is more specific than classical TaqMan® probes (11). A mismatch in the MGB probe is critical for the specific detection of '*Ca.* Phytoplasma mali', for discrimination between '*Ca.* Phytoplasma mali' and '*Ca.* Phytoplasma prunorum' (12).

Here we describe the qPCR protocol for detection of '*Ca.* Phytoplasma mali', '*Ca.* Phytoplasma prunorum', and '*Ca.* Phytoplasma pyri' from fruit trees. The protocol was developed by Nikolić et al. (13) for diagnostic purposes and it is based on TaqMan® MGB real-time PCR, which uses identical primer pairs and different probes. Primers and probes anneal within a variable region of the intergenic spacer region (IGS) between 16S rRNA and 23S rRNA. The qPCR reactions using TaqMan® (hydrolysis) MGB probes, which allow the use of shorter oligonucleotides, are more specific than classical TaqMan® probes. Additionally, the amplification of plant DNA co-extracted with phytoplasma DNA is included in order to check the quality of extracted DNA for each sample. Potential inhibition of PCR is also checked by the analysis of two different dilutions of extracted DNA. This assay allows fast, reliable, sensitive, and specific diagnosis of phytoplasmas within the 16SrX group of fruit trees from the *Rosaceae* family. The use of 384-well plates facilitates the analysis of many samples simultaneously and the limited number of handling steps means the assay could be potentially applied to a high-throughput diagnostics of phytoplasmas in certification programmes or phytoplasma surveys. In this chapter, a step-by-step analysis of the system, together with the precautions that should be taken to avoid contamination of samples and reagents, is described in detail.

2. Materials

2.1. Samples and Controls

1. Extracted DNAs of samples (see Note 1).
2. Tenfold dilutions of extracted DNAs: 5 μL of DNA, 45 μL of sterile nuclease-free water (see Notes 1 and 2).
3. Negative control of isolation (NCI): extraction method control prepared from material known to be uninfected (see Note 3).
4. No template control (NTC): sterile deionized water (the same as used for qPCR master mix preparation) (see Note 4).

Table 1
MGB probes for apple proliferation (AP)-specific, pear decline (PD)-specific, and European stone fruit yellows (ESFY)-specific amplicons

Name	Orientation	Sequence (5′-3′)	Working concentration (μM)
SAD	Forward	TGG TTA GAG CAC ACG CCT GAT	10
SAD	Reverse	TCC ACT GTG CGC CCT TAA TT	10
AP	Probe	*FAM*-CAA AGT ATT TAT CTT AAG AAA ACA AGC T-*NFQ*[a]	2.5
ESFY	Probe	*FAM*-CAA AAT ATT TAT TTT AAA AAA ACA AGC TC-*NFQ*	2.5
PD	Probe	*FAM*-AAT ATT TAT TTT AAA AAA AAG CTC TTT G-*NFQ*	2.5

Primer pairs are identical for all three species-specific assays (SAD) (13)
[a] *NFQ* nonfluorescent MGB quencher

5. Positive controls (PC) (sample containing target sequence): DNA from a known AP-, PD-, and ESFY-infected plant (see Notes 3–5).

2.2. qPCR Reagents

1. 2× Maxima™ qPCR Master Mix (Fermentas) (see Note 6).
2. 2× TaqMan Universal PCR Master Mix (Applied Biosystems) (see Notes 6 and 7).
3. Human 18S rRNA (20×) (Applied Biosystems): 18S rRNA primer-probe mix. The 18S rRNA probe has a VIC fluorescent label (see Notes 6 and 8).
4. Molecular biology grade water (nucleic acid- and nuclease-free) or sterile-filtered double deionized water (see Note 9).
5. Primers and probes (Table 1) (see Note 10).

2.3. Equipment

1. qPCR reaction plates: e.g., 384-well optical reaction plates (Applied Biosystems) (see Notes 6 and 11).
2. Optical adhesive covers (e.g., Applied Biosystems).
3. Centrifuge with a rotor for plates.
4. Microfuge with a rotor for tubes.
5. Vortexer.
6. Real-time thermal cycler: e.g., ABI PRISM® 7900 HT Sequence Detection System (Applied Biosystems) (see Note 6).
7. Software (e.g., SDS 2.3, Applied Biosystems) for fluorescence acquisition and calculation of threshold cycles (quantification cycles; *Cq*) (see Note 6).
8. Two UV chambers (see Notes 12 and 13).
9. Two sets of pipettes.

10. DNase-free tubes, racks for tubes.
11. Microfuge tube opener.
12. Aerosol-barrier pipette tips.

3. Methods

3.1. Setting Up Plates

1. Arrange the samples in the reaction plate according to the sample number and amplicon. For each primer set, include the NTC, NCI, and PC as a part of the sample group. See Fig. 1 for an example of a typical plate plan.

	18S rRNA				AP																			
	1	2	3	4	5	6	7	8	9	10	11	12	13	14	15	16	17	18	19	20	21	22	23	24
A	S1		S1 d		S1		S1 d																	
B	S2		S2 d		S2		S2 d																	
C	S3		S3 d		S3		S3 d																	
D	S4		S4 d		S4		S4 d																	
E	S5		S5 d		S5		S5 d																	
F	S6		S6 d		S6		S6 d																	
G	S7		S7 d		S7		S7 d																	
H	S8		S8 d		S8		S8 d																	
I	S9		S9 d		S9		S9 d																	
J	S10		S10 d		S10		S10 d																	
K	S11		S11 d		S11		S11 d																	
L	S12		S12 d		S12		S12 d																	
M	S13		S13 d		S13		S13 d																	
N	S14		S14 d		S14		S14 d																	
O	NCI		NCI d		NCI		NCI d																	
P	N 1	N 2	P	P	N 1	N 2	P	P																

Fig. 1. Typical plate plan of qPCR for AP phytoplasma. The plate arrangement shown represents an assay in which 14 samples from fruit trees will be assayed for the presence of AP. A similar plate plan for PD and ESFY phytoplasmas can be used. Each DNA sample is tested with a phytoplasma-specific amplicon and an amplicon for an endogenous control (18S rRNA), in separate qPCR reactions. All reactions are performed in two replicate wells in two dilutions (undiluted and tenfold dilution) on the same 384-well plate. *SX (X = 1–14)* extracted DNAs of samples; *SX d* dilutions of extracted DNAs; *NCI* negative control of isolation; *NCI d* dilutions of negative control of isolation; *N 1* no template control—put sterile deionized water in the plate at the start of pipetting (NTC1); *N 2* no template control—put sterile deionized water in the plate at the end of pipetting (NTC2); *P* positive DNA target control (PC).

Table 2
Master mixes for apple proliferation (AP)-specific, pear decline (PD)-specific, and European stone fruit yellows (ESFY)-specific amplicons

Components	Final concentration	Volume per well (μL)
Sterile nuclease-free water		0.84
2× Maxima™ qPCR master mix	1×	5.0
Forward primer SAD (10 μM)	900 nM	0.9
Reverse primer SAD (10 μM)	900 nM	0.9
Probe (2.5 μM)	90 nM	0.36
Total volume		8.0

Table 3
Master mix for 18S rRNA amplicon

Components	Final concentration	Volume per well (μL)
Sterile nuclease-free water		2.5
2× TaqMan Universal PCR master mix	1×	5.0
Human 18S rRNA (20×)	1×	0.5
Total volume		8.0

2. Calculate the volume of components needed for the assay master mixes according to Tables 2 and 3. Include sufficient reagents for the number of samples (undiluted and tenfold diluted) and controls to be tested in duplicate. Include 10% extra volume for loss during pipetting and possible errors.
3. Remove all reagents from freezer and allow them to thaw. Mix all thawed reagents well with the vortexer and spin them briefly (~5 s) in a microfuge.
4. Prepare two test assay master mixes (18S rRNA, and AP, PD, or ESFY) in individual tubes. For each mix, the reagents are added in the order and quantity calculated in step 2. After completion, mix all reagents well with the vortexer and spin them briefly (~5 s) in a microfuge (see Note 14).
5. Take a 384-well qPCR plate. Fill the appropriate number of wells with 8 μL of each master mix according to the plate plan (see Note 15).
6. Add 2 μL of undiluted sample, sample diluted tenfold, and control DNA into each test well. Always start by adding water for the first NTC (NTC1), continue with DNA samples and

controls. Finish with water for the last NTC (NTC2). Using this order of DNA addition facilitates, the identification of any possible source of contamination (see Note 15).

7. After adding DNA, the qPCR plate should be covered with an optical adhesive cover and sealed with a plastic applicator (see Note 16).
8. Centrifuge the plate for 1 min at 1,000 × *g* to ensure that both the reaction mix and the DNA are collected at the bottom of the wells (see Note 17).

3.2. Perform qPCR Run

1. Switch on the thermal cycler in advance, according to the manufacturer's instructions to warm it up and to stabilize temperature and light sources.
2. Transfer the qPCR plate to the real-time thermal cycler and run qPCR using the following universal cycling program (consult the instrument manual for specific instructions, see Note 18):

1 cycle	2 min at 50°C (UNG activation step)
	10 min at 95°C (polymerase activation)
45 cycles	15 s at 95°C (DNA denaturation)
	1 min at 60°C (annealing and extension)

3.3. Analyze Data

1. Analyze and export raw data (see instrument manual for detailed instructions about document setup, baseline, and threshold settings) (see Note 19).
2. Import data into Microsoft Excel or equivalent spreadsheet program with statistical features. For each of the two replicates of a sample, calculate the average *Cq* value. If large differences between the replicates are observed, the analysis should be repeated (see Note 20). All samples with an average *Cq* value lower than the cut-off value should be considered positive for that particular assay (see Note 21).
3. If the target is detected in the extraction blank control (NCI), contamination of reagents or samples during DNA extraction should be suspected. The results of sample analysis are therefore not reliable; the analysis is not valid and should be repeated, including the extraction of DNA.
4. NTC analysis of any assays must not result in a positive qPCR signal; otherwise, contamination during preparation of qPCR should be suspected. The results of the sample analyses are not reliable and the qPCR analysis must be repeated (see Note 22).
5. The presence of all target sequences in the positive controls must be confirmed. If the target is not detected in these controls, the analysis is not valid and errors in the PCR reaction should be considered.

6. The *Cq* of the isolated DNA measured by endogenous controls (e.g., 18S rRNA, *cox*) must be in the predicted range (see Note 23). If not, then the analysis is not valid and some errors in the extraction procedure or the presence of inhibitors should be considered.
7. If a negative result is obtained for a sample with typical symptoms of phytoplasma infection, the sample should be retested to check for the presence of other phytoplasma species (see Note 24).

4. Notes

1. In general, the total amount of DNA added to a PCR reaction should be less than 100 ng. Mix all DNA samples and their dilutions well, and spin them briefly (about 5 s) in a microfuge before use. If DNA samples were stored frozen, they should be allowed to thaw before mixing.
2. All samples should be tested at two dilutions. Testing of dilutions is a PCR inhibition control, checking for the presence of inhibitors in the sample that could influence the efficiency of the qPCR amplification. If there is no inhibition, the shapes of the amplification curves are the same for undiluted and tenfold diluted DNA, and the difference in the *Cq* values for the tenfold diluted DNA is 3.3 cycles higher than the undiluted one with 100% efficiency in amplification. In the case of strong inhibition (e.g., the *Cq* value for the tenfold dilution of 18S rRNA is lower than that of undiluted 18S rRNA), the qPCR reaction may give false negative results. Analyses should be repeated with DNA diluted more. If strong inhibition is still detected, DNA should be extracted using a different method. If no inhibition is detected while using the protocol for many samples of the same type, the use of only one DNA dilution is possible, but in that case we suggest that each sample is tested by qPCR in triplicate instead in duplicate. In our case, when the KingFisher automated procedure described in Chapter 12 of this volume is used for extraction of DNA, tenfold diluted DNA is suitable for testing.
3. The assay is validated only for fruit trees (apples, pears, and stone fruits). When the assay is used for other sample materials, a healthy sample and a spiked healthy sample of the same material must be included in the analysis to check for possible cross-reactions of probes with the new sample material and to prove that detection of the target from the new sample material is possible, respectively.

4. A nontemplate control (NTC) and a positive control (PC) are performed in every qPCR run. NTC contains all reagents used for qPCR, but sterile deionized water is added into the reaction instead of the sample DNA. NTC should be used for the same amplicon as the sample in duplicate, one at the start of pipetting (NTC1) and one at the end (NTC2) for each amplicon.
5. The positive control (PC) is a sample that contains target sequence DNA. To make assays more uniform, we suggest preparing several aliquots of appropriate controls and putting them in the freezer so that a fresh aliquot can be used for every analysis to prevent degradation of DNA by repeated freezing and thawing.
6. Use of names of chemicals or equipment in this chapter implies no approval of them to the exclusion of others. This information is given for the convenience of users and does not constitute an endorsement of the products or services. Equivalent products may be used if they can be shown to lead to the same results.
7. Chemicals from Applied Biosystems for the 18S rRNA TaqMan assay and those of Fermentas were compared. The obtained results were very similar.
8. In order to control the DNA extraction procedure, an endogenous control must be included in the assays. This is preferably a conserved gene, which provides confirmation that the extraction from different plant species has been successful, and it should be present in high copy numbers in the genome in order to increase the sensitivity of the assay. Instead of 18S rRNA, other endogenous controls may be used, e.g., plant cytochrome oxidase (*cox*) (14).
9. Use of molecular biology grade (nucleic acid- and nuclease-free) or sterile-filtered double deionized water in all recipes and protocol steps is essential. Do not use DEPC-treated water because the slightly acidic pH may promote primer degradation. It is easy to contaminate a PCR assay by reusing an old bottle of water, therefore aliquot the water into tubes and then use one tube per assay.
10. Purchase lyophilized oligonucleotides from any commercial source, synthesized at a 25 nmol scale. For these, standard desalting is sufficient and no additional purification is required. Briefly centrifuge the tubes of dried oligonucleotide to get the contents to the bottom. Resuspend the oligonucleotides in molecular biology grade water to 100 μM. Dilute each stock of forward and reverse primers to 10 μM, prepare aliquots of these working solutions and store them at −20°C. When stored properly and subjected to minimal freeze–thaw cycles, a primer stock may last at least 2 years.

Depending on the vendor the probe may be supplied in 100 μM solutions. In this case, dilute a small amount of 100 μM stock to 2.5 μM, prepare aliquots of these working solutions and store them at −20 °C. Avoid repetitive freeze–thaw cycles, and thaw on ice before use to preserve the integrity of the probe. Protect the probe from excessive exposure to light (e.g., put the tubes with probes in dark plastic bags/containers) to prevent photobleaching of the fluorescent dyes and evaporation.

11. Alternatively, you can use 96-well optical reaction plates. If 96-well plates are used, the total reaction volume should be 25 μL (20 μL of master mix and 5 μL of sample DNA).
12. Reaction mixes and the addition of DNA samples should be prepared in separate UV chambers using dedicated laboratory equipment (pipettes, tips, tubes, microfuge tube opener, racks for tubes, lab coat, gloves): first, prepare reaction mixes and load them onto the qPCR plate (no DNA other than primers and probes should be present here); secondly, add DNA of samples in another UV chamber. Both steps should be done in a separate location from the places used for homogenization of the samples and DNA extraction.
13. In all stages of setting up PCR reactions, the following precautions should be taken in order to avoid contamination of samples and reagents: (1) use only filter tips (aerosol barrier tips) and use a fresh tip for each pipetting step; (2) close reagent/sample tubes once desired volume has been aspirated to the tip; (3) always wear gloves and change them if you suspect contamination; (4) use only clean, sterile plasticware; (5) all tubes should be opened by microfuge tube opener.
14. This step may be prepared in advance and the tubes may be put on ice or stored at 4°C for up to a day before preparing the reaction plate.
15. An automated liquid handling system (e.g., Multiprobe® II PLUS EX, PerkinElmer) may be used for pipetting a large number of DNA samples and master mixes onto the 384-well plates. If the plates are pipetted manually, start with pipetting the master mixes onto the bottom of the wells and continue by adding DNA samples and controls to the side walls of the wells. This facilitates visual checking during the addition of samples. Avoid the touching of master mixes inside the tubes with the pipette tip. It is advisable to use multistep pipettors. In this case, use one tip to fill up all wells with the same master mix and one tip for two replicates when adding the DNA of the sample. In the case of liquid handling machines, the pipetting order of master mixes and DNA may be reversed: DNA is applied first (onto the bottom of the wells), the plate is then centrifuged to ensure the DNA is collected at the bottom and then master

mixes are added without touching the well bottoms, followed by plate sealing and centrifugation.

16. Do not touch the optical adhesive cover with fingers.
17. Once prepared, reactions must be processed immediately. If this is not possible, they may be stored in the dark at 2–8°C for a maximum of a few hours.
18. Before starting the run: (1) adjust the reaction volume, if needed; (2) assign samples to locations on the plate; (3) assign detectors to your samples (choose the correct filters according to the probes used in amplicons: FAM for phytoplasma amplicons and VIC for the 18S rRNA amplicon). The temperature ramp rate should be similar to Applied Biosystems 7900 thermal cyclers. The standard mode for temperature rates on 7900HT is 1.6°C/s up and 1.6°C/s down. Do not use faster temperature ramp rates.
19. The transformation of the fluorescence signal into *Cq* data (e.g., Ct-threshold cycle, Cp-crossing point) (15), as well as methods for baseline and threshold settings, vary between instrument models. The specific instrument manual must be consulted. When analyzing the raw data, it is important to adjust the cycle threshold (*Cq*) of the amplification plot to within the geometric (exponential) phase of amplification, preferably at the beginning of the geometric phase. In the log view, this is the linear part of the amplification plot. This is critical for correct quantitative analysis because the geometric phase represents the point of the reaction at which *Cq* is quantitatively related to the amount of initial PCR template. We have experimentally determined that automatic baseline and threshold at 0.065 are usually suitable when using Applied Biosystems 7900 thermal cyclers. Manually check the multicomponent plots for all samples (wells). In wells containing positive samples, the fluorescence of a fluorescent reporter (in this case FAM dye) should be increasing with cycles. A *Cq* value where no increase of FAM fluorescence is observed must not be considered as positive.
20. It is always important to run duplicate or, preferably, triplicate reactions of single dilutions when performing a qPCR assay. If the *Cq* values are very high, the use of replicates is especially important to make sure the reaction is working properly. If data from the replicate reactions vary significantly, the reactions should be repeated, unless high variability in the *Cq* values are attributed to the stochastic effect of target copy distribution in replicate reactions. This occurs when less than ten copies of the target are present in the reaction (16, 17).
21. The cut-off value is determined for each assay in the process of assay validation before the assay is put into routine use. It is the highest *Cq* value still considered to be a positive result.

The following guidelines should be followed for the determination of the cut-off value during validation of the assay. Analyze serial dilutions of known positive and negative samples in different sample types. At least five dilution steps should be analyzed (usually tenfold dilution steps), each containing a sample in five replicate wells. Next, search for the last group of samples in which there is no amplification in some of the replicates. Take the range of the highest *Cq* values observed. Round this *Cq* value up to the next half value and then add 0.5 to this *Cq* value to take into account the difference in threshold chosen between runs. This *Cq* value is the cut-off value and should be determined for each assay at the time of the method validation. The cut-off value for the same assay may vary from laboratory to laboratory for various reasons (e.g., the qPCR machine and software used). Therefore each laboratory should revalidate the cut-off values for any new assay prior to its implementation.

22. If the amplification of target occurs in the NTC, it may be a result of the contamination of reagents, degradation of the primer mix, or the formation of nonspecific PCR products. If contamination of the reagents is suspected, use fresh aliquots of all reaction components. In the case of 18S rRNA, high *Cq* values (>36) for NTC are not unusual: it is a result of DNA traces in bacterial polymerase. Even in cases in which all amplification products, including those that are nonspecific, contribute to the fluorescence signal, they may be considered insignificant if the *Cq* values of the NTC or NCI are higher than the cut-off value.

23. Quality and quantity of isolated DNA greatly influence the detection of phytoplasmas. Monitoring the *Cq* values of endogenous controls (e.g., *cox*, 18S rRNA) allows the estimation of both parameters, thereby increasing the confidence in results. The *Cq* values for 18S rRNA in the DNA extraction procedure used in our laboratory by the KingFisher automated system (*see* Chapter 12 of Mehle et al. (18)) vary from 13 to 25 (while using *cox Cq* values are up to 33). Higher *Cq* values are an indicator of lower amounts of plant DNA and presumably lower amounts of phytoplasma DNA in the sample, which therefore decreases the confidence of results. This is a signal that such samples should be re-extracted and tested again. Occasionally, higher *Cq* values are only an indicator of inhibitors present in the sample, therefore higher dilutions of DNA sample should be retested.

24. '*Ca.* Phytoplasma mali' causes diseases in the genus *Malus* (apple), '*Ca.* Phytoplasma pyri' in the genus *Pyrus* (pear), and '*Ca.* Phytoplasma prunorum' in the genus *Prunus* (stone fruit) (1). Occasionally, these may be identified infecting plants other than the typical host, e.g., '*Ca.* Phytoplasma mali' found in stone fruits and in European and Asian pears (*Py. communis* and

Py. pyrifolia) (1, 19–21). Therefore it is worthwhile including, as an additional test, a universal phytoplasma assay (e.g., assays described in Chapters 21 of, Christensen et al. (22)) in the diagnostic procedure, to confirm the presence of any phytoplasmas and/or to detect other unidentified phytoplasma species. In our lab, we perform three assays simultaneously—each sample is tested with the following assays: specific AP, ESFY, or PD sets of primers/probes, universal phytoplasma UniRNA amplicon (23), and 18S rRNA amplicon as an endogenous control for the quality of DNA extraction. In cases where the universal amplicon gave a positive result but the specific amplicon gave a negative result, we retest such samples using other sets of primers/probes specific for other phytoplasma species.

References

1. Seemüller E, Schneider B (2004) '*Candidatus* Phytoplasma mali', '*Candidatus* Phytoplasma pyri' and '*Candidatus* Phytoplasma prunorum', the causal agents of apple proliferation, pear decline and European stone fruit yellows, respectively. Int J Syst Evol Microbiol 54:1217–1226
2. EPPO (2011) EPPO A2 List of pests recommended for regulation as quarantine pests. http://www.eppo.org/QUARANTINE/listA2.html. Last Accessed 10 June 2011
3. Torres E et al (2005) Real-time PCR for simultaneous and quantitative detection of quarantine phytoplasmas from apple proliferation (16SrX) group. Mol Cell Probes 19:334–340
4. Jarausch W et al (2004) Establishment of a quantitative real-time PCR assay for the quantification of apple proliferation phytoplasmas in plants and insects. Acta Hortic 657:415–420
5. Galetto L, Bosco D, Marzachì C (2005) Universal and group-specific real-time PCR diagnosis of flavescence doreé (16Sr-V), bois noir (16Sr-XII) and apple proliferation (16Sr-X) phytoplasmas from field-collected plant hosts and insect vectors. Ann Appl Biol 147:191–201
6. Yvon M et al (2009) Specific detection and quantification of the phytopathogenic agent '*Candidatus* Phytoplasma prunorum'. Mol Cell Probes 23:227–234
7. Pignatta D et al (2008) A real-time PCR assay for the detection of European stone fruit yellows phytoplasma (ESFYP) in plant propagation material. Acta Hortic 781:499–504
8. Babini AR et al (2008) Investigations with real-time PCR assay on the transmissibility of pear decline phytoplasma (PDP) with dormant buds. Acta Hortic 781:495–498
9. Baric S, Dalla VJ (2004) A new approach to apple proliferation detection: a highly sensitive real-time PCR assay. J Microbiol Meth 57: 135–145
10. Kutyavin IV et al (2000) 3′-Minor groove binder-DNA probes increase sequence specificity at PCR extension temperatures. Nucl Acids Res 28:655–661
11. Yao Y, Nellåker C, Karlsson H (2006) Evaluation of minor groove binding probe and Taqman probe PCR assays: Influence of mismatches and template complexity on quantification. Mol Cell Probes 20:311–316
12. Aldaghi M et al (2007) Development of a new probe for specific and sensitive detection of '*Candidatus* Phytoplasma mali' in inoculated apple trees. Ann Appl Biol 151:251–258
13. Nikolić P et al (2010) A panel of real-time PCR assays for specific detection of three phytoplasmas from the apple proliferation group. Mol Cell Probes 24:303–309
14. Weller SA et al (2000) Detection of *Ralstonia solanacearum* strains with a quantitative, multiplex, real-time, fluorogenic PCR (TaqMan) assay. Appl Environ Microbiol 66:2853–2858
15. Bustin SA et al (2009) The MIQE guidelines: minimum information for publication of quantitative real-time PCR experiments. Clin Chem 55:611–622
16. Morrison TB, Weis JJ, Wittwer CT (1998) Quantification of low-copy transcripts by continuous SYBR Green I monitoring during amplification. Biotechniques 24:954–962
17. Ellison SLR et al (2006) Routes to improving the reliability of low level DNA analysis using real-time PCR. BMC Biotechnol 6:33
18. Mehle N, Nikolić P, Rupar M, Boben J, Ravnikar M, Dermastia M (2012) Automated DNA

extraction for large numbers of plant samples. In: Dickinson M, Hodgetts J (eds) Methods in molecular biology. Springer, New York

19. Lee I-M et al (1995) Detection of multiple phytoplasmas in perennial fruit trees with decline symptoms in Italy. Phytopathology 85:728–735
20. Del Serrone P et al (1998) Occurrence of apple proliferation and pear decline phytoplasmas in diseased pear trees in Hungary. J Plant Pathol 80:53–58
21. Mehle N et al (2007) First report of '*Candidatus* Phytoplasma mali' in *Prunus avium*, *P. armeniaca* and *P. domestica*. Plant Pathol 56:721
22. Christensen NM, Nyskjold H, Nicolaisen M (2012) Real-time PCR for universal phytoplasma detection and quantification. In: Dickinson M, Hodgetts J (eds) Methods in molecular biology. Springer, New York
23. Hren M et al (2007) Real-time PCR detection systems for Flavescence dorée and Bois noir phytoplasmas in grapevine: comparison with conventional PCR detection and application in diagnostics. Plant Pathol 56: 785–796

Chapter 24

Reverse Transcription-PCR for Phytoplasma Detection Utilizing Crude Sap Extractions

Paolo Margaria and Sabrina Palmano

Abstract

Phytoplasmas are routinely detected by nucleic acid-based techniques. These approaches rely on enriched phytoplasma DNA extracts of good quality, following labor intensive and time-consuming purification protocols. Here we describe a very rapid, specific, sensitive, and reliable method for flavescence dorée phytoplasma detection, based on real-time Taqman® reverse transcription-PCR of the 16S rRNA. The protocol is particularly useful for large-scale screening of vineyards and nurseries, pathogen surveys, and field epidemiological studies.

Key words: Crude sap, Grapevine, Real-time TaqMan® reverse transcription-PCR, Ribosomal RNA

1. Introduction

The use of polymerase chain reaction (PCR) to amplify phytoplasma DNA has been a fundamental tool in the detection, characterization, and classification of these uncultivable pathogens. The 16S rRNA gene has been the major target for universal or group specific-primer design, although other genes have been used successfully for the same purpose (1). Due to the low concentration and uneven distribution of phytoplasmas in the plant host (2), nested PCR (3–5) is often necessary to increase sensitivity. Diagnostic assays based on real-time PCR have become common (6), as they achieve high levels of sensitivity, reducing the risk of contamination that characterizes nested PCR assays and avoiding gel-based postamplification analysis. The limitation for many phytoplasma detection methods is not strictly the sensitivity, but the fact that they rely on the need for a relatively pure DNA template whose extraction is expensive and time consuming, making the above protocols unsuitable for high-throughput detection.

Matt Dickinson and Jennifer Hodgetts (eds.), *Phytoplasma: Methods and Protocols*, Methods in Molecular Biology, vol. 938, DOI 10.1007/978-1-62703-089-2_24, © Springer Science+Business Media, LLC 2013

Here we describe a rapid, reliable, and sensitive assay based on a simple total tissue grinding process that avoids the use of hazardous agents normally used in DNA extractions, followed by reverse transcription (RT)-PCR.

PCR-based assays can target either DNA or RNA. As the transcriptome is context-dependent, i.e., the mRNA content varies with physiology or development, the choice for the RT-PCR target must be focused on a constitutively expressed gene or, as in the described protocol, on a ribosomal gene. Targeting the 16S rRNA has the additional advantage of the high rRNA copy number present in a living cell, in contrast to the two copies of the 16S rRNA gene present in the phytoplasma genome (7), resulting in an increased amount of initial template for amplification. Comparison between a diagnostic method that uses nested PCR (on phytoplasma DNA-enriched preparations) and RT-PCR (using crude sap) has been performed and statistical analyses demonstrated substantial agreement between the two sets of results (8). Furthermore, adapting the RT-PCR to the real-time TaqMan® system noticeably increases the sensitivity of the assay (9). Real-time TaqMan® RT-PCR has been successfully applied to the detection of flavescence dorée (FD), one of the most serious diseases of grapevine, associated with a phytoplasma of the 16SrV group, recently proposed as '*Candidatus* Phytoplasma vitis' (10). The disease is subject to quarantine restrictions in Europe. Specific as well as large-scale detection methods are particularly important for field surveys and for control of propagation material in the context of mandatory control strategies required to monitor and limit the spread of the disease. Moreover, the protocol described here may be used in combination with protocols for the simultaneous detection of other grapevine pathogens such as RNA viruses, using the same crude sap as the template in the amplification reaction (9, 11–13). The overall advantages of real-time RT-PCR compared to nested PCR are summarized in Table 1.

2. Materials

Prepare all solutions using ultrapure water (prepared by purifying deionized water to attain a sensitivity of 18 MΩ cm at 25°C) and analytical grade reagents.

2.1. Sample Preparation

1. Sodium carbonate buffer: 15 mM sodium carbonate (Na_2CO_3), 34.9 mM sodium hydrogen carbonate ($NaHCO_3$), pH 9.6. Weigh 1.59 g Na_2CO_3, 2.93 g $NaHCO_3$ and transfer to a 1-L glass beaker. Add about 800 mL water, mix and adjust pH with HCl (see Note 1). Transfer to a 1-L graduated cylinder and add water to a final volume of 1 L. Autoclave and store at room temperature.

Table 1
Comparison of nested PCR and real-time RT-PCR techniques used for diagnosis of flavescence dorée phytoplasma in grapevine samples

Parameter	Nested PCR	Real-time RT-PCR
Execution time	2 days	3–4 h
No. of samples processed/day	*n*	4*n*
Technical difficulty	Medium	Easy
Hazardous chemicals	Chloroform, phenol, β-mercaptoethanol, ethidium bromide, or other DNA stain for electrophoresis	None
Cost	Medium	Medium
Sensitivity	Very high	Very high
Multiple pathogen detection	Phytoplasmas	Viruses, phytoplasmas

2. Grinding buffer: 2% polyvinylpyrrolidone mol. wt. 40,000 (PVP40), 0.2% bovine serum albumin (BSA), 1% sodium metabisulphite ($Na_2S_2O_5$), 0.05% Tween 20 in sodium carbonate buffer (see Note 2).
3. GES buffer: 100 mM glycine, 50 mM NaCl, 1 mM EDTA, 0.5% Triton X-100, pH 9.0. Weigh 1.50 g glycine, transfer to a 500-mL beaker, and add about 100 mL of water. Adjust pH to 9.0 with NaOH. Add 0.58 g NaCl and 0.07 g EDTA. Transfer to a graduated cylinder and add water to a final volume of 200 mL. Autoclave (see Note 3) then cool down and add 1 mL Triton X-100. Store at room temperature.
4. BIOREBA "Universal" extraction bags, 12 × 14 cm (BIOREBA AG, catalogue number 430100).
5. Ice bucket.

2.2. Real-Time RT-PCR

Store all reagents at −20°C, otherwise specified.

1. Buffer A (Applied Biosystems) (see Note 4).
2. dNTPs: prepare a mix containing 2 mM each nucleotide.
3. $MgCl_2$: 25 mM solution.
4. TaqMan® probe: 5′-TTTCGGTATGTAAAGTTCT-3′, marked with FAM-6 (6-carboxyfluorescein) at the 5′ end, and with TAMRA (tetramethylrhodamin) at the 3′ end. Prepare a 10 μM working solution aliquot in sterile distilled water.
5. Forward primer FD-395F: 5′-GCCGCGTGAACGATGAA-3′, 10 μM solution in sterile distilled water.

6. Reverse primer FD-480R: 5′-GAATAACGTCAAGATAGTT TTTCCACT-3′, 10 μM solution in sterile distilled water.
7. Murine leukemia virus (MuLV) reverse transcriptase, 50 U/μL.
8. RNase inhibitor, 20 U/μL.
9. AmpliTaq Gold enzyme 5 U/μL (Applied Biosystems).
10. Sterilized ultrapure water. Store at room temperature.
11. MicroAmp Fast optical 96-well reaction plates (Applied Biosystems, catalogue number 4346906).
12. MicroAmp optical adhesive film (Applied Biosystems, catalogue number 4311971).
13. StepOnePlus real-time PCR system (Applied Biosystems) or other real-time PCR thermal cycler.

3. Methods

Procedures are carried out at room temperature.

3.1. Sample Preparation

1. Prepare fresh grinding buffer for each occasion (see Note 5).
2. Wash grapevine leaf samples with water and dry using paper towels (see Note 6).
3. Sample a leaf fragment from symptomatic regions and from the basal region next to the petiole. In the case of propagation material with few leaves, take samples from the subcortical tissue as well (see Note 7). Sample a total of 0.3–1 g of material depending on the condition and quantity of available tissue. Put each sample in an extraction bag (see Note 8).
4. Add grinding buffer to each bag, using 10 mL per 0.5 g sample.
5. Homogenize the plant samples with a rolling ball or a press (see Note 9).
6. Pipette 100 μL of GES buffer into 0.2 mL PCR tubes or strips. Mix the sap inside the extraction bag and aliquot 8 μL of crude sap into the GES buffer (see Notes 10 and 11). Mix by gentle agitation by hand. Heat at 95°C for 10 min in a thermal cycler (see Note 12). Immediately cool the tubes/strips in ice–water mix (see Note 13).

3.2. Real-Time TaqMan RT-PCR

1. Prepare the PCR mix by adding 10.17 μL water, 2.5 μL buffer A, 2.5 μL dNTPs, 5.5 μL $MgCl_2$, 0.5 μL TaqMan® probe, 0.75 μL forward primer, 0.75 μL reverse primer, 0.1 μL MuLV, 0.1 μL RNase inhibitor, 0.125 μL AmpliTaq Gold per sample (see Note 14). Vortex and aliquot 23 μL to each well in the PCR plate.
2. Aliquot 2 μL of boiled crude sap template to each well (see Note 15). Cover the plate using the adhesive film.

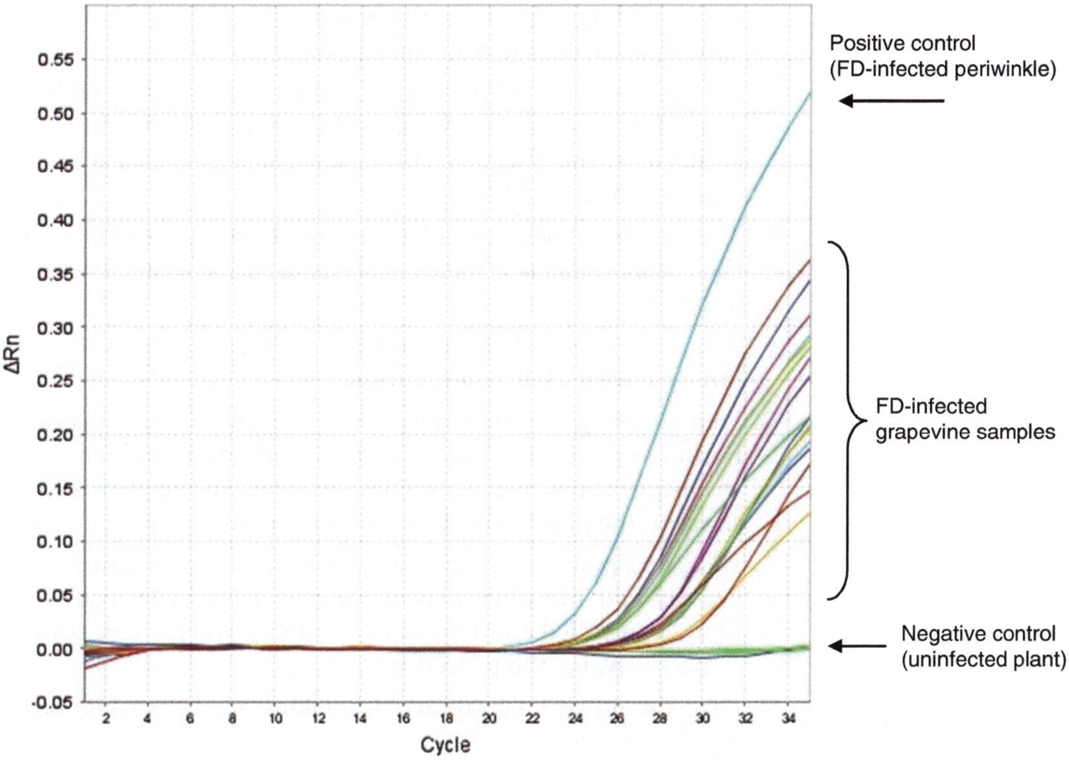

Fig. 1. Amplification plot of grapevine samples, 15 of which were infected with flavescence dorée (FD) phytoplasma. The threshold cycle of the positive control from infected periwinkle is generally lower than those from infected grapevine samples, where the phytoplasma titre is often low. The plot shows ΔRn as a function of the cycle number, where ΔRn is the magnitude of normalized fluorescence signal generated by the reporter in each cycle during the PCR amplification.

3. Set the reporter and quencher in the real-time machine software as FAM and TAMRA, respectively.
4. Set amplification parameters as follows: 52°C for 30 min (reverse transcription), 95°C for 10 min (AmpliTaq activation) and 35 amplification cycles, consisting of 15 s at 95°C and 1 min at 58°C. A representative amplification plot is shown in Fig. 1.

4. Notes

1. The pH of the solution should be very close to 9.6. If necessary, adjust pH to 9.6 using a few drops of 10% HCl.
2. Ensure complete solubilization of PVP40 and BSA, then add Tween 20 only after complete solubilization of the other components, ensuring absence of precipitates or lumps.

3. Avoid adding Triton X-100 before autoclaving, because the detergent causes the solution to boil over.
4. Do not use the Gold Buffer, which is supplied with the AmpliTaq enzyme by Applied Biosystems. Buffer A is only available in the Taqman 1000 Reaction Buffer A Pack (catalogue number 4304441).
5. The grinding buffer must be prepared fresh each time, because the components precipitate when stored for more than a few hours.
6. The washing step helps to reduce treatment residues that may interfere with PCR amplification.
7. Leaf sampling is critical due to the uneven phytoplasma distribution in the host. Choose samples from many different leaves to ensure representative sampling to establish the phytosanitary status of the plant. Avoid sampling from dead or rotten tissues.
8. It is good practice to include in each assay a fresh sample from healthy and FD phytoplasma-infected periwinkle maintained under controlled conditions, as a negative and positive control, respectively, for the procedure.
9. Care should be taken to avoid breaking the bags while grinding the sample.
10. Take the 8 μL from the net side of the bags, opposite to where the leaf sample was placed. In this way, pipetting of leaf residues in GES buffer is minimal.
11. Include a sample consisting of GES buffer at this step as a buffer-only negative control.
12. The boiling step in GES buffer is necessary to release the rRNA and make it available for the RT reaction.
13. Rapid cooling with ice–water is critical to maintain the rRNA in a free and linear state, available for the RT step.
14. Vortex the PCR reagents for 3 s before use to ensure homogenization and avoid precipitates. Spin the enzymes for 3 s at maximum speed in a microfuge. The PCR mix may be prepared during the boiling step to save time.
15. Care should be taken when opening the tube/strips to avoid cross-contamination.

Acknowledgements

P. M. was supported by the grant "Studi sui fattori che favoriscono le epidemie di flavescenza dorata in Piemonte e loro superamento," sub-project C, from Regione Piemonte, Italy (Report number 12851).

References

1. Bertaccini A (2007) Phytoplasmas: diversity, taxonomy, and epidemiology. Front Biosci 12: 673–689
2. Firrao G, Garcia-Chapa M, Marzachì C (2007) Phytoplasmas: genetics, diagnosis and relationships with the plant and insect host. Front Biosci 12:1353–1375
3. Lee I-M et al (1994) Use of mycoplasmalike organism (MLO) group-specific oligonucleotide primers for nested-PCR assays to detect mixed-MLO infections in a single host plant. Phytopathol 84:559–566
4. Gundersen DE, Lee I-M (1996) Ultrasensitive detection of phytoplasmas by nested-PCR assays using two universal primer pairs. Phytopathol Mediterr 35:144–151
5. Clair D et al (2003) A multiplex nested-PCR assay for sensitive and simultaneous detection and direct identification of phytoplasma in the Elm yellows group and Stolbur group and its use in survey of grapevine yellows in France. Vitis 42:151–157
6. Galetto L, Marzachì C (2010) Real-time PCR diagnosis and quantification of phytoplasmas. In: Weintraub PG, Jones P (eds) Phytoplasmas: genomes, plant hosts and vectors. CAB International, Cambridge, pp 1–18
7. Schneider B, Seemüller E (1994) Presence of two sets of ribosomal genes in phytopathogenic mollicutes. Appl Environ Microbiol 60:3409–3412
8. Margaria P et al (2007) Detection of Flavescence doreé phytoplasma in grapevine by reverse-transcription PCR. Plant Dis 91: 1495–1501
9. Margaria P, Turina M, Palmano S (2009) Detection of Flavescence dorée and Bois noir phytoplasmas, *Grapevine leafroll associated virus-1* and *-3* and *Grapevine virus A* from the same crude extract by reverse transcription-RealTime Taqman assays. Plant Pathol 58: 838–845
10. The IRPCM Phytoplasma/Spiroplasma Working Team—Phytoplasma taxonomy group (2004) '*Candidatus* Phytoplasma', a taxon for the wall-less, non-helical prokaryotes that colonize plant phloem and insects. Int J Syst Evol Microbiol 54:1243–1255
11. Osman F, Rowhani A (2006) Application of a spotting sample preparation technique for the detection of pathogens in woody plants by RT-PCR and real-time PCR (TaqMan). J Virol Methods 133:130–136
12. Osman F et al (2007) Real-time RT-PCR (TaqMan®) assays for the detection of *Grapevine Leafroll associated viruses* 1–5 and 9. J Virol Methods 141:22–29
13. Osman F, Rowhani A (2008) Real-time RT-PCR (TaqMan®) assays for the detection of viruses associated with Rugose wood complex of grapevine. J Virol Methods 154: 69–75

Chapter 25

In-Field Diagnostics Using Loop-Mediated Isothermal Amplification

Jenny Tomlinson

Abstract

Loop-mediated isothermal amplification (LAMP) is a method for amplification and detection of target organisms which, unlike polymerase chain reaction, does not require thermal cycling. LAMP assays can be developed in the laboratory for subsequent deployment in the field, where the simplicity of isothermal amplification makes LAMP a suitable method for rapid detection of phytoplasmas with levels of sensitivity and specificity approaching those of more complex and time-consuming laboratory methods.

Key words: Detection, On-site testing, Loop-mediated isothermal amplification, Polymerase chain reaction, Primer design

1. Introduction

Methods for detection of phytoplasmas (and other pathogens) that are based on the polymerase chain reaction (PCR) can offer advantages of sensitivity and specificity over conventional detection methods. However, the equipment required for thermal cycling is relatively complex and expensive. Real-time PCR can be adapted to automated or semiautomated high-throughput testing, but the complexity of the instruments used for real-time PCR is compounded by the requirement to perform fluorescence monitoring concurrently with accurate thermal cycling. These factors limit the routine use of PCR and real-time PCR to relatively well-equipped laboratories. Isothermal amplification methods have been developed as alternatives to PCR which can be performed without thermal cycling equipment. Loop-mediated isothermal amplification (LAMP) is a method by which a target specific region of DNA can be amplified during incubation at a single temperature (1, 2). LAMP uses strand-displacing DNA polymerase and an arrangement

Matt Dickinson and Jennifer Hodgetts (eds.), *Phytoplasma: Methods and Protocols*, Methods in Molecular Biology, vol. 938, DOI 10.1007/978-1-62703-089-2_25,

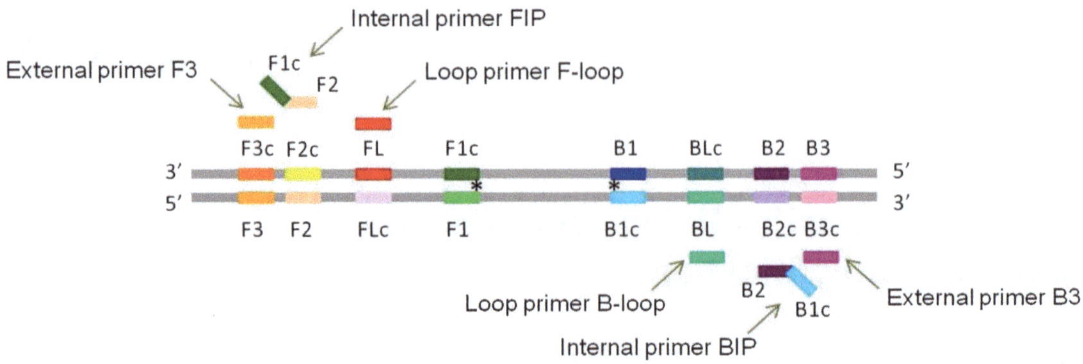

Fig. 1. Arrangement of the primers used for loop-mediated isothermal amplification (LAMP). Internal primers FIP and BIP each target two primer binding sites (F1 and F2, and B1 and B2, respectively). * indicates the positions where the most salient mismatches for target/nontarget discrimination should be placed.

of external and internal primers, plus optional "loop" primers, (3) (Fig. 1) which generate amplification products containing loop regions. The single-stranded loop regions serve as primer binding sites, allowing amplification to proceed without thermal denaturation of the template. Unlike some isothermal amplification methods which can sustain amplification of only short regions of target nucleic acid, LAMP is a highly efficient method for amplification of DNA, with amplicons of typically 200–300 bp in length. LAMP generates products consisting of alternately oriented repeats of the target sequence with a characteristic ladder-like appearance when visualized by gel electrophoresis. LAMP assays have been developed for a range of plant pathogens (4) including several phytoplasmas (5–9) (see Note 1).

This chapter describes how to perform LAMP for phytoplasma detection and how to adapt laboratory-developed LAMP assays for in-field use.

2. Materials

1. LAMP primers (see Notes 2–5, and Fig. 1). Internal primers FIP and BIP should be HPLC purified.
2. *Bst* DNA polymerase (New England Biolabs).
3. $MgSO_4$: 50 mM.
4. Betaine: 5 M.
5. dNTP mix: each at 10 mM.
6. Molecular biology grade water.
7. 1.5 mL Eppendorf tubes (or similar).
8. Vortex mixer.
9. 0.2 mL reaction tubes (or similar).

10. Heated block or waterbath.
11. DNA extracted from sample (see Note 6).
12. Pipettors and filter tips (see Note 7).
13. Agarose.
14. 1× TBE: 89 mM Tris base, 89 mM boric acid, 2 mM EDTA.
15. Ethidium bromide (10 μg/μL). Ethidium bromide is a known mutagen: wear gloves when handling, and dispose of with care.
16. Gel casting tray and comb(s).
17. Gel tank and power pack.
18. 6× gel loading buffer and DNA ladder (any manufacturer). The exact sizes of the bands in the ladder are not critical because LAMP products contain molecules of a range of sizes.
19. UV transilluminator.

3. Methods

3.1. LAMP Reactions

Prepare reactions using dedicated pipettors and tips, and preferably in an area dedicated to setting up reactions, to avoid contaminating reagents (see Note 7).

1. Make up a master mix for the total number of reactions to be carried out plus at least two additional reactions. Add the components shown in Table 1 to a 1.5 mL Eppendorf tube. Betaine and $MgSO_4$ concentrations should be optimized for each assay (see Note 8).
2. Vortex to mix, then distribute 24 μL aliquots into 0.2 mL reaction tubes. Add 1 μL DNA to each reaction.
3. Reactions should be performed in duplicate, and reactions containing water instead of DNA should be included as negative controls in each run.
4. Place reaction tubes into a heated block or waterbath at 65°C for 60 min (see Note 9).
5. The reaction can be terminated by heating to 80°C for 5–10 min to denature the polymerase.

For a discussion of alternative reagents, see Note 10.

3.2. Analysis of LAMP Reaction Products

1. LAMP products can be visualized by agarose gel electrophoresis. For alternative detection methods for use outside the laboratory, see Notes 10–12.
2. To make a 100 mL gel, mix 1.2 g agarose with 100 mL 1× TBE, then microwave for 60–90 s to melt the agarose.

Table 1
Components of a loop-mediated isothermal amplification reaction

Reagent	Starting concentration	Volume per reaction (μL)	Final concentration
Thermopol buffer[a]	10×	2.5	1×
$MgSO_4$[a]	50 mM	0–4	2–10 mM
dNTP mix	10 mM each	3.5	1.4 mM
Primer F3	10 μM	0.5	200 nM
Primer B3	10 μM	0.5	200 nM
Primer FIP	100 μM	0.5	2 μM
Primer BIP	100 μM	0.5	2 μM
Primer F-loop	100 μM	0.25	1 μM
Primer B-loop	100 μM	0.25	1 μM
Betaine	5 M	3–7	0.6–1.4 M
Bst polymerase	8 units/μL	1	0.32 units/μL
Molecular biology grade water		To give a final reaction volume of 24 μL	

[a]10× thermopol buffer contains 20 mM $MgSO_4$

3. When cooled to <60°C, add approximately 5 μL ethidium bromide to give a final concentration of 0.5 μg/mL.
4. Assemble a gel-casting tray with an appropriate comb and slowly pour in the agarose. Leave to set for 30 min, then remove the comb and transfer the gel (still in the casting tray) to the gel tank.
5. Load 5–10 μL of each LAMP product onto the gel mixed with loading buffer in the ratio 5:1. In addition to the LAMP products, load a DNA ladder into one of the lanes.
6. Connect the gel tank to the power pack and run the gel at approximately 110 V for 40–60 min.
7. Examine the gel under UV illumination. LAMP products have a ladder-like appearance (Fig. 2).

4. Notes

1. LAMP can also be used for detection of RNA targets. Use of a thermostable reverse transcriptase such as Thermoscript (Life Technologies) enables both reverse transcription and LAMP to be carried out simultaneously at approximately 65°C.

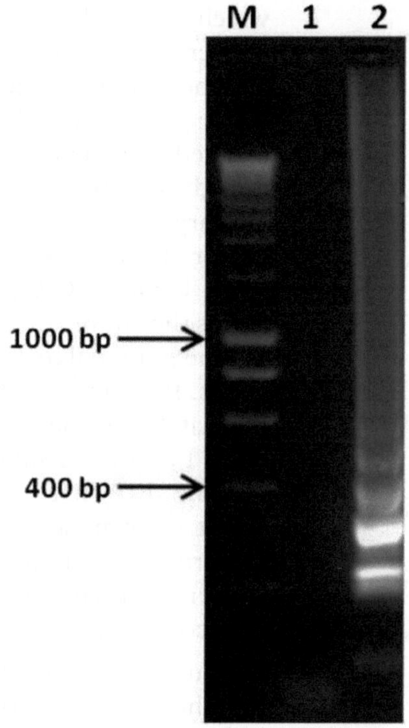

Fig. 2. Appearance of LAMP product when visualized by agarose gel electrophoresis, showing typical ladder-like pattern. *M* size marker; *lane 1* negative reaction; *lane 2* positive reaction.

The polymerase in Isothermal Master Mix (see Note 9) has some reverse transcriptase activity, allowing amplification of RNA without additional reverse transcriptase. For some assays, reaction time and/or assay sensitivity is improved by a short incubation at a lower temperature (approximately 50–55°C) to allow more efficient reverse transcription before increasing the temperature to 65°C for the remainder of the reaction.

2. The layout of primers for a LAMP assay is shown in Fig. 1. Note that the internal primers FIP and BIP each target two primer binding sites (F1c and F2, and B1 and B1c, respectively). Primer design criteria are similar to those for PCR; primers should be designed with reference to alignments of sequences for target and nontarget species, and the physical properties of the primers—specifically melting temperature (determined by length and base composition) and secondary structure—should be taken into account. The melting temperature (T_m) should be approximately 60°C for F3/B3, F2/B2, and F-loop/B-loop and approximately 65°C for F1/B1. Areas of sequence containing substantial secondary structure should be avoided, and primers should ideally be selected which do not form hairpins or primer dimers (self-dimers or heterodimers), especially any involving the 3′ ends of primers,

which could be extended to form amplification artifacts. Primers should be designed to target variation between target and nontarget species, and to mitigate against intraspecific variation. Where variation between target sequences cannot be avoided, degeneracies or synthetic bases such as inosine can be introduced into the primers. For inclusive detection of different targets, primers should be designed to position degeneracies or mismatches away from the 3′ ends of the primer binding sites. Unlike PCR primers, LAMP primers are distributed across the entire amplicon (200–300 bp), a characteristic which has the potential to confer a high degree of specificity to LAMP assays. Primers should be designed to position the most salient mismatches at the 5′ ends of the internal primers FIP and BIP (Fig. 1). For maximum specificity, the 3′ ends of all primers (F3/B3, FIP/BIP, and F-loop/B-loop) should also be used to target discriminatory mismatches if possible. LAMP can be performed without loop primers—or with a single loop primer—if it is not possible to design large enough loop regions to accommodate loop primers (see Note 4).

3. Various free oligonucleotide calculators are available for assessing the properties of primers including T_m and secondary structure (e.g., (10)). Reliable results can be obtained using salt-adjusted T_m values ((Na^+) = 50 mM).

4. LAMP can be performed with external and internal primers only (1), but amplification is usually accelerated by the inclusion of loop primers (3) as this allows all loops in the amplification product to act as primer binding sites (the amplification product contains repeats of the target sequence with alternating orientation, so the internal primers can bind to only half of the loops available). If design constraints limit the options for design of loop primers, assays can be designed without loop primers or with a single loop primer.

5. The following software programs are available for design of LAMP primers: PrimerExplorer (Eiken, Japan) (http://primerexplorer.jp/e/); LAMP Designer (PREMIER Biosoft, USA).

6. The protocol given above is suitable for use in the development of LAMP assays in the laboratory, where conventionally extracted DNA can be tested to establish assay sensitivity and specificity. LAMP assays are more suitable for use outside the laboratory than PCR-based methods and require relatively minor modification for deployment in the field. LAMP assays typically display higher sensitivity than comparable PCR assays for the same targets; LAMP also has the significant advantage of being more tolerant of substances which are found to be inhibitory to PCR. In combination, these features allow the use of more crude DNA extraction methods than are typically used for PCR testing. One method involves preparing samples by shaking in a

buffer and applying the disrupted sample to a lateral flow device (LFD). A section of the LFD membrane is then excised from the device and added directly to the LAMP reaction (11). An alternative method involves preparing a crude macerate in alkaline polyethylene glycol (PEG) buffer and adding a sample of the macerate directly into the reaction tube (6, 12). Such methods are extremely rapid and require no laboratory equipment, making them highly suited to use in the field. Detection methods appropriate to non-laboratory conditions (see Notes 11 and 12), and the use of rapid LAMP reagents (see Note 10), can be combined with a simplified DNA extraction method to create extremely simple workflows which can be performed in the field by nonspecialist operators.

7. LAMP reactions are very efficient and typically generate large amounts of amplification product. As a result of this, care must be taken to prevent contamination of laboratory facilities and equipment with amplification products, because carryover of extremely small amount of amplification products can generate spurious results. Pre- and postamplification steps should be rigorously separated: store and mix reagents in a dedicated area and use only dedicated pipettors and stocks of molecular biology grade water; keep all pipettors and other equipment used for postamplification manipulations (such electrophoresis) separate from clean equipment; minimize as much as possible the opening of reaction tubes after amplification. Some methods for detection or visualization of LAMP products require reaction tubes to be opened after amplification—extreme care should be taken when using these methods. If segregation of pre- and postamplification equipment is not possible, consider using a homogeneous (closed-tube) method for detection of amplification. These include real-time detection of fluorescence or turbidity, direct observation of turbidity (precipitated magnesium pyrophosphate), and colorimetric detection using hydroxy naphthol blue (HNB) (see Note 12). For on-site testing, workflows containing as few steps as possible are less prone to contamination than methods with a greater number of manipulations. Reaction mix should be prealiquoted into individual reaction tubes prior to use in the field.
8. New assays should be optimized in terms of the concentrations of $MgSO_4$ and betaine. An initial test of new primers can be carried out using $MgSO_4$ at 6 mM and betaine at 1.2 M. Test a dilution series of target DNA to establish the limit of detection using the initial reaction composition, then test different $MgSO_4$ and betaine concentrations using DNA concentrations either side of the limit of detection, in order to identify combinations which result in an increase (or decrease) in the sensitivity of the assay. Test $MgSO_4$ concentrations between 4 and 10 mM and betaine concentrations between 0.6 and 1.6 M.

The concentrations of other components (*Bst* polymerase, dNTPs, primers) and the ratio of internal, external, and loop primer concentrations can also be optimized but these generally have a lesser effect on the performance of the assay.

9. Using *Bst* polymerase and the conditions described above, a reaction time of 60 min is likely to be sufficient for most assays, although sensitivity may be slightly improved by increasing this to 90 min for some assays. Prolonged incubation beyond 90 min is unlikely to increase sensitivity and may result in the eventual generation of nonspecific amplification artifacts. A reaction temperature of 65°C can be used for most assays; however, the reaction temperature can be optimized between approximately 62 and 66°C, and for some assays this may increase the speed of the reaction or result in a slight increase in sensitivity.
10. The majority of LAMP methods published to date have reported the use of *Bst* polymerase. An alternative to *Bst* polymerase is Isothermal Master Mix (OptiGene, Horsham, UK), which contains a highly processive strand-displacing polymerase and the intercalating dye EvaGreen for real-time monitoring of the accumulation of amplification products. Isothermal Master Mix typically displays a more rapid amplification (typically 10–30 min) than achieved using *Bst* polymerase.
11. Real-time LAMP using intercalating dyes can be carried out on platforms developed for real-time PCR, but these instruments are typically relatively costly due to their rapid thermal cycling capability, and few are suitable for use in non-laboratory locations. The Genie II instrument (OptiGene, Horsham, UK) is a platform specifically developed for real-time LAMP, and its compact size and portability make this instrument particularly well suited to field use. As real-time detection requires no postamplification steps, the risk of carryover contamination is substantially reduced. When monitored in real time, LAMP produces amplification plots that are similar in appearance to those generated by real-time PCR (Fig. 3a). After amplification on the Genie II, products are typically analyzed by a slow annealing step with concurrent fluorescence monitoring, because the temperature at which the amplification product anneals/melts is consistent and characteristic for each assay (Fig. 3b). Using this analysis, specific amplification products can be distinguished from any nonspecific amplification artifacts. In addition, this approach can allow two or three amplification products to be resolved in the same reaction if the products have sufficiently different annealing/melting temperatures. The ability to resolve multiple products can allow multiple pathogens to be detected in a single reaction; alternatively the pathogen-specific test can be combined with an internal control assay to assist in the interpretation of negative results.

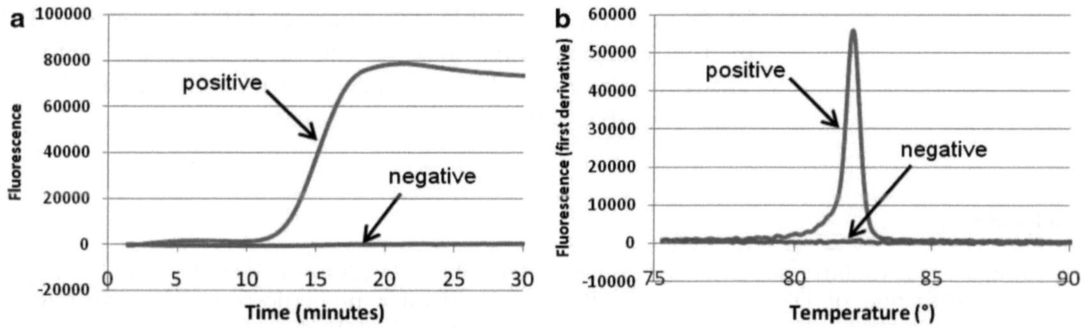

Fig. 3. Typical results of real-time LAMP using fluorescence monitoring. (**a**) Amplification plot showing fluorescence during incubation at 65°C; (**b**) annealing/melt analysis showing the temperature at which the amplification product anneals/melts after amplification.

12. LAMP products can be analyzed in the same ways as PCR products by gel electrophoresis or real-time fluorescent monitoring with an intercalating dye such as SYBR Green. An alternative approach is the detection of labels incorporated during amplification using LFDs (11, 13). Labels can be incorporated using labeled nucleotides, primers, and/or probes; if two different labels are incorporated, the dual-labeled amplification product can be detected in a sandwich immunoassay in an LFD format. This approach is also applicable to other amplification methods including PCR and isothermal methods such as recombinase polymerase amplification (RPA), but is particularly useful for the resolution of mixed LAMP products. Unlike PCR where mixed products can be separated by gel electrophoresis on the basis of size, each LAMP product contains molecules of different sizes (Fig. 2) and has a ladder-like appearance on a gel, making it difficult to resolve mixed products. The incorporation of different labels in each assay allows mixed products to be resolved after amplification.

    The large amount of amplification product typically generated in a LAMP reaction makes LAMP amenable to the use of several additional detection methods which are insufficiently sensitive for use with PCR, and other amplification methods which generate smaller amounts of product. Magnesium pyrophosphate generated as a by-product of polymerase activity reaches sufficiently high levels in LAMP reactions to allow direct observation of a white precipitate in positive reactions, or real-time or end-point turbidity measurement using a simple instrument (14, 15). Similarly, color change methods can be used to distinguish positive and negative reactions, for example SYBR Green—or other intercalating dyes such as PicoGreen—can be added at the end of the reaction and a color change from orange to yellow observed directly (16). HNB can be used as a color-change indicator (17); unlike SYBR Green (which is inhibitory to the LAMP reaction at the levels required for a visible color change), HNB can be included in the reaction mix, making this a closed-tube method. However, the color change achieved with HNB is typically less clear than that achieved using intercalating dyes.

References

1. Notomi T et al (2000) Loop-mediated isothermal amplification of DNA. Nucl Acids Res 28:e63
2. Nagamine K et al (2001) Loop-mediated isothermal amplification reaction using a nondenatured template. Clin Chem 47:1742–1743
3. Nagamine K, Hase T, Notomi T (2002) Accelerated reaction by loop-mediated isothermal amplification using loop primers. Mol Cell Probes 16:223–229
4. Tomlinson J, Boonham N (2008) Potential of LAMP for detection of plant pathogens. CAB Reviews: Perspectives in Agriculture, Veterinary Science, Nutrition and Natural Resources 3, 066, p 1–7
5. Tomlinson JA, Boonham N, Dickinson M (2010) Development and evaluation of a one-hour DNA extraction and loop-mediated isothermal amplification assay for rapid detection of phytoplasmas. Plant Pathol 59:465–471
6. Hodgetts J et al (2011) Development of rapid in-field loop-mediated isothermal amplification (LAMP) assays for phytoplasmas. Bull Insectol 64:S41–S42
7. Bekele B et al (2011) Use of a real-time LAMP isothermal assay for detecting 16SrII and XII phytoplasmas in fruit and weeds of the Ethiopian Rift Valley. Plant Pathol 60:345–355
8. Obura E et al (2011) Detection of phytoplasma by loop-mediated isothermal amplification of DNA (LAMP). J Microbiol Methods 84:312–316
9. Yankey EN et al (2011) Improving molecular diagnostics for the detection of lethal disease phytoplasma of coconut in Ghana. Bull Insectol 64:S47–S48
10. Kibbe WA (2007) OligoCalc: an online oligonucleotide properties calculator. Nucl Acids Res 35:W43–W46
11. Tomlinson JA, Dickinson MJ, Boonham N (2010) Rapid detection of *Phytophthora ramorum* and *P. kernoviae* by two-minute DNA extraction followed by isothermal amplification and amplicon detection by generic lateral flow device. Phytopathology 100:143–149
12. Chomczynski P, Rymaszewski M (2006) Alkaline polyethylene glycol-based method for direct PCR from bacteria, eukaryotic tissue samples, and whole blood. Biotechniques 40: 454–458
13. Kiatpathomchai W et al (2008) Shrimp Taura syndrome virus detection by reverse transcription loop-mediated isothermal amplification combined with a lateral flow dipstick. J Virol Methods 153:214–217
14. Mori Y et al (2004) Real-time turbidimetry of LAMP reaction for quantifying template DNA. J Biochem Biophys Methods 59:145–157
15. Mori Y et al (2001) Detection of loop-mediated isothermal amplification reaction by turbidity derived from magnesium pyrophosphate formation. Biochem Biophys Res Commun 289:150–154
16. Iwamoto T, Sonobe T, Hayashi K (2003) Loop-mediated isothermal amplification for direct detection of *Mycobacterium tuberculosis* complex, *M. avium*, and *M. intracellulare* in sputum samples. J Clin Microbiol 41:2616–2622
17. Goto M et al (2009) Colorimetric detection of loop-mediated isothermal amplification reaction by using hydroxy naphthol blue. Biotechniques 46:167–172

Chapter 26

DNA Bar-Coding for Phytoplasma Identification

Olga Makarova, Nicoletta Contaldo, Samanta Paltrinieri, Assunta Bertaccini, Henriette Nyskjold, and Mogens Nicolaisen

Abstract

Phytoplasma identification has proved difficult due to their inability to be maintained in vitro. DNA barcoding is an identification method based on comparison of a short DNA sequence with known sequences from a database. A DNA barcoding tool has been developed for phytoplasma identification. While other sequence-based methods may be well adapted to identification of particular strains of phytoplasmas, often they cannot be used for the simultaneous identification of phytoplasmas from different groups. The phytoplasma DNA barcoding protocol in this chapter, based on the *tuf* and 16SrRNA genes, can be used to identify the following phytoplasma groups: 16SrI, 16SrII, 16SrIII, 16SrIV, 16SrV, 16SrVI, 16SrVII, 16SrIX, 16SrX, 16SrXI, 16SrXII, 16SrXV, 16SrXX, 16SrXXI.

Key words: 16S rRNA, DNA barcoding, Identification, Quarantine barcode of life, *tuf*

1. Introduction

Phytoplasmas infect over 200 plant species (1). Based on the 16S ribosomal gene (16S rRNA) sequence, over 20 groups have been established within the '*Candidatus* Phytoplasma' taxon (2, 3). Some phytoplasmas are tightly regulated by plant protection authorities as quarantine organisms (grapevine "flavescence dorée," grapevine "bois noir," palm lethal yellowing, apple proliferation, potato stolbur). To date, no protocol for in vitro cultivation of phytoplasmas has been developed, making classical microbiological identification procedures impossible. Moreover, spread of phytoplasma-associated diseases, facilitated by migration of insect vectors due to global climate change and increased international movement of people and goods, may result in emergence of phytoplasma diseases in regions previously atypical for the pathogen,

Matt Dickinson and Jennifer Hodgetts (eds.), *Phytoplasma: Methods and Protocols*, Methods in Molecular Biology, vol. 938, DOI 10.1007/978-1-62703-089-2_26,

leaving plant protection inspectors unprepared for identification based solely on symptomatology. Therefore, efficient protocols for phytoplasma identification are in great demand.

One of the most common phytoplasma identification methods is based on nested amplification of the 1.8 kbp region of the 16S rRNA gene, including flanking regions, using P1 (4) and P7 primers (5), followed by amplification of a 1.2 kbp fragment with R16F2n and R16R2 primers (6). This amplicon is subsequently subjected to RFLP (restriction fragment length polymorphism) analysis. For finer differentiation of phytoplasmas other genes are used, e.g., translation elongation factor *tuf*, protein translocase subunit *secY*, ribosomal protein operon *rp*, 16S–23S rRNA intergenic sequence, and protein translocase subunit *secA* (7–10), but often the primers only amplify specific 16Sr groups so they can be used only for discrimination of phytoplasmas within a particular 16Sr group. Although it is by far the most popular method of phytoplasma identification, RFLP-based methods suffer from several problems: it is not easy to set up, it requires the use of expensive restriction enzymes and toxic polyacrylamide gels, it is necessary to use reference strains for pattern comparison and, finally, it only considers a small fraction (restriction sites) of the available sequence information.

DNA barcoding is a technique where short DNA sequences (DNA barcodes) are used for species identification (11, 12). Briefly, DNA from an unidentified species is extracted, amplified with a set of generic primers, sequenced, and finally the sequence is compared with the sequences from a database of standards to assign the sample to a species. This method is based on several assumptions: (1) the DNA barcode is present in all species; (2) it is possible to amplify the region in all species with just one set of generic primers; (3) sequence differences between species are much greater than differences between individuals of the same species; (4) the barcode is relatively short, so that it can be sequenced quickly; and (5) taxonomy, to which DNA barcodes are linked, is already established. The main advantages of this method are that no morphological traits are needed for identification, DNA can be obtained from various sources including environmental samples and museum specimens, and just one generic assay is used. A limitation is the risk of contamination resulting in false identifications and a variable detection limit, which depends on the DNA concentration in preparations. Under-representation of sequences of particular species in the database may be another problem.

The adoption of DNA barcoding principles to the identification of plant pests and pathogens has been proposed. The QBOL (quarantine barcode of life) project, financed by the EU Commission and aimed at the development of a universal identification system for the main groups of plant pathogens including phytoplasmas, is focused on quarantine organisms (13). Work under this project includes obtaining high quality reference barcode sequences for quarantine and regulated pathogens, development and validation of protocols, establishment of an online sequence database, and a reference collection of pathogen strains.

Several considerations had to be made when developing a DNA barcoding system for phytoplasmas. Firstly, as phytoplasma DNA is present as a minor component in a plant DNA background and secondly, as phytoplasmas often coexist with other bacteria, the primers should not amplify plant DNA or unrelated bacterial DNA. Among many regions tested, two barcodes, based on the *tuf* and 16S rRNA genes, proved to be immune to these problems and hence were selected as final DNA barcodes for phytoplasma identification.

To develop the *tuf* barcode, all phytoplasma *tuf* genes available in the NCBI GenBank were aligned and a conserved region within the *tuf* gene, present in all phytoplasmas, was identified. Combinations of primers (Table 1), which allowed for minor sequence variations between individual groups of phytoplasmas, were designed for the region, and the 420–444 bp barcode was amplified from all phytoplasma strains tested (16SrI, 16SrII, 16SrIII, 16SrIV, 16SrV, 16SrVI, 16SrVII, 16SrIX, 16SrX, 16SrXI, 16SrXII, 16SrXV, 16SrXX, and 16SrXXI groups). The phylogenetic tree constructed from these sequences followed branching patterns of trees constructed using full 16S rRNA sequences, suggesting that the *tuf* barcode enables fine discrimination of phytoplasmas on both group and subgroup levels (our unpublished data).

While the full 16S rRNA gene is commonly used for phytoplasma taxonomy, its use as a barcode is limited by its relatively long length and associated difficulties with sequencing. Nevertheless, it was possible to identify a variable region of approximately 600 bp in the 5′ of the 16S rRNA gene, and its potential for DNA barcoding was explored. The 600 bp amplicon was recovered from all phytoplasma strains tested, and construction of a phylogenetic tree verified its capacity to distinguish most phytoplasmas at the 16Sr group level, suggesting that it may be used for phytoplasma DNA barcoding.

The current DNA barcoding tool for phytoplasma identification, developed within the QBOL project, includes two primer sets for nested amplification of the two DNA barcode regions (*tuf* and 16S rRNA genes), an online sequence database with reference sequences for each DNA barcode for strains covering the most important groups described so far, including quarantine phytoplasmas, and a reference strain DNA gene bank. The identification pipeline (Fig. 1) comprises: DNA extraction from plants suspected to be infected with phytoplasma; nested PCR amplification of DNA barcodes; sequencing; sequence analysis and assembly; and online identification.

Table 1
Primer sequences

Barcode	Direct/nested PCR	Primer	Forward/reverse	Primer mix components	Sequence (5′-3′)	Proportions in mixture
tuf	D	Tuf340	For	Tuf340a Tuf340b	GCTCCTGAAGAAARAGAACGTGG ACTAAAGAAGAAAAAGAACGTGG	Mix 1:1 to 10 μM
tuf	D	Tuf890	Rev	Tuf890a Tuf890b Tuf890c	ACTTGDCCTCTTTCKACTCTACCAGT ATTTGTCCTCTTTCWACACGTCCTGT ACCATTCCTCTTTCAACACGTCCAGT	Mix 1:1:1 to 10 μM
tuf	N	Tuf400	For	Tuf400a Tuf400b Tuf400c Tuf400d Tuf400e	**GTAAAACGACGGCCAGT**GAAACAGA AAAACGTCAYTATGCTCA **GTAAAACGACGGCCAGT**GAAACTTC TAAAAGACATTACGCTCA **GTAAAACGACGGCCAGT**GAAACA TCAAAAAGACAYTATGCTCA **GTAAAACGACGGCCAGT**GAAA CAGAAAAAAGACAYTATGCTCA **GTAAAACGACGGCCAGT**CAAAC AGCTAAAAGACATTATYCTCA	Mix 1:1:1:1:1 to 10 μM
tuf	N	Tuf835	Rev	Tuf835a Tuf835b Tuf835c	*TAATACGACTCACTATAGGG*AACA TCTTCWACHGGCATTAAGAAAGG *TAATACGACTCACTATAGGG*AACAC CTTCAATAGGCATTAAAAAWGG *TAATACGACTCACTATAGGG*AAC ATCTTCTATAGGTAATAAAAAAGG	Mix 1:1:1 to 10 μM
16S rRNA	DN	P1-ATT	For	P1-ATT	AAGAGTTTGATCCTGGCTCAGG	
16S rRNA	D	P7	Rev	P7	CGTCCTTCATCGGCTCTT	
16S rRNA	N	P625	Rev	P625	ACTTAYTAAACCGCCTACRCACC	

Bold indicates M13F tag, italic indicates T7R tag

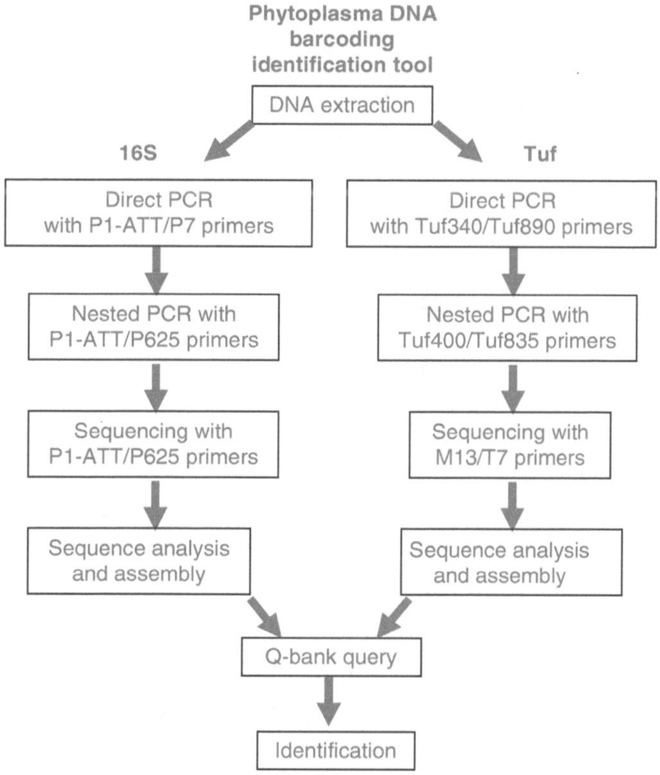

Fig. 1. Phytoplasma identification DNA barcoding flowchart.

It has to be emphasized that although this method is efficient for identification of phytoplasma strains both from lab-maintained collections and field material, it still has some limitations. When mixed infections occur, cloning is required to separate sequences of all phytoplasmas present in the sample before sequencing and analysis.

Finally, it is worth mentioning that while the QBOL project provided the initial platform for the development of the current DNA barcoding phytoplasma identification system, there is a potential for expansion of the DNA sequence database, and therefore the phytoplasma community is encouraged to contribute by submitting new sequences.

2. Materials

2.1. DNA Extraction

1. Phytoplasma-infected plant material (see Note 1).
2. Healthy plant material for negative control.
3. Liquid nitrogen.
4. Tissue homogenizer (e.g., Geno/Grinder 2000, SPEX CertiPrep) or sterile mortar and pestles for plant tissue disruption.

5. 5 mm grinding steel beads and grinding tubes (for Geno/Grinder).
6. Microfuge tubes.
7. Spatulas.
8. DNA extraction kit (e.g., DNeasy, QIAGEN) (see Note 2).
9. Centrifuge with cooling.
10. Heating block.
11. Spectrophotometer.

2.2. PCR

1. Ice.
2. Vortexer.
3. Microfuge for PCR tubes and 1.5 mL tubes.
4. PCR tubes and microfuge tubes.
5. Thermocycler.
6. 10 μM forward primer (Table 1).
7. 10 μM reverse primer (Table 1).
8. 25 mM $MgCl_2$.
9. 10 mM dNTPs mix (2.5 mM each).
10. Milli-Q water.
11. Positive control DNA (20 ng/μL) (see Note 3).
12. DNA from the infected and healthy samples (20 ng/μL).
13. Promega GoTaq DNA polymerase, 5 U/μL.
14. Promega 5× colorless GoTaq Flexi reaction buffer.
15. Fermentas *Taq* DNA polymerase, 5 U/μL (see Note 4), for PCR reactions with P1-ATT/P7 (see Note 5).
16. Fermentas 10× *Taq* reaction buffer with KCl.

2.3. Agarose Gel Electrophoresis

1. Agarose.
2. TAE (Tris/acetate/EDTA) electrophoresis buffer (50× stock): 242 g Tris base, 57.1 mL glacial acetic acid, 100 mL 0.5 M EDTA, double distilled water to 1 L. Adjust the pH to 8.5. To obtain 1× working solution, dilute the stock by 50× with double distilled water.
3. Ethidium bromide.
4. Loading dye (see Note 6).
5. Molecular weight marker (100–1,500 bp range).
6. Gel electrophoresis equipment.
7. Microwave oven.
8. Transilluminator.

2.4. Sequencing

1. M13F and T7R primers for the *Tuf* barcode; P1-ATT and P625 primers for the 16S barcode.

2.5. Software for DNA Sequence Assembly and Alignments

1. CLC DNA Workbench 6 (CLCbio) or any other sequence assembly software.

2.6. Online QBOL DNA Barcoding Database

1. http://www.q-bank.eu/.

3. Methods

See Notes 7 and 8.

3.1. DNA Extraction

1. Put the plant samples into the grinding tubes together with eight steel beads (see Note 9).
2. Submerge the tubes in liquid nitrogen.
3. Disrupt the samples by shaking in Geno/Grinder at the rate of 1,200 strokes per minute for 1 min. Repeat two more times (see Note 10).
4. Use ≤100 mg of the ground plant material for DNA extraction. To make transfer of the plant material easier, briefly freeze the recipient microfuge tube and the spatula in liquid nitrogen. Use a new spatula for each sample to avoid cross-contamination.
5. Proceed with DNA extraction by following the instructions from the kit manual.
6. Elute the DNA from the kit column in 50 μL of elution buffer.
7. Determine the DNA concentration by measuring absorbance with a spectrophotometer.
8. Aliquot the DNA and store at −20°C (see Note 11).

3.2. PCR

See Notes 12 and 13.

3.2.1. Direct PCR

1. Thaw the following reagents at room temperature, gently vortex, and then briefly spin down: reaction buffer, $MgCl_2$, and P1-ATT/P7 primers (1.8 kbp 16S rRNA product) or Tuf340/Tuf890 primers (0.55 kbp *tuf* product) (Figs. 2 and 3). Place reagents on ice.
2. Make calculations for the master mix (Table 2). Remember to include reactions for no template control (NTC) and positive and negative (nontarget) controls.
3. Mix together water, $MgCl_2$, reaction buffer, dNTPs, primers, and *Taq* DNA polymerase (see Note 14). Gently vortex, centrifuge, and aliquot 24 μL of the master mix per reaction into individual PCR tubes and return to ice (see Note 15).

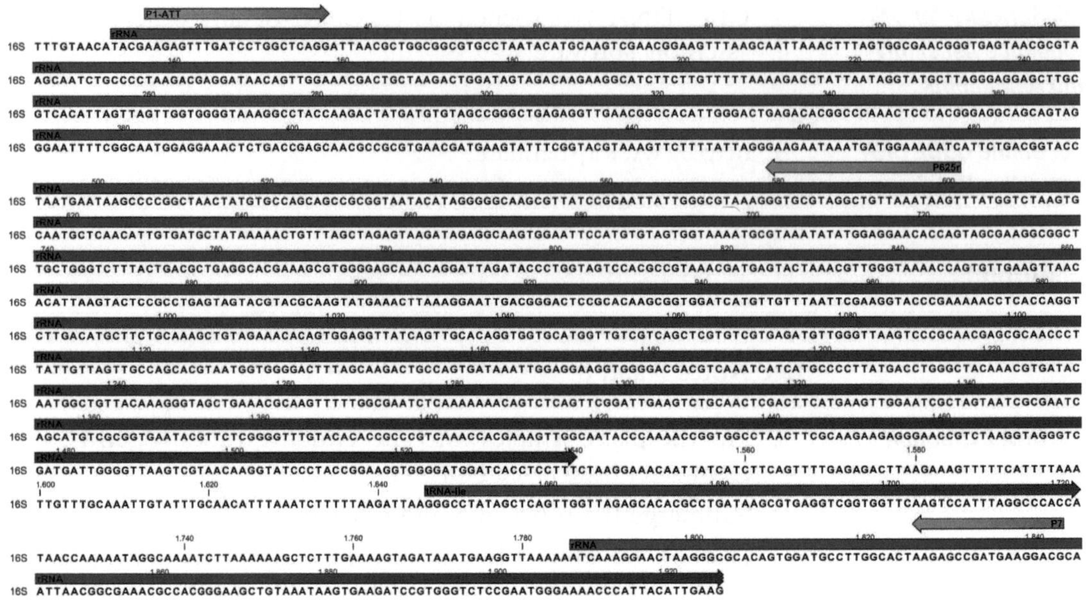

Fig. 2. Phytoplasma 16Sr RNA gene and primers used for the 16Sr RNA barcode amplification.

Fig. 3. Phytoplasma *tuf* gene and primers used for the *tuf* barcode amplification.

Table 2
Master mix for PCR assays

Components	Primer pair	
	P1-ATT/P7	P1-ATT/P625 Tuf340/Tuf890 Tuf400/Tuf835
Milli-Q water	12.6 µL	11.3 µL
PCR reaction buffer	2.5 µL Fermentas 10× PCR buffer with KCl	5 µL Promega 5× PCR buffer
25 mM $MgCl_2$	1.5 µL	1.5 µL
10 mM dNTPs mix	2 µL	1 µL
10 µM forward primer	2.5 µL	2.5 µL
10 µM reverse primer	2.5 µL	2.5 µL
Taq DNA polymerase (5 U/µL)	0.4 µL Fermentas *Taq* DNA polymerase	0.2 µL Promega GoTaq polymerase

Volumes indicated below are for one reaction. The total volume of a single reaction is 25 µL. To prepare master mix, multiply the amounts indicated below by the number of reactions you are going to run plus a couple of extras to allow for pipetting errors

4. Add 1 µL of DNA (20 ng/µL) and mix. Remember to include positive and negative control DNA and a NTC. Briefly spin down and return to ice.
5. Place the PCR tubes into a thermocycler and run a PCR program (conditions in Table 3).
6. Prepare 1:30 dilutions (in water) of the PCR product obtained for nested PCR.
7. Store DNA at −20°C.

3.2.2. Nested PCR

1. Using P1-ATT/P625 primers (0.6 kbp 16S rRNA product) or Tuf400/Tuf835 primers (0.4 kbp *tuf* product), repeat steps 1–3 from Subheading 3.2.1.
2. Add 1 µL of the diluted direct-PCR product (Subheading 3.2.1, step 6), mix, briefly centrifuge, and return to ice.
3. Place the PCR tubes into a thermocycler and run a PCR program (Table 3).
4. Use 2.5 µL of the PCR product for gel electrophoresis.
5. Store the remaining PCR product at −20°C for sequencing.

3.3. Agarose Gel Electrophoresis

1. Prepare a 1 % agarose gel. For 200 mL of 1 % agarose gel, dissolve 2 g of agarose in 200 mL of 1× TAE buffer by heating the mixture in a microwave oven. Let it cool down.

Table 3
Cycling conditions for the PCRs

Step	Temperature (°C)	Time	Number of cycles
P1-ATT/P7			
Initial denaturation	94	5 min	1
Denaturation	94	1 min 30 s	35
Annealing	55	2 min	
Extension	72	3 min	
Final extension	72	7 min	1
Hold	4	As long as required	1
Tuf340/Tuf890; Tuf400/Tuf835; P1-ATT/P625			
Initial denaturation	94	3 min	1
Denaturation	94	15 s	35
Annealing	54/64[a]	30 s	
Extension	72	1 min	
Final extension	72	7 min	1
Hold	4	As long as required	1

[a]64°C for P1-ATT/P625; 54°C for others

2. Pour the agarose into a tray, insert combs, remove any air bubbles, and let it set (approximately 20 min).
3. Place the gel into the electrophoresis chamber and pour 1× TAE buffer over, so the buffer covers it completely.
4. Mix 2.5 μL of PCR product, 2.5 μL of water, and 1 μL of 6× loading dye.
5. Load the samples, NTC, positive and negative controls and 5 μL of the DNA ladder (50 ng/μL) into the wells.
6. Connect the electrodes to the power source and run the gel (see Note 16).
7. Stain the gel with ethidium bromide for 10 min, and wash in water for 10 min.
8. Analyze the gel on a transilluminator by visualizing DNA under UV light.
9. Based on the positive and negative controls, choose samples with the correct amplicon size for sequencing.

3.4. Sequencing

1. Prepare samples according to the sequencing company's requirements (see Note 17).
2. Use M13F and T7R primers for sequencing the *tuf* barcode (see Note 18) and P1-ATT and P625 for the 16S rRNA barcode (Table 1). Always sequence amplicons with both the forward and reverse primers.

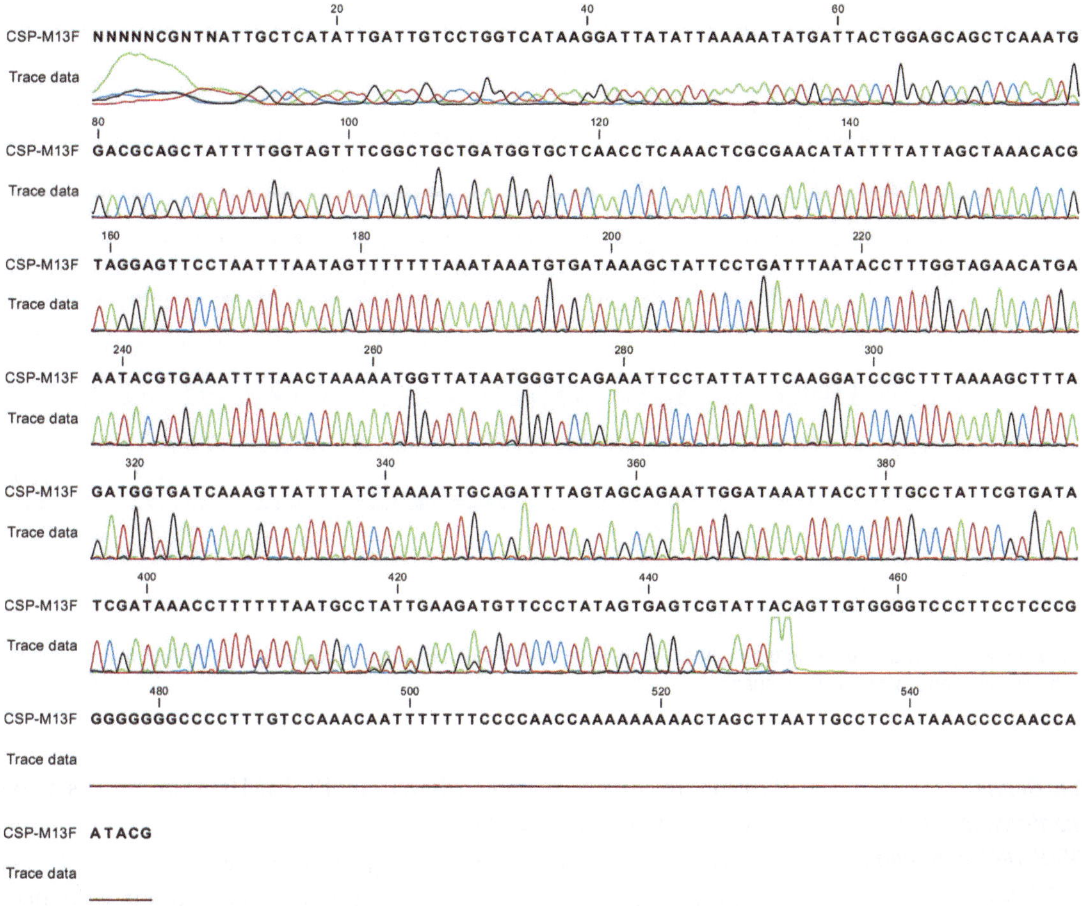

Fig. 4. A good quality sequence chromatogram. Single peaks indicate absence of contamination in the sample and a good quality sequence.

3.5. DNA Sequence Analysis and Assembly

1. Sequencing companies usually return the sequencing results in both plain text and chromatogram formats (Figs. 4 and 5). Carefully check chromatograms for multiple overlapping peaks (see Note 19).
2. Assemble both (forward and reverse) readings into one consensus sequence. We use CLC workbench for DNA assembly. However, other sequence assembly software packages may be used.
3. Identify primer regions and remove primer sequences and any flanking sequences.
4. Carefully solve nucleotide conflicts (if there are any) by vigilant examination of the chromatograms and manual editing of the text.
5. Save the final consensus sequence.

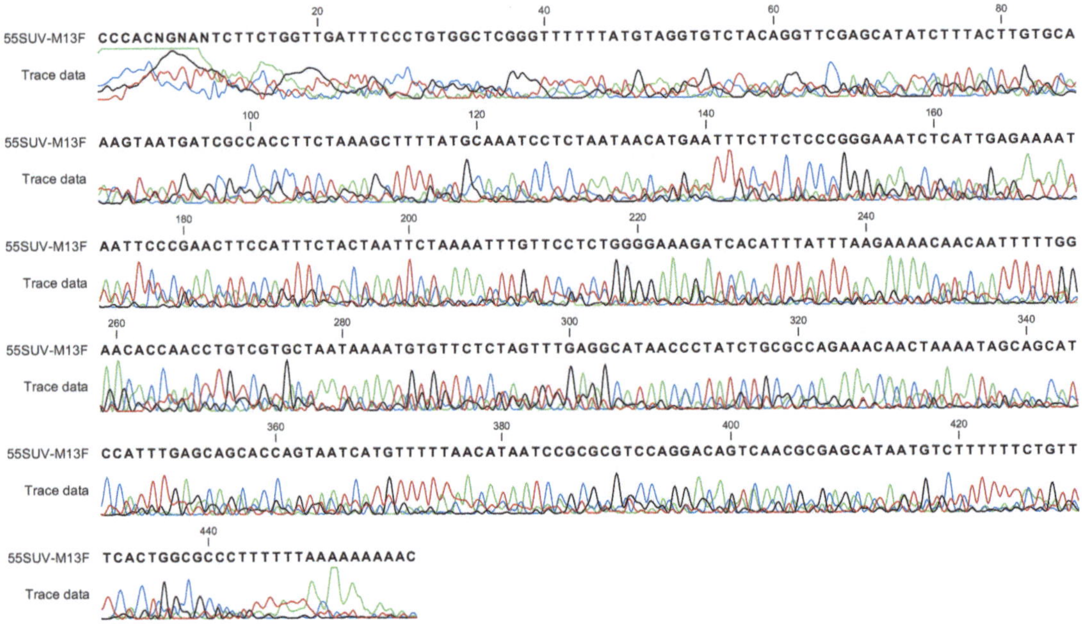

Fig. 5. A poor quality sequence chromatogram. Multiple peaks indicate presence of contamination and/or mixed infection in the sample and a poor quality sequence.

3.6. Online Identification Using QBOL DNA barcoding Database

1. Go to http://www.q-bank.eu/BioloMICSSequences.aspx?expandparm=f (Fig. 6).
2. Paste your sequence into the query window (Fig. 7) and start the alignment. A BLAST search against all sequences deposited in the QBOL database will be performed (see Note 20).
3. View the results (Fig. 8).

4. Notes

1. Use phloem tissue or midribs from fully expanded leaves for DNA extraction. If possible use plant material showing symptoms. Optimization of the DNA extraction protocol may be necessary for "difficult" plant species, because suboptimal DNA extraction will lead to decreased detection.
2. We use the DNeasy Extraction kit (QIAGEN), but alternative protocols may be used for phytoplasma DNA extraction, including those detailed in Chapters 12 and 13 of this book.
3. This can be any phytoplasma DNA sample which has previously been shown to be infected. Alternatively a reference strain may be obtained from the QBOL GenBank phytoplasma strain depository (http://www.q-bank.eu/).

Fig. 6. Q-bank phytoplasma identification web page.

4. In our experience, Fermentas *Taq* DNA polymerase works better for amplification of longer PCR products, such as the full-length 16S rRNA region.
5. This is a modified version of the previously published P1 primer (2). The nucleotides corresponding to the 3′ end of this primer vary between strains. To accommodate for differences between the strains, the last three nucleotides (ATT) were removed.
6. Commercially available loading dyes can be purchased from a number of manufacturers. However, make sure that you use the same loading buffer for the DNA ladder and for loading

Fig. 7. Submission of a query to the online DNA barcodes database. The online tool performs a BLAST search against the DNA barcodes database. Various parameters regarding the alignment method may be adjusted.

samples, so they will migrate at the same rate and allow correct size estimation of amplicons.

7. Following good laboratory practice is essential for successful phytoplasma identification using this method. Take extreme care to not cross-contaminate your samples, especially when performing nested PCR assays. Always use filter tips and change gloves whenever contamination is suspected. It is also advisable to spatially separate the areas where DNA samples are processed from where PCR reactions are set up, as failure to do so may contribute to increased risk of cross-contamination.
8. Some techniques and reagents used in this protocol may pose health hazards. Always wear proper protective clothing and conform to the local safety regulations. Familiarize yourself with the MSDS for all reagents before use.

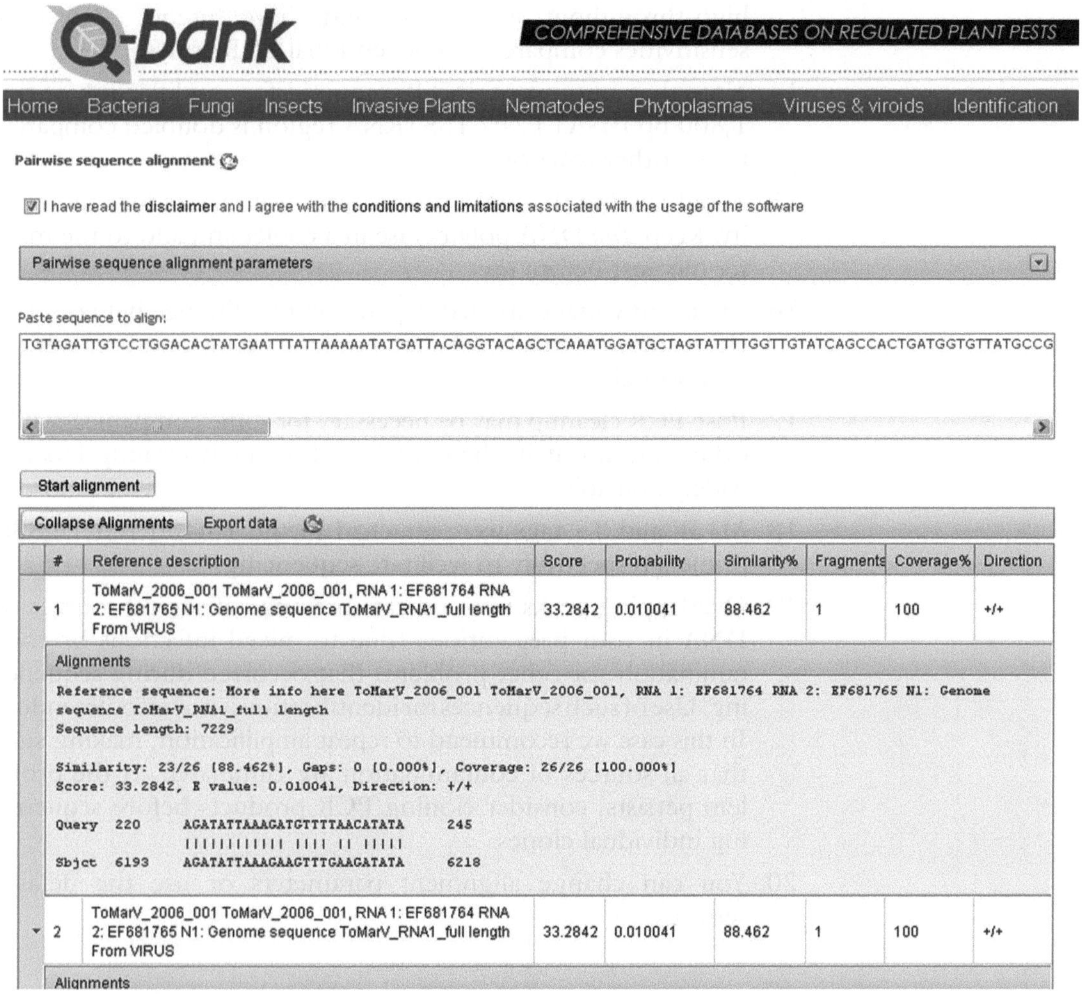

Fig. 8. Example of identification results.

9. You may use mortars and pestles for homogenization instead but ensure they are thoroughly sterilized between uses.
10. Remember to re-freeze the samples/tubes briefly between the grinding steps, as it is important that the samples do not thaw.
11. Storing DNA in small aliquots will help avoid repeated freeze–thaw cycles, which have a negative effect on DNA preservation.
12. To ensure greater reliability of the method, we strongly advise simultaneous use of two barcodes (both *tuf* and 16SrRNA) using the nested PCR assays. However, the use of the Tuf400/Tuf835 barcode in a direct PCR assay alone is often sufficient for phytoplasma identification.
13. Real-time TaqMan PCR methods (14) may be used beforehand. These are easily performed, can be set up for

high-throughput screening, and have higher detection sensitivities compared to conventional PCR assays.

14. Note that amount of dNTPs required for amplification of the 1,800 bp P1-ATT/P7 16S rRNA region is doubled compared to the other reactions.
15. Avoid heating the DNA polymerase, as it may reduce its activity. Keep *Taq* DNA polymerase in a cooler and add to the master mix just before use.
16. Time and voltage applied depend on the distance between the electrodes. Normally 5 V/cm is adequate for good separation in most gels.
17. Post-PCR cleanup may be necessary for some companies, while others include it in their services. Post-PCR cleanup kits are widely available.
18. M13F and T7 tags were attached to the Tuf400 and Tuf835 primers respectively to facilitate sequencing.
19. Overlapping peaks may indicate presence of several sources of DNA in your preparations (due to mixed infections or contamination) or other problems that occurred during sequencing. Use of such sequences for identification is not recommended. In this case we recommend to repeat amplification, making sure that all sources of contamination are eliminated. If the problem persists, consider cloning PCR products before sequencing individual clones.
20. You can change alignment parameters or use the default settings.

References

1. Lee I-M, Davis RE, Gundersen-Rindal DE (2000) Phytoplasma: phytopathogenic mollicutes. Annu Rev Microbiol 54:221–255
2. The IRPCM Phytoplasma/Spiroplasma Working Team—Phytoplasma taxonomy group (2004) '*Candidatus* Phytoplasma', a taxon for the wall-less, non-helical prokaryotes that colonize plant phloem and insects. Int J Syst Evol Microbiol 54:1243–1255
3. Wei W et al (2007) Computer-simulated RFLP analysis of 16S rRNA genes: identification of ten new phytoplasma groups. Int J Syst Evol Microbiol 57:1855–1867
4. Deng S, Hiruki C (1991) Amplification of 16S rRNA genes from culturable and nonculturable mollicutes. J Microbiol Methods 14:53–61
5. Smart CD et al (1996) Phytoplasma-specific PCR primers based on sequences of the 16S–23S rRNA spacer region. Appl Environ Microbiol 62:2988–2993
6. Gundersen DE et al (1996) Genomic diversity and differentiation among phytoplasma strains in 16S rRNA groups I (aster yellows and related phytoplasmas) and III (X-disease and related phytoplasmas). Int J Syst Bacteriol 46:64–75
7. Hodgetts J et al (2008) Phytoplasma phylogenetics based on analysis of *secA* and 23S rRNA gene sequences for improved resolution of candidate species of '*Candidatus* Phytoplasma'. Int J Syst Evol Microbiol 58:1826–1837
8. Lee I-M, Zhao Y, Bottner KD (2006) SecY gene sequence analysis for finer differentiation of diverse strains in the aster yellows phytoplasma group. Mol Cell Probes 20: 87–91
9. Botti S, Bertaccini A (2003) Variability and functional role of chromosomal sequences in 16SrI-B subgroup phytoplasmas including aster yellows and related strains. J Appl Microbiol 94:103–110

10. Marcone C et al (2000) Classification of aster yellows-group phytoplasmas based on combined analyses of rRNA and tuf gene sequences. Int J Syst Evol Microbiol 50:1703–1713
11. Hebert PDN et al (2004) Ten species in one: DNA barcoding reveals cryptic species in the neotropical skipper butterfly *Astraptes fulgerator*. PNAS USA 101:14812–14817
12. Hebert PDN, Ratnasingham S, deWaard JR (2003) Barcoding animal life: cytochrome *c* oxidase subunit 1 divergences among closely related species. Proc R Soc Lond B (Suppl) 270:S96–S99
13. Bonants P et al (2010) QBOL: a new EU project focusing on DNA barcoding of Quarantine organisms. EPPO Bull 40:30–33
14. Christensen NM et al (2004) Distribution of phytoplasmas in infected plants as revealed by real-time PCR and bioimaging. Mol Plant Microbe Interact 17:1175–1184

Chapter 27

Phylogenetic Analyses of Phytoplasmas Based on Whole-Genome Comparison

Hiromi Nishida

Abstract

It is no longer as difficult to determine genomic DNA sequences of uncultured bacteria as it once was, due to the development of DNA sequencing technology. It is likely that the number of whole-genome sequences of phytoplasmas will increase. In this chapter, two major strategies of whole-genome comparison studies, viz. gene content and orthologous protein sequence comparisons, are described. In general, horizontal gene transfer has greater influence on gene content-based phylogenetic analysis than orthologous protein sequence-based analysis. However, horizontal gene transfer has occurred rarely during the evolution of *Mollicutes*. Thus, the two phylogenetic topologies of the *Mollicutes* based on the two different strategies are similar.

Key words: Gene content comparison, *Mollicutes*, Mycoplasma, Orthologous protein sequence comparison

1. Introduction

When the number of substitutions at the evolutionary point of divergence is insufficient to show the relationships between organisms using a single gene comparison-based analysis, then comparison of single gene sequences does not have sufficient resolution to show the exact phylogenetic relationships. In contrast, whole-genome comparison is a powerful tool that enables greater accuracy in phylogenetic evaluation than that based on a single gene sequence comparison (1, 2). There are two major strategies of whole-genome

Matt Dickinson and Jennifer Hodgetts (eds.), *Phytoplasma: Methods and Protocols*, Methods in Molecular Biology, vol. 938, DOI 10.1007/978-1-62703-089-2_27,

comparison studies, viz. gene content and orthologous protein sequence comparisons (3). When the phylogenetic relationship among mycoplasmas on the basis of 143 orthologous protein sequence comparison was studied, the phylogenetic resolution was much greater than that achieved by 16S rRNA sequence comparison (4). In addition, genes evolving under *Ureaplasma*-specific selection from the 143 orthologues were detected (5).

2. Materials

2.1. Sequence Database

A genome includes a lot of function-unknown genes. Some of the function-unknown genes encode proteins conserved between different organisms. Therefore, we should select orthologous genes based not on annotated function but on the encoded amino acid sequence. The choice of orthologous gene set depends on the relationships of the test organism to other organisms with genome data already available for comparison purposes. For example, the number of orthologous genes among phytoplasmas is more than that among the *Mollicutes.* Microbial genome database for comparative analysis (MBGD, http://mbgd.nibb.ac.jp/) (6) is useful for selecting orthologous genes among different bacterial genomes. On this web site, we can select organisms from more than 1,000 bacterial species and create a table of orthologues from the selected species.

2.2. Software

Molecular Evolutionary Genetics Analysis (MEGA) version 5 (http://www.megasoftware.net/) (7) is a very useful software package for constructing multiple sequence alignments, distance matrix-based phylogenetic trees, maximum parsimony phylogenetic trees, maximum likelihood phylogenetic trees, etc.

Statistical Computing R (http://www.r-project.org/) and Microsoft Excel are useful for fundamental operations (e.g., construction and rearrangement of matrix data).

3. Methods

3.1. Creating Table of Orthologues from Selected Bacterial Genomes

1. Select organisms in "Organism Selection" on the MBGD web site. For example, in order to create the orthologue table of the *Mollicutes*, select the 27 *Mollicutes* (*Acholeplasma laidlawii*, '*Candidatus* Phytoplasma asteris' AYWB, '*Ca.* Phytoplasma asteris' OY, '*Candidatus* Phytoplasma australiense', '*Candidatus* Phytoplasma mali', *Mycoplasma agalactiae* 5632,

M. agalactiae PG2, *M. arthritidis*, *M. capricolum*, *M. conjunctivae*, *M. crocodyli*, *M. gallisepticum*, *M. genitalium*, *M. hominis*, *M. hyopneumoniae* 232, *M. hyopneumoniae* 7448, *M. hyopneumoniae* J, *M. mobile*, *M. mycoides*, *M. penetrans*, *M. pneumoniae*, *M. pulmonis*, *M. synoviae*, *Mesoplasma florum*, *Ureaplasma parvum* ATCC 27815, *U. parvum* ATCC 700970, and *U. urealyticum*) from the list of bacterial species.

2. After the selection, click "Execute Clustering" on the web site.
3. Change from 3 (default) to 2 as "Minimum cluster size" in "Search conditions."
4. Next, click "Show cluster table." After that, click "Save selected table."
5. Finally, download the orthologue table to a personal computer (see Note 1). Open the table as a Microsoft Excel file. In the case of the 27 *Mollicutes*, 1776 gene clusters were obtained. We can choose the evolutionary lineage-specific genes on the basis of the orthologue table. For example, in the case of the 27 *Mollicutes*, 27 of the 1776 gene clusters are phytoplasma-specific (Table 1).

3.2. Gene Content Comparison-Based Phylogenetic Analysis

1. Construct a gene presence/absence data matrix (for example, in the case of the 27 *Mollicutes*, 1776 gene clusters × 27 organisms), according to the orthologue table obtained in Subheading 3.1. In the gene presence/absence data matrix, change presence to "A" and absence to "T" (see Note 2). According to the A/T data matrix, make an input file for multiple alignment in MEGA. This input file is recognized as a nucleotide alignment data file in MEGA.
2. Select "Construct/Test Neighbor-joining tree" in "Phylogeny." After selecting "Construct/Test Neighbor-joining tree," select the number of bootstrap replications (e.g., 1,000) and select "No. of differences" in "Model/Method."
3. After that, click "Compute." The neighbor-joining tree of the 27 *Mollicutes* is shown in Fig. 1 (see Note 3).

3.3. Orthologous Protein Sequence Comparison-Based Phylogenetic Analysis

1. Extract proteins shared among all the selected organisms as orthologous proteins, according to the orthologue table obtained in Subheading 3.1.
2. Open the MEGA software; click "Edit/Build Alignment" on "Align."
3. Select "Create a new alignment" and then click "OK."
4. Copy the FASTA-format file of multiple amino acid sequences and paste it in "Alignment Explorer" in MEGA.

Table 1
Phytoplasma-specific genes

Ca. Phytoplasma mali	*Ca.* Phytoplasma australiense	*Ca.* Phytoplasma asteris AYWB	*Ca.* Phytoplasma asteris OY	Function	Gene name
ATP_00104	PAA_0768	AYWB_052	PAM_720	Malate/citrate symporter	*citS*
ATP_00420	PAA_0361	AYWB_617 AYWB_220 AYWB_179 AYWB_284 AYWB_077	PAM_571 PAM_074 PAM_388 PAM_645 PAM_523 PAM_550 PAM_570 PAM_686 PAM_341 PAM_334 PAM_375 PAM_545	Replicative DNA helicase	*dnaB*
ATP_00420	PAA_0361	AYWB_220 AYWB_179 AYWB_284 AYWB_046 AYWB_338 AYWB_389 AYWB_078 AYWB_617	PAM_335 PAM_571 PAM_074 PAM_388 PAM_645 PAM_523 PAM_550 PAM_570 PAM_686	Replicative DNA helicase	*dnaB*
ATP_00415	PAA_0360	AYWB_172 AYWB_086 AYWB_616 AYWB_219 AYWB_178 AYWB_285 AYWB_048	PAM_075 PAM_387 PAM_644 PAM_524 PAM_572 PAM_685 PAM_549 PAM_569	DNA primase	*dnaG*
ATP_00068	PAA_0755 PAA_0756	AYWB_529	PAM_191	ABC-type peptide/nickel transport system substrate-binding protein	*dppA*
ATP_00397	PAA_0127 PAA_0258 PAA_0723 PAA_0815 PAA_0316 PAA_0409 PAA_0790	AYWB_618 AYWB_045 AYWB_354 AYWB_206 AYWB_380 AYWB_359 AYWB_194 AYWB_233	PAM_503 PAM_566 PAM_397 PAM_536 PAM_700 PAM_760 PAM_702 PAM_323 PAM_363 PAM_511 PAM_320 PAM_360 PAM_040 PAM_656 PAM_408	DNA-directed RNA polymerase specialized sigma subunit	*fliA*

ATP_00382	PAA_0136	AYWB_096	PAM_658	Probable ATP-dependent Zn protease	*hflB*
ATP_00412	PAA_0385 PAA_0395 PAA_0600 PAA_0314 PAA_0648	AYWB_214 AYWB_373 AYWB_204 AYWB_371	PAM_326 PAM_366	ATP-dependent Zn protease	*hflB*
ATP_00222	PAA_0779	AYWB_018	PAM_030	Cation transport ATPase	*mgtA*
ATP_00048	PAA_0627	AYWB_121	PAM_612	Phosphatidylserine decarboxylase	*psd*
ATP_00047	PAA_0626	AYWB_120	PAM_613	CDP-diacylglycerol-serine *O*-phosphatidyltransferase	*pssA*
ATP_00450	PAA_0769	AYWB_051	PAM_721	NAD-dependent malic enzyme	*sfcA*
ATP_00398	PAA_0724 PAA_0791 PAA_0814 PAA_0317 PAA_0408	AYWB_209 AYWB_150 AYWB_PIII07 AYWB_PI05 AYWB_PII04 AYWB_PIV05 AYWB_232 AYWB_378 AYWB_193 AYWB_271 AYWB_004	PAM_062 PAM_701 PAM_410 PAM_564 PAM_324 PAM_364 PAM_655 PAM_041 PAM_512 PAM_396 PAM_726 PAM_319 PAM_359 PAM_535 PAM_699 PAM_759	Single-stranded DNA binding protein	*ssb*
ATP_00403	PAA_0304 PAA_0365	AYWB_222 AYWB_282 AYWB_074 AYWB_181	PAM_552 PAM_344 PAM_688	Thymidylate kinase	*tmk*
ATP_00006 ATP_00492	PAA_0481	AYWB_624	PAM_092	ABC-type Mn/Zn transport system, periplasmicMn/Zn-binding protein	*znuA*
ATP_00402	PAA_0328 PAA_0399	AYWB_182 AYWB_223	PAM_071 PAM_391 PAM_648 PAM_520 PAM_543 PAM_345	Putative phage-related protein (endonuclease)	

(continued)

Table 1 (continued)

Ca. Phytoplasma mali	*Ca.* Phytoplasma australiense	*Ca.* Phytoplasma asteris AYWB	*Ca.* Phytoplasma asteris OY	Function	Gene name
ATP_00093	PAA_0619	AYWB_113	PAM_622	Similar to phage-related protein predicted endonuclease	
ATP_00421	PAA_0354 PAA_0292 PAA_0721	AYWB_384 AYWB_614 AYWB_286 AYWB_176 AYWB_363 AYWB_216	PAM_407 PAM_525 PAM_386 PAM_573 PAM_682 PAM_684 PAM_643 PAM_338	Hypothetical protein	
ATP_00263	PAA_0283 PAA_0461 PAA_0305 PAA_0366 PAA_0442 PAA_0728	AYWB_239	PAM_554	Hypothetical protein	
ATP_00404	PAA_0410 PAA_0338 PAA_0388	AYWB_212 AYWB_376	PAM_562	Hypothetical protein	
ATP_00037	PAA_0298 PAA_0357	AYWB_019	PAM_477 PAM_031	Hypothetical protein	
ATP_00265	PAA_0138	AYWB_016	PAM_027	Hypothetical protein	
ATP_00129	PAA_0817	AYWB_559	PAM_160	Hypothetical protein	
ATP_00098	PAA_0533	AYWB_132	PAM_595	Hypothetical protein	
ATP_00144	PAA_0475	AYWB_141	PAM_606	Hypothetical protein	
ATP_00229	PAA_0522	AYWB_557	PAM_162	Hypothetical protein	
ATP_00280	PAA_0177	AYWB_543	PAM_177	Hypothetical protein	

5. Construct each multiple alignment of the amino acid sequences of proteins shared among all the selected organisms using CLUSTAL W (8) or MUSCLE (9) of MEGA.
6. Concatenate all multiple alignments of orthologous proteins as an input file in MEGA. This input file is recognized as a protein alignment data file in MEGA.
7. Select "Construct/Test Maximum likelihood tree," "Construct/Test Neighbor-joining tree," or "Construct/Test Parsimony tree" in "Phylogeny."

4. Notes

1. In the table of orthologues, blank means a lack of the gene; multiple genes in a cell mean existence of multiple genes with sequence similarity. If a gene product consists of multiple similar regions to different gene products, the gene is classified into different gene clusters. In order to select orthologues from multiple genes, we should construct and evaluate the phylogenetic tree for the multiple genes.
2. In the input file consisting of A/T characters, any combinations (for example, G/C, G/A, etc.) can be available.
3. In the *Mollicutes* phylogenetic analysis, gene content comparison-based phylogenetic relationships (Fig. 1) is similar to orthologous protein sequence comparison-based phylogenetic relationships (4). This strongly suggests that horizontal gene transfer was rare during the *Mollicutes* evolution. Interestingly, it has reported that phytoplasma genomes contain potential mobile units and prophage sequences (10–12). It has been hypothesized that phytoplasmas have a unique mechanism of recombination using potential mobile units to adjust to the diverse environments of plants and insects (11). Considering that *Acholeplasma* lacks the prophage sequences, it has been suggested that ancient bacteriophage attack triggered phytoplasma evolution (12).

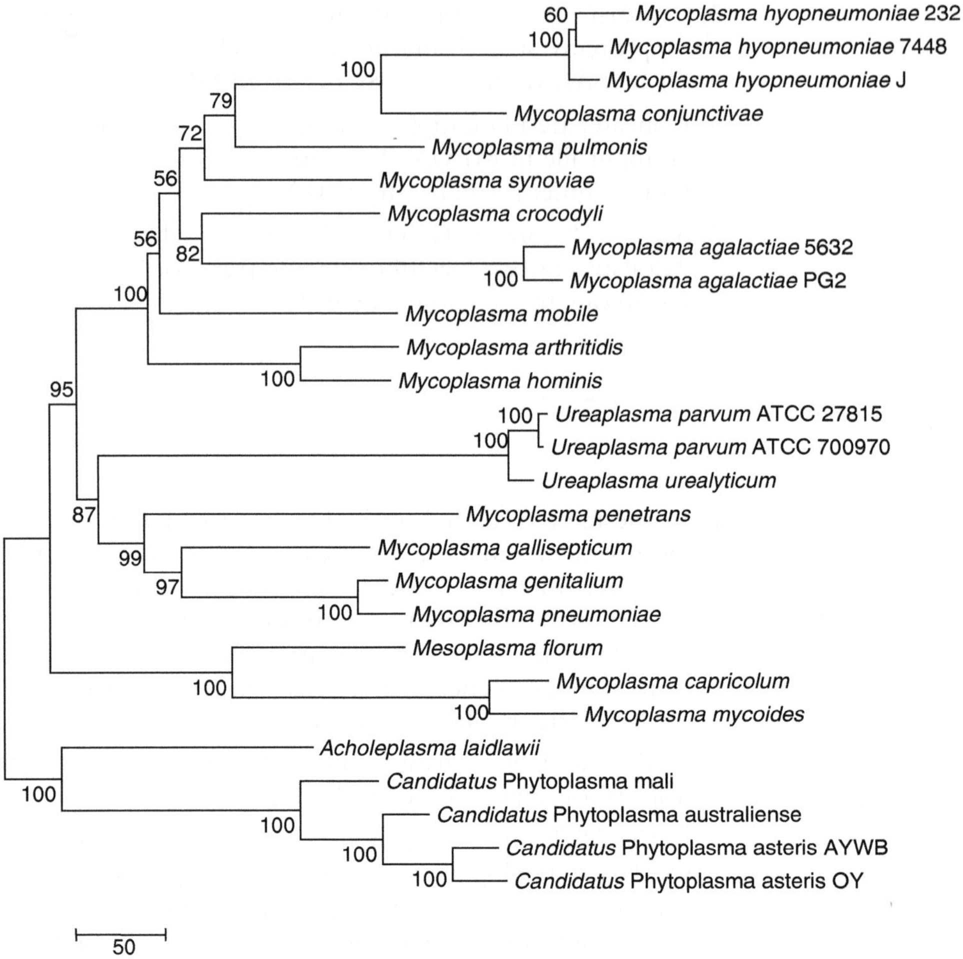

Fig. 1. Phylogenetic relationships on the basis of gene content comparisons among 27 *Mollicutes*. The orthologue cluster analysis among the 27 bacteria was performed using the microbial genome database (MBGD) (6). The analysis produced a gene presence/absence data matrix (1776 gene clusters × 27 organisms), which was used to generate the distance matrix between all pairs of the 27 bacterial species. On the basis of the distance matrix, a neighbor-joining tree was reconstructed using MEGA software (7). The bootstrap was performed with 1,000 replicates. The bar indicates a 50-gene difference.

References

1. Alam MT et al (2010) Genome-based phylogenetic analysis of *Streptomyces* and its relatives. Mol Phylogenet Evol 54:763–772
2. Oshima K et al (2011) Unique evolution of *Symbiobacterium thermophilum* suggested from gene content and orthologous protein sequence comparisons. Int J Evol Biol 2011:376831
3. Wolf YI et al (2002) Genome trees and the tree of life. Trends Genet 18:472–479
4. Oshima K, Nishida H (2007) Phylogenetic relationships among mycoplasmas based on the whole genomic information. J Mol Evol 65:249–258
5. Oshima K, Nishida H (2008) Detection of the genes evolving under *Ureaplasma*-specific selection. J Mol Evol 66:529–532
6. Uchiyama I, Higuchi T, Kawai M (2010) MBGD update 2010: toward a comprehensive resource for exploring microbial genome diversity. Nucl Acids Res 38:D361–D365
7. Tamura K et al (2011) MEGA5: molecular evolutionary genetics analysis using maximum

likelihood, evolutionary distance, and maximum parsimony methods. Mol Biol Evol 28: 2731–2739

8. Thompson JD, Higgins DG, Gibson TJ (1994) CLUSTAL W: improving the sensitivity of progressive multiple sequence alignment through sequence weighting, position-specific gap penalties and weight matrix choice. Nucl Acids Res 22:4673–4680
9. Edgar RC (2004) MUSCLE: multiple sequence alignment with high accuracy and high throughput. Nucl Acids Res 32:1792–1797
10. Dickinson M (2010) Mobile units of DNA in phytoplasma genomes. Mol Microbiol 77: 1351–1353
11. Bai X et al (2006) Living with genome instability: the adaptation of phytoplasmas to diverse environments of their insect and plant hosts. J Bacteriol 188:3682–3696
12. Wei W et al (2008) Ancient, recurrent phage attacks and recombination shaped dynamic sequence-variable mosaics at the root of phytoplasma genome evolution. PNAS USA 105: 11827–11832

Chapter 28

The *i*PhyClassifier, an Interactive Online Tool for Phytoplasma Classification and Taxonomic Assignment

Yan Zhao, Wei Wei, Ing-Ming Lee, Jonathan Shao, Xiaobing Suo, and Robert E. Davis

Abstract

The *i*PhyClassifier is an internet-based research tool for quick identification and classification of diverse phytoplasmas. The *i*PhyClassifier simulates laboratory restriction enzyme digestions and subsequent gel electrophoresis and generates virtual restriction fragment length polymorphism (RFLP) patterns. Based on RFLP pattern similarity coefficient scores, the *i*PhyClassifier gives instant suggestions on group and subgroup classification status of the phytoplasma strains under study. The *i*PhyClassifier also aligns the query sequences with that of reference strains of all previously described '*Candidatus* Phytoplasma' species, calculates sequence similarity scores, and assigns the phytoplasmas under study into respective '*Ca.* Phytoplasma' species as related strains according to the guidelines set forth by the Phytoplasma Taxonomy Group of the International Research Program on Comparative Mycoplasmology. Additional functions of the *i*PhyClassifier include delineation of potentially new phytoplasma groups and subgroups as well as new '*Ca.* Phytoplasma' species. This chapter describes the program components, the operational procedure, and the underlying principles of the *i*PhyClassifier operation. The chapter also provides hints on how to interpret the results.

Key words: '*Candidatus* Phytoplasma' species assignment, *i*PhyClassifier, Phytoplasma classification, Restriction fragment length polymorphism, Similarity coefficient

1. Introduction

Despite numerous efforts, attempts to cultivate phytoplasmas in cell-free media have not been successful; therefore, for phytoplasmas, measurable phenotypic characters suitable for conventional microbial characterization and classification remain essentially inaccessible. Owing to advances in molecular biology and DNA amplification

Matt Dickinson and Jennifer Hodgetts (eds.), *Phytoplasma: Methods and Protocols*, Methods in Molecular Biology, vol. 938, DOI 10.1007/978-1-62703-089-2_28, © Springer Science+Business Media, LLC 2013

technology, phytoplasmal genes with different degrees of nucleotide sequence conservation can be readily amplified from small amounts of DNA extracted from samples of infected materials (Chapters 14, 15, 16, and 17 of this book). As a result, genetic relatedness and phylogenetic relationships of diverse phytoplasmas can be determined through molecular analyses of conserved genes, forming a basis for phytoplasma molecular taxonomy and classification. Genes encoding 16S ribosomal RNAs (rRNAs) are highly conserved in nucleotide sequence across the phytoplasma clade, yet contain ample information for differentiation of a wide array of phytoplasma strains. Therefore, 16S rRNA genes have served as the primary character for phytoplasma molecular taxonomy and classification. The genus level taxon '*Candidatus* Phytoplasma' and existing phytoplasma classification schemes were established based on 16S rRNA gene sequences. Currently, there are two widely accepted phytoplasma classification schemes. One is based on phylogenetic analysis of 16S rRNA gene sequences (1–5) and another is based on restriction fragment length polymorphism (RFLP) analysis of a 1.2 kb polymerase chain reaction (PCR)-amplified 16S rRNA gene fragment (6–9). While both schemes can reliably classify diverse phytoplasmas into groups, the latter offers a mechanism, by distinguishing subtle RFLP pattern differences, to identify and differentiate distinct subgroup lineages among phytoplasmas within individual groups. With periodic update and recent development of computer-simulated RFLP analysis, the PCR-RFLP based classification scheme has delineated 31 phytoplasma groups and more than 100 subgroups (10–17). Typically, PCR-RFLP analysis of phytoplasmal 16S rRNA segments is done in the absence of complete nucleotide sequence information. Currently, as sequence information has become readily available, does RFLP analysis still remain a useful tool for phytoplasma identification and classification? We believe the answer is positive. First, the established phytoplasma 16S rRNA RFLP patterns have served as standard keys for phytoplasma strain identification and classification and have become authoritative expositions for research in the field. Second, although pairwise sequence comparison and phylogenetic analysis can be used to assess the relationships among phytoplasma strains, the outcomes (percentage sequence similarity or tree topology) do not directly reveal "visible" markers provided by RFLP analysis. While RFLP analysis remains a valuable tool for studying phytoplasma diversity and classification, the method by which RFLP analysis is carried out has evolved. The introduction of virtual RFLP analysis represents a technological advance in phytoplasma identification and classification. The *i*PhyClassifier brings the 16S rRNA gene sequence-based phytoplasma research toolbox to the internet. The group/subgroup classification function of the *i*PhyClassifier was devised according to the PCR-RFLP analysis based phytoplasma classification scheme (8,14), and the taxonomic assignment function was designed following the guidelines set by the International Committee of Systematic Bacteriology Subcommittee for the Taxonomy of

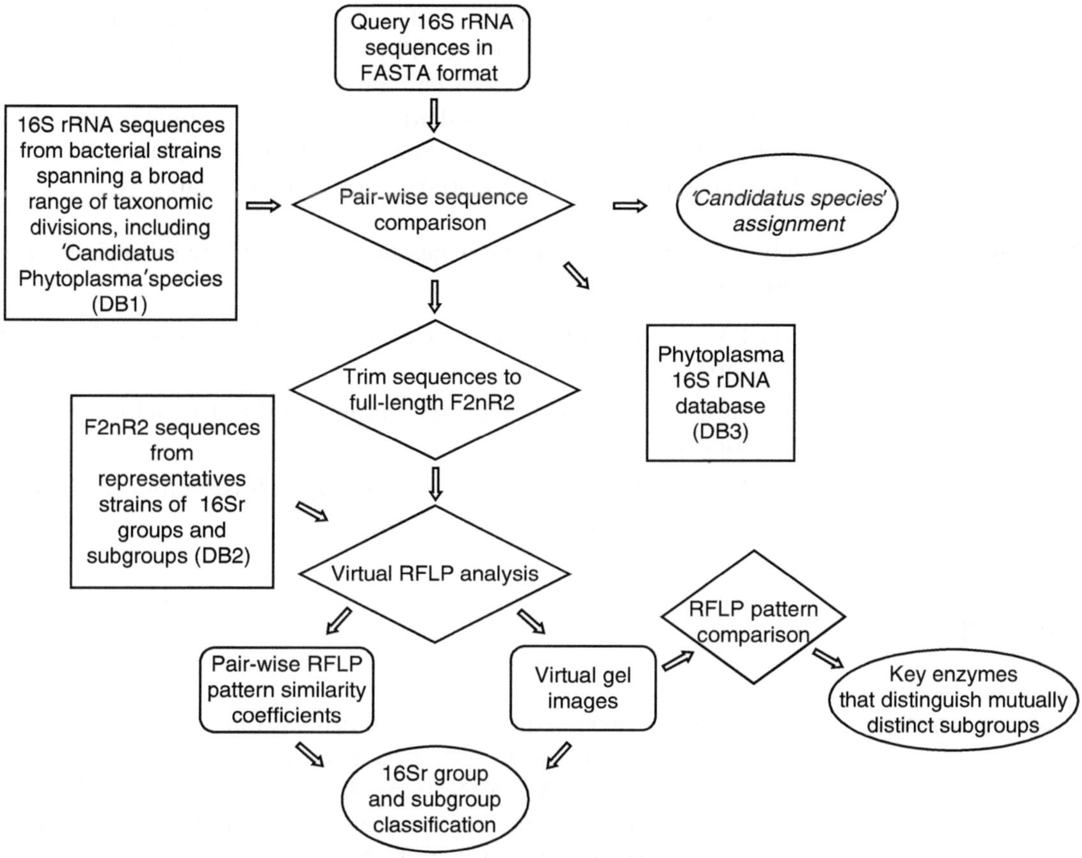

Fig. 1. Diagrammatic representation of the operational process of the *iPhyClassifier*. *Rectangles* represent input and output files; *squares* represent databases; *diamonds* represent computational operations; and *ovals* represent recommendations on tentative '*Candidatus* Phytoplasma' species assignment, 16Sr group/subgroup classification status, and key restriction enzymes that distinguish mutually distinct phytoplasma subgroup lineages. DB1, a set of full- or near-full-length 16S rRNA sequences from reference strains of all formally described '*Ca.* Phytoplasma' species, reference strains of proposed (11,18) but yet to be formally described '*Ca.* Phytoplasma' species, and type strains of other bacterial species spanning a broad range of taxonomic divisions; DB2, a set of F2nR2 sequences from representative strains of all previously delineated phytoplasma 16Sr groups and subgroups; and DB3, a set of F2nR2 sequences compiled from all phytoplasma 16S rRNA sequences currently deposited in the GenBank.

Mollicutes, the International Research Program for Comparative Mycoplasmology, Phytoplasma/Spiroplasma Working Team—Phytoplasma Taxonomy Group (18) (see Note 1). The complete operational process of the *i*PhyClassifier is outlined in Fig. 1.

2. Materials

2.1. System Requirement and Point of Access

The *i*PhyClassifier can be operated from any computer with an internet connection regardless of its platform (PC, Mac, Unix, or Linux) and underlying operating system. No special hardware or software other than a web browser is required. Although all web browsers including Mozilla Firefox, Internet Explorer, and Google Chrome work well with the *i*PhyClassifier, virtual gels are best viewed using Mozilla Firefox. The *i*PhyClassifier can also be operated from mobile devices such as iPad, iPhone, and Google phone. The web address of the *i*PhyClassifier server is http://plantpathology.ba.ars.usda.gov/cgi-bin/resource/iphyclassifier.cgi.

2.2. Program Components

The current version of *i*PhyClassifier comprises the following three program modules:

1. PM1: sequence similarity search and similarity score calculation module (see Note 2).
2. PM2: intelligent sequence trimming and virtual RFLP analysis module (see Note 3).
3. PM3: virtual electrophoretic gel image plotting module (PM3) (see Note 4).

3. Methods

3.1. Query Sequence Input

1. An ideal query sequence is a full or near full-length 16S rRNA. As a minimum, it should encompass the entire F2nR2 region of the 16S rRNA (see Note 5). Query sequences should contain no ambiguous base symbols and should be compiled in FASTA format (19) (see Note 6).
2. The query sequence input panel is located at the left side of the *i*PhyClassifier web page.
3. A query sequence can either be uploaded as a precompiled file from the user's computer to the *i*PhyClassifier web server or directly typed or pasted into the query sequence input window. Multiple query sequences can be entered at once in Multi-FASTA format (a concatenated list of sequences in FASTA format).

3.2. Function Selection and Query Submission

1. The function selection and query submission panel is located at the right side of the *i*PhyClassifier web page.
2. Select one or more functions in a single operation by checking the box(es) to the left of the desired function(s). The current version of the *i*PhyClassifier can perform the following functions:

   - '*Candidatus* Phytoplasma' species assignment.
   - 16Sr group/subgroup classification.
   - RFLP similarity coefficient calculation.
   - Virtual gel image generation.
   - RFLP pattern comparison.

3. For RFLP similarity coefficient calculation, users may define the maximum length (0, 1, 2, 3, 4, 5, or 6 nucleotides) by which two DNA fragments can differ yet still be considered as representing the same size during pattern comparison. The default deviation value is five nucleotides.
4. Users can specify which restriction enzyme patterns and phytoplasma group are to be compared with the query strain(s) by using the drop-down menus next to the RFLP pattern comparison function (see Note 7).
5. After selecting desired functions and appropriate parameters, click the "submit query" button to start the *i*PhyClassifier operation.

3.3. Program Output and Interpretation of Results

3.3.1. "Candidatus Phytoplasma" Species Assignment

1. The first step of the 'Ca. Phytoplasma' species assignment operation is to perform pairwise comparisons between each query sequence and the sequences in the database DB1 (see Note 8) for rapid identification of the query's phylogenetically closest neighbors, using the BLAST algorithm (20).
2. If none of the '*Ca*. phytoplasma species' in DB1 appears among the top 50 hits returned from the BLAST search, or one or more '*Ca*. phytoplasma species' appear(s) among the top hits but share(s) ≤91% sequence similarity with the query, the operation will abort, warning that the query sequence may not be of phytoplasmal origin.
3. If at least one '*Ca*. phytoplasma species' is among the top hits returned from the BLAST search and shares ≥92% sequence similarity with the query, the operation will proceed and the query sequence will be fed into the cLUSTALw program (21) for global alignment with all phytoplasma sequences in the database DB1Phy (a subset of DB1) and for sequence similarity score calculation (see Note 9).
4. Depending on the top sequence similarity score, the following examples illustrate types of output statement.

Scenario 1: *the sequence similarity score between the query and the top hit is 100%*. The query 16S rRNA sequence is identical to that of '*Ca*. Phytoplasma name' reference strain (GenBank accession number). The phytoplasma under study is either the '*Ca*. Phytoplasma name' reference strain or a closely related strain that cannot be distinguished by 16S rRNA sequence alone.

Scenario 2: *the sequence similarity score between the query and the top hit is >97.5% but <100%*. The query 16S rRNA sequence shares similarity with that of the '*Ca*. Phytoplasma name' reference strain (GenBank accession number). The phytoplasma under study is a '*Ca*. Phytoplasma name'-related strain.

Scenario 3: *the sequence similarity score between the query and the top hit is >91% but equal to or less than 97.5%*. The query 16S rRNA sequence shares less than 97.5% similarity with that of any previously recognized '*Ca.* Phytoplasma' species reference strain. The most closely related phytoplasma is '*Ca.* Phytoplasma name' (GenBank accession number), whose 16S rRNA sequence shares (n)% similarity with the query. The phytoplasma under study may therefore represent a new '*Ca.* Phytoplasma' species (18).

3.3.2. 16Sr Group/Subgroup Classification

According to conventions (see Note 10) and depending on the similarity coefficient score, the following examples illustrate types of output statement.

Scenario 1: *the similarity coefficient between the query and the top hit equals 1.00*. The virtual RFLP pattern derived from the query 16S rRNA F2nR2 fragment is identical to the reference pattern of 16Sr group (group letter, e.g., 16SrI), subgroup (subgroup letter, e.g., A) (GenBank accession number). The phytoplasma under study is a member of (e.g., 16SrI-A).

Scenario 2: *the similarity coefficient between the query and the top hit is >0.97 but <1.00*. The virtual RFLP pattern derived from the query 16S rRNA F2nR2 fragment is most similar to the reference pattern of the 16Sr group (group letter, e.g., 16SrII), subgroup (subgroup letter, e.g., B) (GenBank accession number), with a pattern similarity coefficient value of (e.g., 98%). The phytoplasma under study is a variant of (e.g., 16SrII-B).

Scenario 3: *the similarity coefficient between the query and the top hit is >0.85 but equal to or less than 0.97*. The virtual RFLP pattern derived from the query 16S rRNA F2nR2 fragment is different from the reference patterns of all previously established 16Sr groups/subgroups. The most similar is the reference pattern of 16Sr group (group letter, e.g., 16SrIII), subgroup (subgroup letter, e.g., C) (GenBank accession number), with a similarity coefficient value of (e.g., 95%). This strain may represent a new subgroup within the 16Sr group (16SrIII in this example). It is recommended that the user compare virtual RFLP patterns and identify the key enzymes that distinguish the new subgroup pattern from previously recognized group/subgroup patterns. An actual laboratory restriction digestion with the key enzyme(s) may help confirm the new subgroup pattern.

Scenario 4: *the similarity coefficient between the query and the top hit is equal to or less than 0.85*. The virtual RFLP pattern derived from the query 16S rRNA F2nR2 fragment is different from the reference patterns of all previously established 16Sr groups/subgroups. The most similar is the reference pattern of the 16Sr group (group letter), subgroup (subgroup letter) (GenBank accession number), with a similarity coefficient value of (e.g., 0.74). This strain may represent a new group. It is recommended that the user compare virtual RFLP patterns and identify the key enzymes that distinguish the new subgroup pattern from previously recognized group/subgroup patterns. An actual laboratory restriction digestion with the key enzyme(s) may help confirm the new group pattern.

4. Notes

1. It is well known that many phytoplasmas have two ribosomal RNA operons, *rrn*A and *rrn*B. While the two *rrn* operons may be identical in some phytoplasma strains, apparent interoperon sequence heterogeneity exists in other strains (22–27). Sequence variations between heterogeneous *rrn* operons may affect restriction sites in the 16S rRNA gene F2nR2 region and, as a result, a phytoplasma may exhibit two different virtual 16Sr RFLP pattern types. It is therefore important to distinguish between subgroup pattern type(s) and final subgroup designation, and to avoid erroneous assignment of the same strain into two different 16Sr subgroups. In this regard, the *i*PhyClassifier uses a three-letter subgroup designation (12), where the first and second letters (in parenthesis) denote the RFLP pattern types of *rrnA* and *rrnB*, respectively, and the third letter designates the 16Sr subgroup. For example, paulownia witches'-broom (PaWB) phytoplasma, a member of the previously delineated subgroup 16SrI-D (7), possesses two sequence-heterogeneous rRNA operons displaying two different 16Sr RFLP pattern types, characteristic of subgroups 16SrI-B and 16SrI-D, respectively; therefore, the subgroup status of PaWB was redesignated as 16SrI-(B/D)D (12). Similarly, classification of clover phyllody (CPh) phytoplasma, a member of subgroup 16SrI-C, was recently redesignated as 16SI-(R/S)C (16). The *i*PhyClassifier is currently being modified to include superimposed subgroup patterns for those phytoplasma strains whose *rrnA* and *rrnB* exhibit two different F2nR2 RFLP pattern types.
2. PM1 compares a query (input) sequence with 16S rRNA sequences from bacterial strains spanning a broad range of taxonomic divisions, and determines whether or not the query sequence is of phytoplasmal origin. If the query sequence is from a phytoplasma, the program will align the query sequence with the 16S rRNA sequences from the reference strains of all previously described '*Ca.* Phytoplasma' species, calculate sequence similarity scores and identify the closest phylogenetic neighbor of the phytoplasma under study.
3. PM2 automatically trims a query phytoplasma 16S rRNA sequence to the full F2nR2 region, performs computer-simulated

RFLP analysis based on the recognition sequences of a set of 17 restriction enzymes (*Alu*I, *Bam*HI, *Bfa*I, *Bst*UI, *Dra*I, *Eco*RI, *Hae*III, *Hha*I, *Hinf*I, *Hpa*I, *Hpa*II, *Kpn*I, *Sau*3AI, *Mse*I, *Rsa*I, *Ssp*I, and *Taq*I), compares the resulting RFLP profiles with that of the reference strain of each previously delineated phytoplasma group/subgroup, and calculates pairwise RFLP pattern similarity coefficients according to the following formula (7,28):

$$F = 2N_{xy} / (N_x + N_y),$$

in which *x* and *y* are two given strains under investigation, *Nx* and *Ny* are the total number of DNA fragments (bands) resulting from digestions by 17 enzymes in strains *x* and *y*, respectively, and *Nxy* is the number of fragments shared by the two strains. The similarity coefficient values are used for determination of the group/subgroup affiliation status of the phytoplasma under study and for delineation of new phytoplasma groups and subgroups.

4. PM3 mimics laboratory agarose gel electrophoretic separation of DNA fragments and generates virtual gel images resulting from *in silico* digestions of the query 16S rRNA sequence and the 16S rRNA sequences from the reference strains of previously delineated subgroups.
5. Typically, the F2nR2 region is about 1.25 kb in length and is bounded by the two conserved sequence blocks corresponding to the annealing sites for phytoplasma-universal 16S rRNA primer pair R16F2n (5′-GAAACGACTGCTAAGACTGG-3′) and R16R2 (5′-TGACGGGCGGTGTGTACAAACCCCG-3′) (29).
6. Because the operation of the *i*PhyClassifier is solely dependent upon sequence information, any error in a query sequence which misrepresents the phytoplasma strain under study could result in erroneous group/subgroup classification and '*Candidatus* species' assignment. While sequence errors may arise at various stages including PCR amplification and DNA sequencing, they usually occur randomly and can be minimized by use of sample replications. To ensure credible results, we highly recommend that consistent sequence data from at least two independent samples, i.e., from two or more infected plants or insect individuals, be obtained. If only one infected plant or insect sample is available for study, mutually consistent sequence data from at least two separate PCRs must be obtained.
7. Both virtual gel image generation function and RFLP pattern comparison function produce images mimicking agarose gel electrophoresis results from PCR-RFLP analysis. While the first function generates gel images resulting from *in silico* digestions of individual query sequence(s) by all 17 individual enzymes, the second function compares gel images resulting

from *in silico* digestions of query sequence(s) and reference sequences from previously delineated group/subgroup representatives. The latter helps identify key restriction enzymes that distinguish different group and subgroup patterns.

8. DB1 is a database that contains full- or near-full-length 16S rRNA from reference strains of previously described '*Ca.* Phytoplasma' species, reference strains of other proposed (11,18) but yet to be formally described '*Ca.* Phytoplasma' species, and all type strains of other named prokaryotic species.
9. In accordance with the convention on 16S rRNA gene sequence-based prokaryotic species delineation (30,31), *i*PhyClassifier implements the guidelines of the IRPCM Phytoplasma Taxonomy Group (18) and presets 97.5% 16S rRNA gene sequence similarity as the cutoff value for new '*Candidatus* Phytoplasma' species recognition.
10. The phytoplasma group/subgroup classification scheme implemented in the *i*PhyClassifier is based on RFLP analysis of a defined set of 17 restriction enzymes on the F2nR2 fragment of phytoplasmal 16S rRNA (7,12). If a query sequence does not fully encompass the entire F2nR2 region, the *i*PhyClassifier operation will produce inaccurate RFLP patterns and wrong classification. A similarity coefficient value of 0.97 was defined as the threshold point for delineation of a new subgroup RFLP pattern type within a given group. Thus, if the virtual RFLP pattern derived from a 16S rRNA F2nR2 fragment of a phytoplasma strain under study has a similarity coefficient of 0.97 or lower compared to all subgroup reference strains in a given group, a new subgroup pattern type is recognized. On the other hand, in agreement with all previously designated groups, a similarity coefficient value of 0.85 or less with all previously recognized subgroups signals a possible new 16Sr group. RFLP patterns that have a similarity coefficient of 0.99 or 0.98 to the reference pattern type of the designated representative member in a given subgroup are considered as variants of the standard pattern type.

References

1. Kuske CR, Kirkpatrick BC (1992) Phylogenetic relationships between the western aster yellows mycoplasmalike organism and other prokaryotes established by 16S rRNA gene sequence. Int J Syst Bacteriol 42:226–233
2. Namba S et al (1993) Phylogenetic diversity of phytopathogenic mycoplasmalike organisms. Int J Syst Bacteriol 43:461–467
3. Gundersen DE et al (1994) Phylogeny of mycoplasmalike organisms (phytoplasmas): a basis for their classification. J Bacteriol 176:5244–5254
4. Schneider B et al (1995) Phylogenetic classification of plant pathogenic mycoplasma-like organisms or phytoplasmas. In: Razin R, Tully JG (eds) Molecular and diagnostic procedures in mycoplasmology. Academic, San Diego, pp 369–380
5. Seemüller E et al (1998) Current status of molecular classification of the phytoplasmas. J Plant Pathol 80:3–26

6. Lee I-M et al (1993) Universal amplification and analysis of pathogen 16S rDNA for classification and identification of mycoplasma-like organisms. Phytopathology 83:834–842
7. Lee I-M et al (1998) Revised classification scheme of phytoplasmas based on RFLP analyses of 16S rRNA and ribosomal protein gene sequences. Int J Syst Bacteriol 48:1153–1169
8. Lee I-M, Davis RE, Gundersen-Rindal DE (2000) Phytoplasma: phytopathogenic mollicutes. Annu Rev Microbiol 54:221–255
9. Schneider B et al (1993) Classification of plant-pathogenic mycoplasma-like organisms using restriction-site analysis of PCR-amplified 16S rDNA. Microbiology 139:519–527
10. Lee I-M et al (2006) '*Candidatus* Phytoplasma americanum', a phytoplasma associated with a potato purple top wilt disease complex. Int J Syst Evol Microbiol 56:1593–1597
11. Wei W et al (2007) Computer-simulated RFLP analysis of 16S rRNA genes: identification of ten new phytoplasma groups. Int J Syst Evol Microbiol 57:1855–1867
12. Wei W et al (2008) Automated RFLP pattern comparison and similarity coefficient calculation for rapid delineation of new and distinct phytoplasma 16Sr subgroup lineages. Int J Syst Evol Microbiol 58:2368–2377
13. Zhao Y et al (2009) '*Candidatus* Phytoplasma tamaricis', a novel taxon discovered in witches'-broom-diseased salt cedar (*Tamarix chinensis* Lour.). Int J Syst Evol Microbiol 59:2496–2504
14. Zhao Y et al (2009) Construction of an interactive online phytoplasma classification tool, *i*PhyClassifier, and its application in analysis of the peach X-disease phytoplasma group (16SrIII). Int J Syst Evol Microbiol 59: 2582–2593
15. Jomantiene R et al (2010) Onion is host for two phytoplasma lineages, subgroups 16SrI-A and 16SrI-(B/L)L, in Lithuania: a *Hinf*I site revealed a SNP marking divergent branches of evolution. J Plant Pathol 92:461–470
16. Jomantiene R et al (2011) Phytoplasmas infecting sour cherry and lilac represent two distinct lineages having close evolutionary affinities with clover phyllody phytoplasma. Eur J Plant Pathol 130:97–107
17. Lee I-M et al (2011) '*Candidatus* Phytoplasma costaricanum', a new phytoplasma associated with a newly emerging disease in soybean in Costa Rica. Int J Syst Evol Microbiol. doi:10.1099/ijs.0.029041-0
18. The IRPCM Phytoplasma/Spiroplasma Working Team—Phytoplasma taxonomy group (2004) '*Candidatus* Phytoplasma', a taxon for the wall-less, non-helical prokaryotes that colonize plant phloem and insects. Int J Syst Evol Microbiol 54:1243–1255
19. Pearson WR, Lipman DJ (1988) Improved tools for biological sequence comparison. PNAS USA 85:2444–2448
20. Altschul SF et al (1990) Basic local alignment search tool. J Mol Biol 215:403–410
21. Thompson JD, Higgins DG, Gibson TJ (1994) CLUSTALW: improving the sensitivity of progressive multiple sequence alignment through sequence weighting, position-specific gap penalties and weight matrix choice. Nucl Acids Res 22:4673–4680
22. Schneider B, Seemüller E (1994) Presence of two sets of ribosomal genes in phytopathogenic mollicutes. Appl Environ Microbiol 60:3409–3412
23. Liefting LW et al (1996) Sequence heterogeneity in the two 16S rRNA genes of Phormium yellow leaf phytoplasma. Appl Environ Microbiol 62:3133–3139
24. Davis RE, Sinclair WA (1998) Phytoplasma identity and disease etiology. Phytopathology 88:1372–1376
25. Davis RE, Dally EL (2001) Nonfunctional tRNA gene in an unusual example of rRNA interoperon sequence heterogeneity in phytoplasma. Phytopathology 91:S21
26. Davis RE et al (2003) Differential amplification of sequence heterogeneous ribosomal RNA genes and classification of the '*Fragaria multicipita*' phytoplasma. Microbiol Res 158:229–236
27. Jomantiene R et al (2002) New group 16SrIII phytoplasma lineages in Lithuania exhibit interoperon sequence heterogeneity. Eur J Plant Pathol 108:507–517
28. Nei M, Li W-H (1979) Mathematical model for studying genetic variation in terms of restriction endonucleases. PNAS USA 76:5269–5273
29. Gundersen DE, Lee I-M (1996) Ultrasensitive detection of phytoplasmas by nested-PCR assays using two universal primer pairs. Phytopathol Mediterr 35:144–151
30. Murray RGE, Schleifer KH (1994) Taxonomic notes: a proposal for recording the properties of putative taxa of prokaryotes. Int J Syst Bacteriol 44:174–176
31. Stackebrandt E, Goebel BM (1994) Taxonomic note: a place for DNA-DNA reassociation and 16S rRNA sequence analysis in the present species definition in bacteriology. Int J Syst Bacteriol 44:846–849

Chapter 29

Phytoplasma Proteomic Analysis

Xianling Ji and Yingping Gai

Abstract

Proteome analysis is becoming a powerful tool in the functional characterization of organisms, and takes a broad, comprehensive, systematic approach to understanding biology. Following the sequencing of the phytoplasma genomes, the next step is to characterize the expressed proteome of phytoplasmas to acquire the verification and functional annotation of all predicted genes and their protein products. Here, we describe the protocol of mulberry dwarf phytoplasma purification, phytoplasma protein extraction and separation by SDS-PAGE, in-gel tryptic digestion of the proteins, separation of the digested peptides by liquid chromatography, and identification of the peptides by mass spectrometry. The protocol described here is also applicable to the analysis of other phytoplasma proteomes.

Key words: In-gel tryptic digestion, LC-MS/MS, SDS-PAGE, Shotgun proteomics

1. Introduction

Phytoplasmas have been associated with diseases in several 100 plant species (1). The inability to culture phytoplasmas in vitro has hindered their characterization at the molecular level (2). Following the sequencing of phytoplasma genomes, the next step is the verification and functional annotation of all predicted genes and their protein products. Therefore, it is necessary to characterize the expressed proteome of phytoplasmas. On a proteome scale, protein expression can be analyzed by high-throughput proteomic technologies. With the development of proteomics, more and more techniques such as two-dimensional polyacrylamide gel electrophoresis (2-DE), matrix-assisted laser desorption ionization-time of flight-mass spectrometry (MALDI-TOF-MS), liquid chromatography-tandem mass spectrometry (LC-MS/MS), and surface-enhanced laser desorption/ionization-time of flight-mass spectra (SELDI-TOF-MS) are used to profile proteins (3). Compared with traditional proteomic methods such as 2-DE,

Matt Dickinson and Jennifer Hodgetts (eds.), *Phytoplasma: Methods and Protocols*, Methods in Molecular Biology, vol. 938, DOI 10.1007/978-1-62703-089-2_29, © Springer Science+Business Media, LLC 2013

shotgun proteomics refers to the direct analysis of complex protein mixtures to rapidly generate a global profile of the protein complement within the mixture. Shotgun proteomics serves as a powerful tool to separate and identify proteins from complex protein mixtures and possesses the virtues of saving time and labor, and having high efficiency (4). Here, we performed LC-MS/MS to characterize the expressed proteome of mulberry dwarf (MD) phytoplasma using shotgun proteomics. The procedures for MD phytoplasma proteome analysis using shotgun proteomics comprise five steps: (1) phytoplasma purification; (2) phytoplasma protein extraction and separation by SDS polyacrylamide gel electrophoresis (SDS-PAGE); (3) in-gel tryptic digestion of the proteins; (4) separation of the digested peptides by liquid chromatography; and (5) identification of the peptides by mass spectrometry.

2. Materials

Prepare all solutions using ultrapure water (prepared by purifying deionized water to attain a conductivity of 18 MΩ cm at 25°C). Use HPLC-grade solvents and highest possible grade reagents. Prepare and store all reagents at room temperature (unless indicated otherwise).

2.1. Mulberry Dwarf Phytoplasma Purification

1. Phosphate-buffered saline (PBS): 137 mM NaCl, 2.7 mM KCl, 4.3 mM Na_2HPO_4, 1.4 mM KH_2PO_4, pH 7.0. Add about 800 mL water to a 1-L glass beaker. Weigh 8 g NaCl, 0.2 g KCl, 1.54 g $Na_2HPO_4 \cdot 12H_2O$, and 0.19 g KH_2PO_4. Mix and adjust pH to 7.0 with HCl. Make up to 1 L with water. Autoclave and store at 4°C.
2. Isolation buffer: 0.3 M glycine, 10 mM $MgCl_2$, 80 U gentamicin, pH 7.4 (see Note 1).
3. Enzyme solution: Dissolve 2.0 g cellulase, 0.8 g hemicellulase and 3.0 g tartaric acid enzyme in 1 L isolation buffer (see Note 2). Sterilize by filtering through a 0.20 μm pore size filter. Store at 4°C.
4. Suspending medium: 0.3 M glycine, 10 mM $MgCl_2$, 0.3 M mannitol, 20 mM Na_2SO_4, 80 U gentamicin, pH 7.4. Store at 4°C.
5. Stock isotonic Percoll (SIP) solution: Add 9 parts (v/v) of Percoll to 1 part (v/v) of sucrose (2.5 M) in water (see Note 3).
6. Dilution buffer: 0.25 M sucrose, 10 mM MOPS pH 7.0.
7. Suspending medium: 0.3 M mannitol, 20 mM MOPS pH 7.0. Store at 4°C.

8. Lysis buffer: 30 mM Tris, 2 M thiourea, 7 M urea, 4% (w/v) CHAPS, 5% (v/v) β-mercaptoethanol (see Note 4).
9. 2-D Quant kit (GE Healthcare Life Sciences).
10. 100-mesh (100 mesh/in.) screen.
11. Cheesecloth (sterilized).

2.2. SDS Polyacrylamide Gel Electrophoresis

1. Acrylamide solution: 30% (w/v) acrylamide, 0.8% (w/v) bis-acrylamide. Weigh 30 g of acrylamide monomer and 0.8 g bis-acrylamide, and transfer to a 100 mL graduated cylinder containing about 40 mL of water. Mix for about 30 min with magnetic stirring apparatus. Make up to 100 mL with water and pass through filter paper. Store at 4°C, in a bottle wrapped with aluminum foil.
2. Resolving gel buffer: 1.5 M Tris–HCl, pH 8.8. Add about 800 mL water to a 1-L glass beaker. Weigh 181.7 g Tris and transfer to the beaker. Mix and adjust pH to 8.8 with HCl. Make up to 1 L with water. Store at 4°C.
3. Stacking gel buffer: 1.0 M Tris–HCl, pH 6.8. Weigh 121.1 g Tris and prepare a 1 L solution as in **step 2**. Store at 4°C.
4. 10% (w/v) Sodium dodecyl sulfate (SDS): Add about 80 mL water to a 200-mL glass beaker, and weigh 10 g SDS. Mix and adjust pH to 7.2 with HCl. Adjust volume to 100 mL with water.
5. Ammonium persulfate: 10% (w/v) solution in water (see Note 5).
6. *N*, *N*, *N'*, *N'*-tetramethyl-ethylenediamine (TEMED), store at 4°C.
7. Isobutanol.
8. SDS-PAGE running buffer: 25 mM Tris, 0.192 M glycine, 0.1% SDS. Weigh 3.03 g Tris and 14.4 g glycine, mix, add 10 mL of 10% SDS solution, and make up to 1 L with water.
9. 5 × SDS-PAGE loading buffer: 10% (w/v) SDS, 250 mM Tris–HCl (pH 6.8), 0.5% (w/v) bromophenol blue, 50% (v/v) glycerol, 5% (v/v) β-mercaptoethanol.
10. Coomassie blue staining buffer: 0.1% (w/v) Coomassie blue R-250, 40% (v/v) methanol, 10% (v/v) acetic acid.
11. Destaining solution: 5% (v/v) ethanol, 10% (v/v) acetic acid.

2.3. In-Gel Trypsin Digestion

Prepare freshly immediately before the digestion process.

1. 25 mM NH_4HCO_3 in 50% (v/v) acetonitrile: Weigh 98.75 mg NH_4HCO_3, add 25 mL of acetonitrile, and make up to 50 mL with water.
2. 25 mM NH_4HCO_3 in water: Weigh 98.75 mg NH_4HCO_3 and make up to 50 mL with water.

3. 10 mM dithiothreitol (DTT) in 25 mM NH_4HCO_3: Weigh 7.71 mg DTT and make up to 5 mL with 25 mM NH_4HCO_3.
4. 55 mM iodoacetamide in 25 mM NH_4HCO_3: Weigh 50.86 mg iodoacetamide and make up to 5 mL with 25 mM NH_4HCO_3.
5. Trypsin solution: 12.5 μg/mL trypsin in 25 mM NH_4HCO_3.
6. 5% (v/v) Trifluoroacetic acid in 50% (v/v) acetonitrile.

2.4. LC-MS/MS

Degas all buffers prior to use.

1. Calibration solutions: The calibration mixture consists of caffeine, MRFA (methionine–arginine–phenylalanine–alanine), and Ultramark 1621 (Thermo Finnigan) dissolved in water/methanol/acetonitrile/acetic acid (25:25:49:1, v/v/v/v) solution. To prepare the calibration solution, follow the manufacturers' instructions (see Note 6).
2. Bovine serum albumin (BSA) tryptic digest standard: BSA digest standards (Protea Biosciences, Inc, USA) are diluted to 10 fmol in 0.1% (v/v) formic acid in water prior to use following the manufacturers' instructions.
3. Peptide solvent solution: water/acetonitrile (95:5, v/v) containing 0.2% (v/v) formic acid. 5 mL acetonitrile and 200 μL formic acid in 95 mL water (LC-MS grade water).
4. Mobile-phase solution A: 0.1% (v/v) formic acid. Add 100 μL formic acid to 100 mL water (LC-MS grade water).
5. Mobile-phase solution B: 0.1% (v/v) formic acid in 84% (v/v) acetonitrile. Add 100 μL formic acid to 84 mL acetonitrile, make up to 100 mL with water (LC-MS grade water).

2.5. Data Processing and Analysis

1. BioWorks software v.3.2.
2. UniProtKB database, http://www.uniprot.org/.
3. Build summary software, v.2.6.5, http://www.proteomics.ac.cn/software/proteomicstools/index.htm.

3. Methods

Carry out all procedures at room temperature unless otherwise specified.

3.1. Mulberry Dwarf Phytoplasma Purification

1. Prepare fresh soft twig tissue from the diseased mulberry plants (see Note 7). Cut 200 g of fresh tissue into pieces, and macerate the pieces in 500 mL cold PBS with a food processor at low speed for 1 min.

2. Rinse the macerated tissue through a 100-mesh (100 mesh/in.) screen, and then squeeze through three layers of cheesecloth. Retain the vascular tissues (see Note 8).
3. Grind the retained vascular tissues in 200 mL of isolation buffer with a mortar and pestle until no vascular fibers are discernible (see Note 9).
4. Mix the sap with 200 mL of enzyme solution and maintain at 35°C for 2 h.
5. Squeeze the sap through three layers of sterilized cheesecloth, and collect the filtrate.
6. Centrifuge at 12,000 × *g* for 1 h at 4°C, collect the precipitate and resuspend with 50 mL of suspending medium.
7. Centrifuge at 5,000 × *g* for 15 min at 4°C, and decant the supernatant to a new tube.
8. Centrifuge at 12,000 × *g* for 1 h at 4°C, resuspend the pellet in 5 mL of suspending medium and mix using a pipette or vortex mixer.
9. Centrifuge at 2,000 × *g* for 5 min at 4°C, and decant supernatant (preparation) to a new tube.
10. Dilute the SIP solution with dilution buffer to give 15, 30, and 50% Percoll solutions.
11. Prepare step gradients as follows: 15% (11 mL), 30% (11 mL), and 50% (6 mL). Layer the solutions into 25 × 89 mm ultraclear centrifuge tubes successively with the densest layer at the bottom and the least dense layer at the top (see Note 10).
12. Load 0.8 mL of preparation onto the top of the discontinuous gradient (see Note 11).
13. Centrifuge at 20,000 × *g* for 20 min at 4°C (see Note 12).
14. Collect the fraction near the interface between the 30 and 50% Percoll layers.
15. Dilute the collected fraction immediately with suspending medium and centrifuge at 100,000 × *g* for 2 h at 4°C.
16. Collect the pellet that aggregated above the surface of the hard cushion of Percoll. Resuspend the pellet in 0.5 mL of lysis buffer, and keep it on ice for 10 min (see Note 13).
17. Sonicate the pellet fraction in lysis buffer at 20 kHz (Model VC750, Bibra Cell™ Sonics and Materials Inc.) at 30% amplitude, 2 s pulse-on, 2.5 s pulse-off, for a total of 5 min in an ice–water bath (see Note 14).
18. Centrifuge the lysates at 16,000 × *g* for 30 min at 20°C, and decant the supernatant to a new tube.
19. Quantify protein using the 2-D Quant kit, following the manufacturers' instructions. Freeze the protein samples immediately in liquid nitrogen and store at −80°C.

3.2. SDS Polyacrylamide Gel Electrophoresis

1. Prepare a 12% acrylamide resolving gel about 10 cm × 10 cm × 1.5 mm according to the recipe as follows: mix 8 mL of acrylamide solution, 0.2 mL of 10% SDS, 5 mL of 1.5 M Tris-Cl (pH 8.8) and 6.6 mL of distilled water in a glass beaker. Add 200 μL of 10% ammonium persulfate and 10 μL TEMED, and gently swirl.
2. Pour the acrylamide solution between the glass plates immediately and gently overlay with isobutanol. Allow the gel to polymerize for at least 1 h (see Note 15).
3. Rinse the isobutanol from the surface with copious amounts of water, and drain excess water.
4. Prepare a stacking gel using the following recipe: mix 0.83 mL of acrylamide solution, 50 μL of 10% SDS, 0.63 mL of 1.0 M Tris-Cl (pH 6.8), and 3.4 mL of distilled water in a glass beaker. Add 50 μL of 10% ammonium persulfate and mix. Add 5 μL TEMED and mix.
5. Pour the stacking gel solution between the glass plates and carefully insert a clean comb.
6. Leave the stacking gel to polymerize for 30–60 min, and then remove comb carefully (see Note 16).
7. Add 5 × SDS-PAGE loading buffer to aliquots of the prepared protein samples at a ratio of 1–5 μL sample. Heat at 95°C for at least 5 min. Centrifuge the heated samples at 10,000 × *g* for 30 s to bring down the condensate (see Note 17).
8. Load samples (50–60 μg protein/lane) and the protein standards (2 μg of markers/lane).
9. Electrophorese at 12 mA/gel until the sample has entered the resolving gel, and then continue at 18 mA/gel till the dye front has reached the bottom of the gel.
10. Following electrophoresis, gently pry apart the gel plates with a spatula. Rinse the gel with deionized water, and then immerse in the Coomassie blue staining buffer for 40–60 min.
11. Destain the gel in destaining solution with gentle agitation until its background becomes colorless and leaves protein bands colored blue, changing the destaining solution several times at 10-min intervals.
12. Scan the gel (see Note 18).

3.3. In-Gel Trypsin Digestion

1. Excise the stained protein bands from the polyacrylamide gel. Cut each entire lane into several slices based on band intensity, regardless of the Coomassie staining profile. Cut each slice into 1 × 1 mm pieces and place into a 1.5 mL microfuge tube (see Note 19).
2. Rinse the gel pieces with 400 μL of HPLC-grade water for 15 min at room temperature on a shaker.

3. Add 400 μL of 25 mM NH_4HCO_3 in 50% acetonitrile and incubate for 15 min at room temperature on a shaker. Discard the supernatant. Repeat this step once or twice (see Note 20).
4. Add 200 μL of acetonitrile for 5 min at room temperature on a shaker. Discard the supernatant. Dry the pieces in a Speedvac for 5 min.
5. Add 100 μL (or enough to cover) of 10 mM DTT in 25 mM NH_4HCO_3. Incubate at 60°C for 1 h. Discard the supernatant. Let the tube cool to room temperature.
6. Add the same volume of 55 mM fresh iodoacetamide in 25 mM NH_4HCO_3. Incubate in the dark at room temperature for 45 min with occasional vortex. Discard the supernatant.
7. Wash the gel pieces with 400 μL of 25 mM NH_4HCO_3 for 15 min at room temperature on a shaker. Discard the supernatant.
8. Wash the gel pieces with 400 μL of 25 mM NH_4HCO_3 in 50% acetonitrile for 15 min at room temperature on a shaker. Discard the supernatant.
9. Add 300 μL of acetonitrile for 5 min at room temperature on a shaker. Discard the supernatant.
10. Dry the gels in a Speedvac for 5 min.
11. Add 20–50 μL (or enough for gel to absorb) of 12.5 μg/mL of modified trypsin in 25 mM NH_4HCO_3 and incubate for 15 min at room temperature until the trypsin solution has been absorbed by the gel pieces, remove and discard excess solution (see Note 21).
12. Add a small amount of 25 mM NH_4HCO_3 without trypsin, aiming to use the minimum volume possible while still keeping the gel pieces immersed throughout the digestion time. Incubate at 37°C for 12–16 h.
13. Extract the gel pieces twice with 50 μL (or enough to cover) of 50% acetonitrile/5% trifluoroacetic acid for 60 min each time at room temperature. Repeat this step once. Pool two extracts and dry in a Speedvac.

3.4. LC-MS/MS

Peptide desalting and reverse phase separation of peptides is performed using an Ultimate 3000 Nano/Capillary LC System or similar system. The LC system is coupled to a linear trap quadrupole (LTQ) mass spectrometer or similar system. We use the following method as described with some modifications (5–7).

1. Operate the LTQ FT mass spectrometer in standard data dependent acquisition mode controlled by Xcalibur 1.4 software.
2. Acquire survey scans (m/z range 400–2,000) on the FT-ICR instrument at a resolution of 100,000 at m/z 400 and one microscan per spectrum.

3. The top 2–5 most abundant multiply charged ions with a minimal intensity at 1,000 counts are subject to MS/MS in the linear ion trap at an isolation width of 2 Thomson.
4. Perform precursor activation with an activation time of 30 ms and set the activation Q at 0.25 and the normalized collision energy to 35%. Set the dynamic exclusion width at ±5 ppm with two repeats and a duration of 30 s.
5. Regulate the automatic gain control target value at $5e^5$ for FT and $1e^4$ for the ion trap, with maximum injection time at 1,000 ms for FT, and 200 ms for the ion trap, respectively.
6. Calibrate the instrument using calibration solutions according to the manufacturers' instructions.
7. Check the performance of the instruments with multiple LC-MS/MS analyses of BSA tryptic digest standard.
8. Redissolve peptide mixtures with 30 μL of peptide solvent solution, and centrifuge at 10,000 × *g* for 5 min.
9. Equilibrate the RP column for 10 min with solution A at 200 μL/min before loading the sample.
10. Load 30 μL of each peptide sample using with the autosampler and desalt on a trap column with a flow of 25 μL/min of solution A for 5 min.
11. Switch the valve to make the trap column online with the analytical column.
12. Start MS acquisition.
13. Start a gradient elution of the peptides on the analytical column at a flow rate of 300 nL/min and separated using the following gradient: (1) from 10 to 40% solution B for 40 min, (2) from 40 to 90% solution B for 10 min, and (3) the remaining percentage of the elution solution is made of solution A (see Note 22).

3.5. Data Processing and Analysis

1. Process MS data using BioWorks v.3.2 (Thermo Electron) to give a peak list files and submit to a Mascot v2.0 (Matrix Science) server for iterative searching against the protein database from UniProtKB (http://www.uniprot.org/) with keyword "Phytoplasma" for peptide and protein identifications.
2. Perform the searches using the "trypsin enzyme" parameter of the software. Methionine oxidation is specified as a variable modification and cysteine carboxamido methylation is specified as the fixed modification.
3. Carry out the peptide database search with the following parameters: parent peptide ion mass-to-charge variation 10 ppm, fragment ion mass variation 0.7 Da, maximum missed cleavage site of trypsin 1.

4. Filter the SEQUEST results using the following parameters:

   Xcorr > 1.55 and ΔCn > 0.15 for the peptides in +1 charged state.

   Xcorr > 1.85 and ΔCn > 0.16 for the peptides in +2 charged state.

   Xcorr > 2.6 and ΔCn > 0.22 for the peptides in +3 charged state.

5. Combine all output results using in-house Buildsummary software (Version: 2.6.5, http://www.proteomics.ac.cn/software/proteomicstools/index.htm).

4. Notes

1. Gentamicin is used to prevent bacterial infections. Other antibiotics also can be used.
2. These enzymes may be substituted by snailase, also known as helicase, this being a mixture of more than 30 enzymes, including cellulose, hemicellulase, seminase, etc. extracted from the crop and digestive tract of snails.
3. Percoll is supplied sterile and can be resterilized by autoclaving at 120°C for 30 min without any change in properties. Autoclaving of Percoll solutions must be carried out in the absence of salts or sucrose. When autoclaving undiluted Percoll, it is recommended that contact with air be minimized to avoid particle aggregation at the Percoll/air interface. This can be accomplished by using a narrow-necked bottle when autoclaving. If particles form, they may be removed by low speed centrifugation. If any significant evaporation occurs during autoclaving, the volume should be replenished with sterile water so that the density is not affected.
4. Other detergents such as Triton X-100, NP-40 and other nonionic or zwitterionic detergents can be used instead of CHAPS.
5. It is best to prepare this fresh each time.
6. Use only glass pipettes or stainless steel syringes when measuring glacial acetic acid. Using plastic pipette tips causes contamination of the acid stock solution that can introduce contaminants into the calibration solution.
7. The diseased plants should exhibit characteristic symptoms, such as small leaves, shortening of internodes, stunting of plants, excessive proliferation of shoots, formation of witches' broom, etc.

8. In plants, phytoplasmas remain restricted to the phloem tissue, so the materials used should be the vascular tissues of twigs.
9. Grinding should be done gently and slowly, to avoid destroying to the phytoplasmas.
10. The gradient is prepared by layering progressively less dense Percoll solutions upon one another; therefore the first solution applied is the 50% Percoll solution. A steady application of the solutions yields the most reproducible gradient therefore the setup described as follows is recommended. First, an ultra-clear centrifuge tube is held upright in a tube stand. Next, a 200 μL pipette tip is placed on the end of a 1,000 μL pipette tip. Both snugly fitting tips are held steady by a clamp stand and the end of the 200 μL tip is allowed to make contact with the inside wall of the tube. Now Percoll solution can be placed inside the 1,000 μL tip and gravity will feed the solutions into the tube slowly and steadily. If the solutions fail to feed down through the tips and into the tube, a small amount of positive air pressure can be applied to the 1,000 μL tip to start the flow. This is done by gently tapping on the wide end of the tip. Preformed gradients can be stored for weeks without a change in gradient shape, provided that the gradient is sterile and is not physically disturbed.
11. The tube should be handled and loaded into the rotor very carefully after the sample is applied to the top of the gradient. Centrifugation should begin as soon as possible after loading.
12. All centrifugation procedures require a balanced rotor, therefore another tube containing precisely the same mass must be generated. In practice this means two gradients must be prepared although the second gradient need not contain an experimental sample but could contain the same volume of water as the volume of sample.
13. This process should be done at as low a temperature as possible and with a minimum of heat generation.
14. In order to obtain maximal coverage of the proteome, it is essential to make sure that the proteins located in cellular membranes and nuclei are released and solubilized. This is easily solved by sonicating cell lysates as indicated.
15. This overlay prevents contact with atmospheric oxygen (which inhibits acrylamide polymerization) in addition to helping to level the resolving gel solution.
16. Try to prepare it about 18–24 h ahead of time to allow maximal polymerization.
17. Centrifuging the samples prior to the run helps remove insoluble components which could produce streaks in the protein lanes when stained.

18. Sandwich the gel between two pieces of acetate film washed in dH_2O, and put the sandwich on a scanner. Print out the scanned image and mark on it where bands were cut. Return the gel to the water-filled dish until ready to excise the bands.
19. Use a fresh scalpel to cut up each slice. Transfer the gel pieces to the tube using this same scalpel. Try to work as quickly as possible, because the gel becomes stickier as it dries out.
20. The gel pieces should shrink and look opaque in this step. If the band pieces are still blue, repeat this step once more.
21. Trypsin is not stable especially when pure and in nonacidic condition, so it should be diluted down from stock solution just before use. Keep the volume small: use *just* enough trypsin solution to cover all the gel pieces.
22. Due to the complexity of the sample, a 50-min or longer gradient is recommended to achieve a good separation of the peptides and minimize the problem of MS undersampling. Samples should be analyzed in duplicate and blank runs be employed to minimize the carryover between injections.

Acknowledgement

This work was funded by the National Natural Science Foundation of China and Science Foundation for the Excellent Youth Scholars of Shandong province of China (grant nos. 30972366, 31070573, 31100478, BS2009NY024, and BS2010NY015).

References

1. Garnier M et al (2001) Mycoplasmas, plants, insect vectors: a matrimonial triangle. C R Acad Sci III 324:923–928
2. Christensen NM et al (2005) Phytoplasmas and their interactions with hosts. Trends Plant Sci 10:526–535
3. Wu K, Zhang Y (2007) Clinical application of tear proteomics: present and future prospects. Proteomics Clin Appl 1:972–982
4. Wu SL et al (2003) Evaluation of shotgun sequencing for proteomic analysis of human plasma using HPLC coupled with either ion trap or fourier transform mass spectrometry. J Proteome Res 2:383–393
5. Brosch M et al (2008) Comparison of Mascot and X!Tandem performance for low and high accuracy mass spectrometry and the development of an adjusted Mascot threshold. Mol Cell Proteomics 7:962–970
6. Collins MO et al (2008) Phosphoproteomic analysis of the mouse brain cytosol reveals a predominance of protein phosphorylation in regions of intrinsic sequence disorder. Mol Cell Proteomics 7:1331–1348
7. Collins MO, Yu L, Choudhary JS (2008) Analysis protein complexes by 1D-SDS-PAGE and tandem mass spectrometry. Protoc Exch. doi: 10.1038/nprot.2008.123

Chapter 30

Preparation of Phytoplasma Membrane Recombinant Proteins

Luciana Galetto, Majid Siampour, and Cristina Marzachì

Abstract

The study of phytoplasma membrane protein interactions with host cell proteins is crucial to understand the life cycle of these unculturable microorganisms within their hosts. A step-by-step protocol for the heterologous expression of phytoplasma membrane proteins in *Escherichia coli* is described, together with a procedure to purify a suitable quantity of fusion antigen for application in the study of phytoplasma–host interactions.

Key words: Antigenic membrane protein, *Escherichia coli*, Histidine-tagged fusion protein, Immunodominant membrane protein, Prokaryotic expression vector

1. Introduction

The cell membrane is the boundary of phytoplasma cell, the place where direct interactions with the host environment first occur. Screening of the available phytoplasma genomes (1) has indicated that several proteins have transmembrane domains and therefore have a putative location in the cell membrane. As phytoplasma membrane proteins are in direct contact with host cells, they are likely to be involved in specific interactions with them.

The most abundant phytoplasma membrane proteins are classified into three types based on their location within the chromosome and their predicted transmembrane structure (2). A role in determining insect vector specificity has been suggested for the major antigenic membrane protein (Amp) of '*Candidatus* Phytoplasma asteris' onion yellows (3) as well as chrysanthemum yellows strains (4). Other putative membrane-associated proteins with predicted extracellular domains are present in the phytoplasma genome and may also be involved in interaction with the host.

Matt Dickinson and Jennifer Hodgetts (eds.), *Phytoplasma: Methods and Protocols*, Methods in Molecular Biology, vol. 938, DOI 10.1007/978-1-62703-089-2_30, © Springer Science+Business Media, LLC 2013

The sequences of some of these genes (such as *amp* and *imp*) are under positive selection (5–7), indicating a possible role of these proteins in the evolution of phytoplasmas. Several transporter systems are present in the phytoplasma genomes sequenced to date (8–11); these are also located within the cell membrane and project hydrophilic solute-binding domains towards the phytoplasma cytoplasm or the host cell environment. These membrane proteins are also putative partners for the interaction with host proteins.

As phytoplasmas cannot be grown in vitro and no transformation protocol is available for these bacteria, the expression of phytoplasma membrane genes in heterologous systems is a prerequisite for further studies on the interaction between the phytoplasma and the plant and vector hosts. Several phytoplasma membrane proteins have been expressed as fusion antigens in heterologous systems. Here we report on the major steps required for the production of phytoplasma membrane recombinant proteins, starting from the identification of the putative transmembrane domain of the target gene, the design of primers for the directional cloning of a PCR product into an appropriate expression vector, the selection of colonies expressing the recombinant protein, the pilot experiment to study a time-course of expression and solubility of the fusion protein, and large-scale polyhistidine-tagged target fusion protein expression and purification under denaturing as well as native conditions (Fig. 1).

It has been reported that the expression of full-length phytoplasma membrane proteins, such as Imp and Amp, induces a

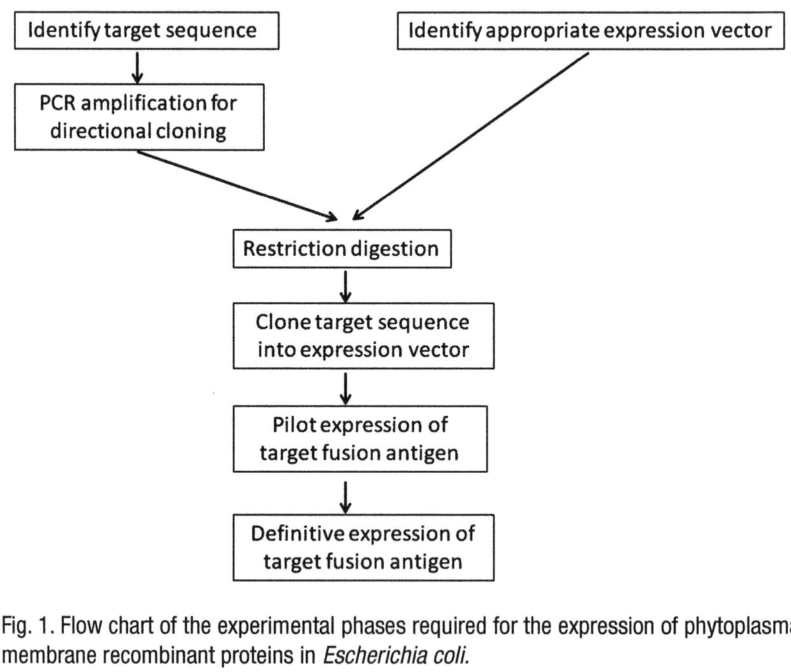

Fig. 1. Flow chart of the experimental phases required for the expression of phytoplasma membrane recombinant proteins in *Escherichia coli*.

significant delay in the growth of the transformed *Escherichia coli*, while deletion of the transmembrane domain has no effect (6, 12). For this reason we suggest the first step is to identify the transmembrane domain of the target protein and, if possible, according to future application of the fusion antigen, to remove it from the expression construct. Several expression vectors are available, and we describe here the pRSET system (Invitrogen), based on pUC-derived expression vectors designed for high-level protein expression and purification of genes expressed in *E. coli*. The target DNA insert must be positioned downstream and in frame with a sequence that codes an N-terminal polyhistidine fusion peptide. This functions both as a metal-binding domain for fusion protein purification and as a tag to detect the fusion antigen. Expression of the target gene is controlled by the phage T7 promoter that drives the expression of gene 10, following induction with isopropyl β-d-thiogalactoside (IPTG). T7 RNA polymerase specifically recognizes this promoter. To reduce the basal level of T7 RNA polymerase and therefore allow a tight control over the expression of fusion antigens that may potentially harm *E. coli*, the BL21 (DE3) pLysS strain of *E. coli* is used for expression. This strain carries the DE3 bacteriophage lambda lysogen with the *lacI* gene, T7 RNA polymerase gene, and a small portion of the *lacZ* gene. The lac repressor specifically represses the expression of T7 RNA polymerase. Addition of IPTG reinstates the expression of T7 RNA polymerase. When the T7 RNA polymerase reaches a sufficient concentration in the cell, it binds to the T7 promoter and expression of the target phytoplasma gene starts. Expression conditions vary depending on the nature of the protein, and therefore a pilot study to optimize the expression of the fusion protein and to clarify whether it is present in the soluble protein fraction or in the inclusion bodies is also addressed. The last step of the protocol describes the procedure for a large-scale purification of the target fusion protein.

2. Materials

Prepare all routine buffers, media, and solutions according to (13) using sterile double distilled H_2O (ddH_2O). Wear gloves when working with acrylamide and other hazards. Dispose of all wastes according to disposal regulations.

2.1. General Equipment

1. Refrigerated microcentrifuge.
2. Centrifuge and rotor for large volume samples.
3. Vortexer.
4. Shaker and thermal shaker.

5. Fume cupboard.
6. Laminar flow hood.
7. Sonicator.
8. Chromatography columns.
9. Agarose gel electrophoretic apparatus.
10. Electrophoretic apparatus for SDS-polyacrylamide gel electrophoresis (SDS-PAGE), e.g., Miniprotean 3 Cell (Bio-Rad).
11. Incubator for 1.5–2.2 mL tubes.
12. Thermal cycler.

2.2. PCR Amplification of Target Gene and Restriction Digestion

1. *Taq* DNA polymerase and 10× reaction buffer.
2. 2 mM dNTP solution.
3. 25 mM $MgCl_2$.
4. Target-specific primers with appropriate restriction enzyme recognition sites at their 5′-ends.
5. Agarose.
6. Tris-borate-EDTA buffer (TBE).
7. Ethidium bromide for gel staining.
8. Kit to purify PCR product, e.g., Geneclean Turbo Kit (MP Biomedicals).
9. Appropriate restriction digestion enzymes together with their concentrated reaction buffers.

2.3. Cloning of PCR Product into an Expression Vector

1. T4 DNA ligase and reaction buffer.
2. *E. coli* strain DH5α competent cells.
3. S.O.C. broth (e.g., Invitrogen).
4. *E. coli* strain BL21 (DE3) pLysS competent cells (Invitrogen).
5. Primers based on the vector sequence. For the application described in this protocol, the T7 promoter primer and the T7 reverse primer (Invitrogen) are used.
6. *Taq* DNA polymerase and 10× reaction buffer.
7. 2 mM dNTPs.
8. 25 mM $MgCl_2$.
9. 50% glycerol solution in ddH_2O (for long storage of cell culture at −80°C).
10. Plasmid purification kit, e.g., FastPlasmid Mini Kit (5 Prime, Eppendorf S.R.L.), but other kits or the procedures described in (13) may also be used.

2.4. Antibiotics, Enzymes, and Media

1. Ampicillin: 100 mg/mL in ddH_2O filter sterilized with a 0.22 μm filter. Store stock solution at −20°C.

2. Chloramphenicol: 25 mg/mL in ethanol (100%). Store stock solution at −20°C.
3. LB medium: 10 g Bacto tryptone, 5 g yeast extract, 5 g NaCl in 950 mL ddH_2O; adjust pH to 7.0 with 1 M NaOH and add ddH_2O to 1 L. Sterilize by autoclaving at 120°C for 20 min.
4. LB agar: Combine 10 g Bacto tryptone, 5 g yeast extract, 5 g NaCl with 950 mL ddH_2O, mix and stir until clear, adjust pH to 7.0 with 1 M NaOH, then add 15 g agar. Bring volume to 1 L with ddH_2O, and sterilize by autoclaving at 120°C for 20 min.
5. Isopropyl-β-D-thiogalactopyranoside, 1 M (IPTG): 238 mg IPTG in 1 mL ddH_2O, filter sterilize with a 0.22 μm filter, store in aliquots at −20°C. (For 100 mM IPTG, dissolve 238 mg IPTG in 10 mL ddH_2O, filter sterilize with a 0.22 μm filter, and store in aliquots at −20°C.)
6. LB amended with ampicillin (100 μg/mL) and chloramphenicol (25 μg/mL): To prepare 5 plates, add 100 μL of ampicillin stock solution and 100 μL of chloramphenicol stock solution to 100 mL molten LB agar (<50°C), mix by hand, pour plates, let them cool to room temperature, seal plates with Parafilm, and store them at 4°C for up to 1 month. To prepare 1 L of LB medium, add 1 mL of ampicillin stock solution and 1 mL of chloramphenicol stock solution to 1 L of LB medium, stir and use immediately.
7. Lysozyme powder. Store at −20°C.

2.5. Protein Extraction Under Denaturing Conditions

1. Basic lysis buffer: 100 mM NaH_2PO_4, 10 mM Tris–HCl, pH 8.0.
2. Resuspension buffer (R): 8 M urea in basic lysis buffer, pH 8.0.
3. Washing buffer 1 (W1): 8 M urea in basic lysis buffer, adjust pH to 6.3 using 1 M HCl.
4. Washing buffer 2 (W2): 8 M urea in basic lysis buffer, adjust pH to 5.9 using 1 M HCl.
5. Elution buffer (E): 8 M urea in basic lysis buffer, adjust pH to 4.5 using 1 M HCl.
6. Ni-NTA agarose slurry: supplied as 50% slurry in ethanol (QIAGEN).

2.6. Protein Extraction Under Native Conditions

1. Basic buffer: 50 mM NaH_2PO_4, 300 mM NaCl, pH 8.0. Dissolve 6.9 g of $NaH_2PO_4 \cdot H_2O$ and 17.53 g NaCl in ddH_2O. Adjust to pH 8.0 using NaOH and bring to a final volume of 1 L with ddH_2O.

2. Lysis buffer (buffer A): 10 mM imidazole, 0.2% Tween 20 in basic buffer, pH 8.0. Dissolve 0.7 g imidazole and 2 mL Tween 20 in a final volume of 1 L basic buffer.
3. Washing buffer (buffer B): 20 mM imidazole in basic buffer, pH 8.0.
4. Elution buffer (buffer C): 250 mM imidazole in basic buffer, pH 8.0.
5. Ni-NTA agarose slurry: supplied as 50% slurry in ethanol (QIAGEN); wash it twice with ddH_2O to remove ethanol and equilibrate it three times with buffer A prior to use.

2.7. SDS-Polyacrylamide Gel Electrophoresis

1. 40% Acrylamide solution (29:1 acrylamide: bis-acrylamide).
2. 1.5 M Tris–HCl, pH 8.8.
3. 10% Sodium dodecyl sulphate (SDS) in ddH_2O.
4. 10% Ammonium persulphate in ddH_2O.
5. *N, N, N′, N′*-tetramethylethylenediamine (TEMED), store at 4°C.
6. *tert*-Butanol.
7. 1.5 M Tris–HCl, pH 6.8.
8. Laemmli buffer (5× stock): 25 mM Tris, 250 mM glycine, 0.1% SDS. Prepare 2 L of a 5× stock solution by dissolving 30 g Tris, 188 g glycine, 10 g SDS in 2 L ddH_2O, sterilize by autoclaving at 120°C for 20 min. To prepare 1 L of 1× Laemmli buffer, mix 200 mL 5× stock solution and 800 mL ddH_2O.
9. Fixative solution for Coomassie staining: 40% ethanol, 10% acetic acid.
10. Coomassie stock solution: 10% H_3PO_4, 10% $(NH_4)_2SO_4$, 0.12% Coomassie brilliant blue G-250.
11. Coomassie working solution: 20% methanol, 80% Coomassie stock solution prepared just before use.
12. SDS sample buffer (3× stock): 15% β-mercaptoethanol, 29% glycerol, 138 mM Tris–HCl pH 6.8, 0.01% bromophenol blue. To prepare 25 mL, mix 7.25 mL of glycerol, 3.45 mL of 1 M Tris–HCl pH 6.8, one spatula tip (a pinch) of bromophenol blue, and 9 mL of ddH_2O. Store at −20°C in 850 μL aliquots. When needed, thaw at room temperature, add 150 μL β-mercaptoethanol, and store at 4°C.
13. Protein molecular weight marker: prestained SDS-PAGE broad range standards.

3. Methods

3.1. Identification of Putative Transmembrane Domains in the Sequence of the Target Protein

1. Copy and paste the target protein sequence into software to predict the putative transmembrane domain (see Note 1).
2. Identify the amino acids involved in the putative transmembrane domains of the target protein (Fig. 2).
3. Go back to the DNA sequence of the protein and exclude the part of the sequence encoding the putative transmembrane domains from the target sequence for cloning in the expression vector.

3.2. Primer Design and Directional Cloning of PCR Product into an Expression Vector

1. Insert restriction site recognition sequences at the 5′-ends of the forward and reverse primers according to the polylinker map of the selected expression vector. Check that the addition of the restriction site at the 5′-end maintains the coding

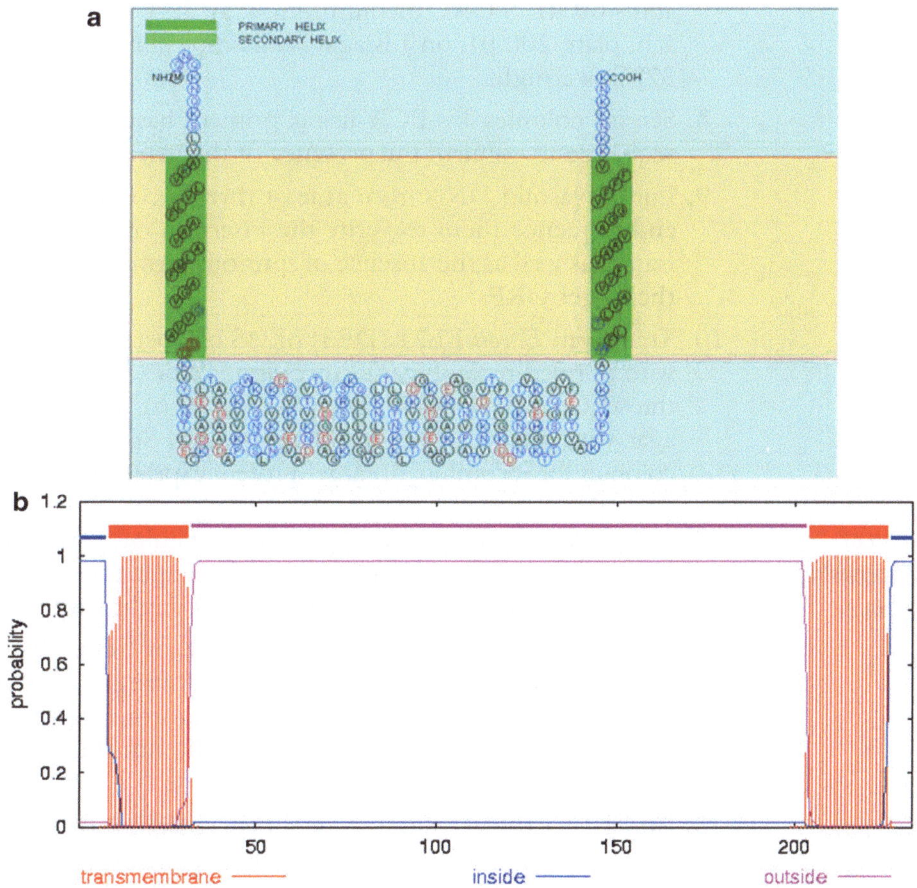

Fig. 2. TMHMM (**a**) and SOSUI (**b**) membrane localization prediction of '*Ca.* Phytoplasma asteris' chrysanthemum yellows strain phytoplasma Amp protein.

sequence of your target gene in frame with the open reading frame (ORF) of the selected expression vector (see Note 2).

2. Perform a PCR using these primers, and check the expected amplicon size by agarose gel electrophoresis.
3. Extract the amplicon from the agarose gel, clean it up with a purification kit of your choice, and quantify it.
4. Digest the purified amplicon (about 1 μg) and the expression vector (about 200 ng) with the two selected restriction enzymes (see Note 3).
5. Purify the double-digested amplicon and expression vector with a purification kit of your choice, and quantify the amount of DNA.
6. Ligate the double-digested amplicon and expression vector together (see Note 4).
7. Transform 100 μL of competent *E. coli* DH5α cells by adding 10 μL of ligation mix, mix gently by tumbling, store on ice for 30 min, incubate at 42°C for 45 s, return to the ice for 2 min, add 400 μL S.O.C. medium, grow at 37°C with shaking for 1 h, plate 200 μL on LB agar with ampicillin, and incubate at 37°C overnight.
8. Screen colonies by PCR using primers based on the vector sequence to confirm the presence of the insert (see Note 5).
9. Purify plasmid DNA from at least three recombinant colonies and sequence them to verify the insertion of the amplicon in frame, as well as the absence of spurious termination codons in the target ORF.
10. Transform *E. coli* BL21 (DE3) pLysS competent cells (200 μL) with purified recombinant expression vector (50 ng) containing the sequenced, in-frame, insert (see Note 6). Plate cells on LB agar with ampicillin and chloramphenicol. Select at least three colonies for recombinant target protein expression (see Note 7).

3.3. Selection of Colonies Expressing the Recombinant Protein

1. Inoculate 2 mL of LB medium containing ampicillin and chloramphenicol (in 15 mL tubes) with single transformed bacterial colonies, and incubate at 37°C overnight with shaking.
2. The following morning, inoculate 5 mL of fresh LB medium (in 50 mL tubes) with 500 μL of the overnight bacterial culture. Incubate the tubes at 37°C with vigorous shaking (above 200 rpm) until the optical density at 600 nm (OD_{600}) reaches 0.4–0.7.
3. Add IPTG to each tube to final concentration of 1 mM (i.e., 5 μL of 1 M IPTG stock in 5 mL of LB medium).
4. Let the cultures grow for 3–5 h.

5. Transfer 4 mL of each culture (in two successive steps) into a 2.2 mL microfuge tube and centrifuge to harvest the bacterial cells (15,000 × *g* for 1 min, at room temperature).
6. Discard the supernatant, add 300 μL buffer R (resuspending buffer), then lyse the bacterial cells by pipetting or gently vortexing. Lysis is complete when the solution becomes translucent.
7. Centrifuge the lysate for 10 min at 15,000 × g and transfer the cleared supernatant containing total solubilized bacterial proteins (cleared lysate) into a fresh 1.5 mL microfuge tube.
8. Add 50 μL of 50% Ni-NTA agarose slurry (see Note 8) to each tube containing the cleared lysate and mix for 20 min at room temperature.
9. Collect the Ni-NTA agarose by centrifugation (6,000 × *g*, 1 min). Discard the supernatant (keep 10 μL of the supernatant for checking via SDS-PAGE).
10. Wash the Ni-NTA agarose pellet twice with 300 μL of wash buffer (buffer W1) followed by centrifugation at 6,000 × *g* for 1 min, discard the supernatant (keep 10 μL of the supernatant for checking via SDS-PAGE).
11. To elute the polyhistidine-tagged fusion protein from Ni-NTA agarose, add 30 μL of elution buffer (buffer E), and gently mix by pipetting. Centrifuge at 6,000 × *g* for 1 min and carefully transfer the supernatant containing the fusion protein into a fresh tube. Repeat the elution step once more.
12. Run the eluted protein (10 μL) in SDS-PAGE alongside appropriate protein molecular weight markers.
13. Incubate the polyacrylamide gel in 50 mL fixative solution for 5 min with gentle agitation. Wash the gel twice for 5 min with ddH_2O, then stain overnight in a freshly made Coomassie working solution (see Note 9). Destain the gel with several changes of ddH_2O, before observing the protein profiles to confirm the expression of the target fusion protein as well as its molecular weight (see Note 10). Identify colonies expressing the fusion protein with the highest efficiency for the use in the next steps, and store the bacterial colonies at –80°C (see Note 5).

3.4. Pilot Experiment to Study Time-Course of Expression and Solubility of Fusion Protein

1. Inoculate 10 mL of LB medium containing ampicillin (100 μg/mL) and chloramphenicol (25 μg/mL) with the bacterial clone expressing the most fusion target protein (identified in Subheading 3.3 step 10), and incubate it at 37°C overnight with shaking.
2. The following day, inoculate 90 mL of pre-warmed (37°C) LB medium (in a flask) with 10 mL of overnight culture (see Note 11). Grow with vigorous shaking (above 200 rpm) to reach an OD_{600} of 0.4–0.7 (typically about 2 h).

3. Take two culture samples (5 mL each), name them A_0 and B_0 (uninduced control samples, see Note 12), centrifuge at 4,000×*g* for 15 min (4°C) to pellet the cells, and store at −20°C until use.
4. Add IPTG to a final concentration of 1 mM to induce fusion protein expression, and incubate the culture at 37°C with vigorous shaking (above 200 rpm).
5. Prepare a 15% SDS-PAGE gel (see Note 13) and store it at room temperature in Laemmli buffer.
6. Take two samples (5 mL each) 1 h after induction with IPTG, name them A_1 and B_1. Centrifuge to collect the bacterial cells (4,000×*g* for 15 min at 4°C) and store the pellets at −20°C. Let the culture continue to grow, collecting samples (5 mL each) at 1 h intervals and naming them A_2...A_n and B_2...B_n (see Note 14). Pellet the cells (4,000×*g* for 15 min at 4°C) and store the pellets at −20°C.
7. For the time-course analysis, thaw the frozen cell pellets from samples A_0 to A_n on ice for 20 min. Total proteins are extracted under denaturing conditions: add 2.5 mL of buffer R to the thawed pellets, gently vortex to lyse cells until the solution becomes translucent, centrifuge at 15,000×*g* for 10 min (at room temperature), and transfer the supernatant (cleared lysate) into a fresh tube (see Note 15). Take 5 μL of cleared lysate from the different times after IPTG induction (A_0–A_n), add 2 μL of 3× SDS sample buffer, boil for 5 min and save for step 9.
8. To examine the solubility of the target fusion antigen, thaw the bacterial pellets (B_0 to B_n) on ice (about 20 min), add 5 mL of buffer A to each tube, add lysozyme to a final concentration of 1 mg/mL, then incubate on ice for 30 min. Lyse the bacterial cells by sonication (10 s, with power setting at 200 W, repeated six times) and store samples on ice between each sonication burst (avoid foaming of the samples). Centrifuge the samples at 15,000×*g* for 10 min at 4°C, and collect the insoluble (pellet) and the soluble phases (supernatant) of the lysate. Resuspend the pellet in 3× SDS sample buffer, boil for 5 min, and save10 μL samples for step 9. Transfer 10 μL of the soluble phase proteins to a fresh microfuge tube, add 3 μL of 3× SDS sample buffer, boil for 5 min, and save for step 9.
9. Prepare 15% acrylamide gels for SDS-PAGE. Load the boiled samples of A_0–A_n and B_0 to B_n insoluble and soluble phases, alongside an appropriate protein molecular weight marker (see Note 16).
10. Run the SDS-PAGE acrylamide gels, and stain with Coomassie blue (Fig. 3).

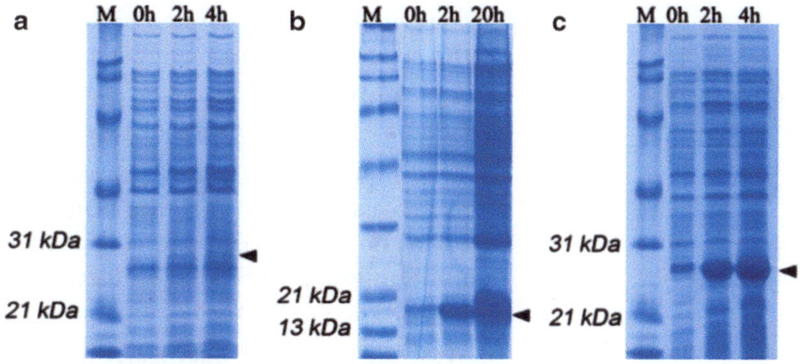

Fig. 3. SDS-polyacrylamide gel electrophoresis (SDS-PAGE) gels of total proteins from time-course expression of three different phytoplasma membrane fusion constructs (**a**, **b** and **c** respectively). *Black arrows* indicate the expected position of each fusion protein. *M* prestained SDS-PAGE broad range standard; 0 h, sample collected before isopropyl β-D-thiogalactoside (IPTG) induction; 2, 4, 20 h, samples collected at 2, 4, and 20 h after IPTG induction.

11. Evaluate the protein expression level and its solubility and determine which conditions are optimal for best expression of your construct based on the specific requirements for future applications (see Note 16).

3.5. Large-Scale Polyhistidine Target Fusion Protein Expression

1. Inoculate 50 mL of LB medium containing ampicillin (100 μg/mL) and chloramphenicol (25 μg/mL) with the selected bacterial clone expressing the fusion target protein with the highest efficiency. Incubate the culture at 37°C overnight with shaking.
2. Inoculate 1 L of fresh LB medium with the overnight culture and incubate at 37°C with shaking (above 200 rpm) (see Note 11).
3. When OD_{600} reaches 0.4–0.7, induce the culture with IPTG to a final concentration of 1 mM. Let the culture grow for additional 4 h (for purification under native condition) or 6 h (for purification under denaturing condition).
4. Centrifuge the culture (4,000 × *g* for 30 min at 4°C) to collect the bacterial cells.
5. Weigh the pellet obtained. Store the bacterial pellet at -20°C overnight before proceeding to preparative purification of polyhistidine fused target protein.

3.6. Large-Scale Polyhistidine Target Fusion Protein Purification

3.6.1. Denaturing Conditions

1. Thaw the bacterial cell pellet on ice for 30 min, and rub it against the wall of the tube with a glass rod to avoid clump formation while resuspending.
2. Add 1–5 mL of lysis buffer R per g of cells (see Note 17), gently vortex and let the lysis proceed at least for 30 min at room temperature.

3. Centrifuge the cell lysate (15,000 × *g* for 30 min) and transfer the supernatant to a fresh 50 mL tube.
4. Add 2 mL of 50% Ni-NTA agarose slurry to the cleared lysate (see Note 18). Gently shake the mixture for 1 h at 4°C.
5. Transfer the mixture into a chromatography column (in one or in successive steps) with the bottom cap closed. When loaded, open the bottom cap and collect the flow-through containing nonbinding proteins (see Note 19). Store the flow-through fraction at −20°C for future analysis by SDS-PAGE.
6. Wash the Ni-NTA agarose column three times with 2 mL of washing buffer 1 (buffer W1); each time take one sample for future SDS-PAGE analysis.
7. Wash the Ni-NTA agarose column twice with 1 mL of washing buffer 2 (buffer W2); each time take one sample for SDS-PAGE analysis.
8. Elute the target fusion protein from the Ni-NTA agarose by adding 0.5 mL elution buffer (buffer E) five times. Add 3 μL of 3× SDS sample buffer to a 10 μL sample of the eluate, then boil for 5 min and load onto an acrylamide gel for SDS-PAGE to measure the purification quality and quantity of target fusion protein (Fig. 4). Include on this gel samples of the flow-through and washes collected during the elution procedure (steps 5–7), to trace the possible release of the target protein in each step.

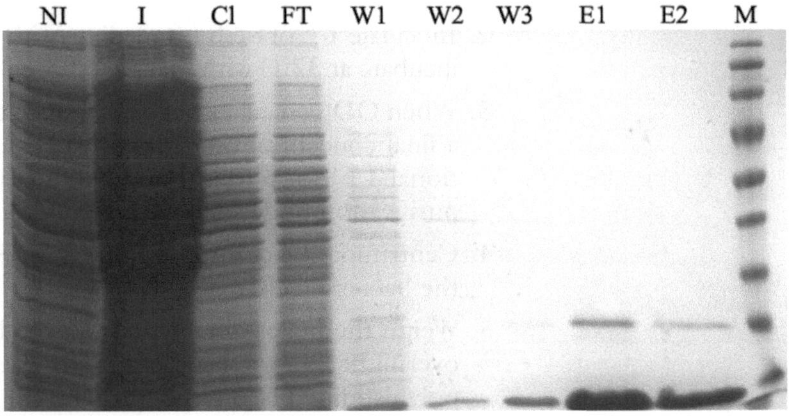

Fig. 4. SDS-PAGE gel of samples from different steps of the expression and purification under denaturing conditions of a phytoplasma membrane protein. *NI* total proteins from uninduced bacterial culture; *I* total proteins from induced bacterial culture; *Cl* total proteins from cleared lysate of induced bacterial culture; *FT* proteins in flow-through following loading of Ni-NTA agarose slurry incubated with resuspended induced bacterial culture; *W1*, *W2*, *W3* proteins in three successive washing steps of the Ni-NTA agarose column; *E1*, *E2*, purified proteins in the two successive elutions from the Ni-NTA agarose column; *M* molecular weight standard markers.

3.6.2. Native Conditions

1. Thaw the bacterial cell pellet on ice for 30 min, and rub it against the wall of the tube with a glass rod to avoid clump formation in resuspending step.
2. Resuspend cells in 20 mL of lysis buffer A (see Note 20) and keep it on ice.
3. Add lysozyme to a final concentration of 1 mM. Slowly mix the solutions by inverting the tube. Incubate on ice for about 30 min.
4. Sonicate the lysate on ice until the lysis is complete (see Note 21).
5. Centrifuge the lysate at 15,000 × *g* for 20 min at 4°C to pellet the cell debris and insoluble proteins in inclusion bodies.
6. Transfer the cleared lysate into a fresh 50 mL tube. Add 1 mL of 50% Ni-NTA agarose slurry equilibrated in buffer A. Gently mix on a shaker for 30 min at 4°C.
7. Load the mixture into a chromatography column with the bottom closed.
8. Remove the bottom cap to elute the unbound proteins. Take a sample of flow-through (10 μL) for SDS-PAGE analysis.
9. Wash the column three times with 2 mL of washing buffer (buffer B). Take samples of each washing (10 μL) for SDS-PAGE analysis.
10. Elute the target protein from the Ni-NTA agarose with 0.5 mL elution buffer (buffer C). Repeat elution four times. Take samples (10 μL) of each elution to confirm quantity and quality of target polyhistidine-tagged protein in SDS-PAGE.

4. Notes

1. Suggested software: TMHMM (http://www.cbs.dtu.dk/services/TMHMM/); SOSUI (http://bp.nuap.nagoya-u.ac.jp/sosui/sosui_submit.html) (14).
2. To maintain the frame you can add one or two nucleotides, checking that a stop codon is not inserted. You can also add few nucleotides to the 5′-ends of both primers to enhance the subsequent restriction digestion.
3. To check for complete digestion of your DNA, set up one control digestion of any covalently closed circular plasmid for each enzyme by digesting with the same enzyme unit/template quantity ratio as in your amplicon/vector digestions. Incubate the digestions at the appropriate temperature for 1 h, then check for linearization of the control plasmid by agarose gel electrophoresis. If digestion is complete then proceed to successive steps with both amplicon and expression vector.

4. For the ligation reaction, use a 3:1 insert to vector ratio, according to the following $ng_i = 3ng_p \times (bp_i/bp_p)$, where ng_i = ng of double-digested and purified amplicon (insert), ng_p = ng of double-digested and purified expression vector (plasmid), bp_i = insert size in base pairs, bp_p = vector size in base pairs. Set up a ligation reaction on ice, by adding 3 units of T4 DNA ligase in the appropriate buffer to the correct amount of double-digested and purified amplicon and expression vector, in a final volume of 10 μL. Also set up a control ligation in which the digested amplicon DNA is replaced with sterile ddH_2O. Incubate both reactions at 4°C overnight.
5. When screening colonies after transformation, ensure the time of the amplification step in the PCR is long enough to ensure amplification of your target amplicon. Streak the selected recombinant colonies on a LB agar/ampicillin master plate. When the sequence of the target insert is available and confirmed as correct, store the corresponding colony in glycerol at −80°C: to do so, grow the transformed colony in 2 mL of LB medium with ampicillin overnight, then mix 700 μL of overnight culture with 300 μL of 50% glycerol, mix thoroughly, drop in liquid nitrogen until frozen, and store at −80°C.
6. Transform *E. coli* BL21 (DE3) pLysS competent cells by adding 1 μL 50 ng/μL purified recombinant expression vector, mix gently by tumbling, store on ice for 30 min, incubate at 42°C for 45 s, transfer back to ice for 2 min, add 400 μL SOC medium, and incubate at 37°C with shaking for 1 h.
7. Select extra recombinant colonies and streak them on an LB agar master plate containing ampicillin and chloramphenicol until the results of the selection of colonies expressing the recombinant protein are available.
8. To equilibrate the Ni-NTA resin, centrifuge the 50% slurry at 6,000 × *g* for 1 min. Remove the supernatant and add 1 volume of reaction buffer (buffer R or A for denaturing or native conditions, respectively). Mix gently then centrifuge it again. Discard the supernatant, wash again once more, and finally make a 50% slurry in the appropriate reaction buffer (buffer R or A).
9. For minigels, 20 mL of freshly prepared Coomassie working solution is usually enough. This solution must be disposed of as toxic waste. After destaining, gels can be stored at 4°C, sealed in a plastic bag in the presence of 1% of acetic acid.
10. The predicted molecular weight of the target protein must be increased to account for the polyhistidine tail as well as other vector amino acids still present in the fusion construct.
11. The flask should have enough free space in it for good aeration of the culture; generally, the maximum ratio of liquid/flask volume required for good aeration is considered to be 0.3.

The basic steps in successful protein expression (membrane or non-membrane proteins) in *E. coli* are choosing the proper expression system, checking for correct and in-frame integration of the insert, and achieving suitable yield of cells expressing the target protein with an efficient expression level in bacterial cells, possibly in native form. The extraction method is also important to ensure complete lysis of the cells, and purification of the target tagged protein from background host proteins. Failure to comply with any of these steps may result in an unsuccessful expression experiment. Therefore, it is recommended to include a negative (uninduced bacterial culture) and a positive control (known to express a fusion protein upon induction) in parallel to the expression system under study, to monitor the expression level of the target protein, its quality at each step, and to locate specific problems.

12. Samples named A are devoted to analysis of the time-course expression of the target protein, and those named B are devoted to the determination of the fusion protein solubility. Over-expressed membrane proteins are believed to associate with hydrophobic domains of other host proteins, forming insoluble inclusion bodies that cannot be purified under native conditions. A rapid rate of protein expression may also cause improper folding, especially when the expressed protein contains cysteine residues able to form disulphide bridges (15), which may be followed by the formation of insoluble inclusion bodies. It is however possible to solubilize most inclusion bodies with strong denaturing buffers and to purify the proteins in denatured form. Induction at lower temperatures (18–30°C) may decrease the expression level and aid the correct folding of the fusion protein, thereby preventing its precipitation in inclusion bodies. The presence of non-denaturing detergents (e.g., 0.1–0.25% Triton X-100) in the extraction buffer may also be helpful to recover some fusion membrane proteins in their native form by reducing their hydrophobic associations. Although it is preferable to express and purify proteins in their native soluble form, there are some techniques available for renaturation of proteins purified under denaturing conditions (16).
13. To prepare a 15% acrylamide minigel, mix together the following ingredients: 1.9 mL of 40% acrylamide mix (29:1), 1.7 mL of ddH_2O, 1.3 mL of 1.5 M Tris–HCl (pH 8.8), 50 μL of 10% SDS, 50 μL of 10% ammonium persulphate, and 5 μL of TEMED. Pour the mix between the two glass plates of the electrophoretic apparatus, overlay with 100 μL *tert*-butanol, and wait for polymerization. Wash *tert*-butanol away with ddH_2O, and pour on top of the 15% running gel about 2.5 mL of 5% acrylamide stacking gel prepared by mixing 313 μL of

40% acrylamide mix (29:1), 1.6 mL of ddH_2O, 315 µL of 1.5 M Tris–HCl (pH 6.8), 25 µL of 10% SDS, 25 µL of 10% ammonium persulphate, and 2.5 µL of TEMED. Insert the plastic comb and wait for polymerization. Remove the comb and wash the wells with ddH_2O.

14. An induction period of 5 h is usually enough, but for some phytoplasma membrane proteins up to 8 h induction may be required. It must be noted that even in successful expression systems, over-expression of foreign genes may be toxic for *E. coli* and this results in a slower rate of growth of the induced host cells compared to the uninduced control. A minor delay in cell growth of induced *E. coli* cells in comparison to uninduced ones can therefore be considered as an indication of successful protein expression. If the cell growth is hindered to a considerable degree, as is common when expressing phytoplasma membrane proteins, the yield of bacterial cells may be poor, resulting in very low levels of fusion protein expression. If this is the case, there are some techniques to improve cell growth as well as the efficiency of the fusion protein expression, such as inducing cells with lower concentrations of IPTG in the expression culture (0.1 mM) and growing bacteria at lower temperatures (20–30°C). Also, it may be helpful to grow the bacterial cells to a high density before adding IPTG, to compensate for the negative effects of recombinant protein expression on bacterial cell growth and propagation. These techniques could be combined, and must be optimized for best results in a pilot experiment. In some cases, the induced bacterial culture grows at a rate comparable to the uninduced control, but the expression level is very low (e.g., 0.1 mg/L for a phytoplasma membrane protein cloned in the pQE vector). In such cases, bacterial cells must be monitored for probable rapid protein degradation and plasmid instability. Rapid protein degradation may occur when the size of the fusion protein is less than 10 kDa. Expression of small proteins may be improved by creating a fusion with carrier proteins such as GST or DHFR. If degradation of the fusion protein happens after cell lysis, it is recommended to include protease inhibitors in the buffers and to work on ice. Also the sequence of the target protein must be checked to exclude the presence of signal sequences causing the improper transcription and translation of the fusion protein resulting in its degradation. Changing the expression system and choosing protease-deficient *E. coli* host strains are also suggested, if problems persist. Plasmid instability, in which the bacterial cells lose the recombinant plasmid (and consequently also the antibiotic resistance), is a problem in some expression systems. As ampicillin is not a stable antibiotic and is rapidly depleted, bacterial cells lacking the expression construct have a better chance to multiply in the

culture medium than those harboring the recombinant plasmid. To test for plasmid instability, induced *E. coli* BL21 (DE3) pLysS cells transformed with the expression construct must be plated on LB agar plates with and without ampicillin. The ratio between the number of colonies grown in the presence and in the absence of the antibiotic determines the plasmid stability (from 0 = highly unstable to 1 = stable) (17). If plasmid instability is a problem, use of a higher concentration of ampicillin (200 μg/mL) or the application of more stable alternative antibiotics in the growth medium is recommended.

15. The supernatant (cleared lysate) contains the total bacterial proteins including the target fusion protein.
16. Run the cleared lysate (A_0–A_n) on 15% acrylamide gel SDS-PAGE. After Coomassie staining, compare the amount of target fusion protein expression at different times after induction. Electrophoretic separation on a parallel gel of cleared lysates from B_0–B_n samples allows the identification of the optimum induction time to achieve the greatest amount of the expressed target fusion protein in soluble native form. These results are crucial to inform the best conditions to set up a large-scale preparation of the target fusion phytoplasma protein. It is generally believed that a good expression is reached when a satisfying amount of protein is expressed and purified in native form. It is important to note that, if the expression yield of the system is low (e.g., <1 mg/L fusion target protein from the bacterial culture), especially when extracting the fusion protein under native conditions, comparison between protein expression levels under different conditions using small-scale analysis of crude protein extract may not be feasible. In fact, poorly expressed proteins may be masked by host cell proteins. If this is the case, it is recommended to move directly to large-scale expression of the fusion protein followed by histidine-tagged protein purification using the Ni-NTA agarose matrix. In a small-scale experiment (under denaturing conditions) and in comparison with protein standards, we obtained an expression level of a phytoplasma membrane protein of about 5 mg/L of culture. Under native conditions, however, much less recombinant protein could be purified because the majority of it formed insoluble inclusion bodies. The volume of culture required for large-scale experiments depends on the expression level of the target protein and also on the amount of purified fusion protein required for successive applications.
17. The required volume of buffer depends on the expression level of target fusion protein and on the quantity of collected cells. Higher expression levels require a larger amount of lysis buffer, so (for example) if the expression level is >10 mg fusion target protein per L culture then the bacterial pellet obtained from

1 L of culture may be resuspended in 100 mL lysis buffer (10× concentration). Lower expression levels require a higher concentration factor and a larger culture size. Under our experimental conditions, when the expression level of target protein was 5 mg/L (and ~12 g of wet cells) a concentration factor of 40× was applied, and the bacterial pellet from 1 L culture was resuspended in 25 mL of lysis buffer. For lower expression levels, it is recommended to dissolve the bacterial pellet at a concentration up to 50× (20 mL of lysis buffer for 1 L culture).

18. The amount of Ni-NTA agarose needed is dependent on the protein expression level. The binding capacity of 50% Ni-NTA agarose to polyhistidine-tagged proteins is normally 2–5 mg/mL depending on the nature of the protein. Under our experimental conditions, for expression levels of 5 mg/L (in the case of polyhistidine-tagged phytoplasma membrane proteins), 2 mL of 50% Ni-NTA agarose was enough to bind 6 mg of polyhistidine-tagged protein.
19. Do not allow Ni-NTA agarose to dry out at any stage.
20. Because the bacterial pellet was obtained from 1 L of culture, a dilution factor of 50 is suggested for resuspension in lysis buffer. The volume of lysis buffer depends on the quantity of protein expressed in soluble native form and also on the wet weight of collected bacterial cells. Generally 1–5 mL lysis buffer per g of bacterial cells is required. The amount of target protein in the soluble phase is considerably lower than that of total expressed fusion protein, especially when membrane proteins are expressed, and therefore it is recommended to keep the lysate more concentrated (concentration factor between 50 and 100). In our experimental conditions, the amount of expressed phytoplasma membrane protein in the soluble phase was about 1 mg/mL for 12 g of cells, from 1 L of culture.
21. Generally six sonication rounds of 15 s each at 200 W with a sonicator equipped with a probe is enough. Hold the probe inside the lysate to avoid foaming. If the lysate is too viscous, the sonication protocol was not sufficient to shear genomic material. To remove this genomic material, use of RNaseA and DNase I at final concentrations of 10 and 5 μg/mL of lysate, respectively, while incubating on ice for 15 min is recommended. Viscous lysate hinders or slows the run of the chromatographic flux.

Acknowledgements

L.G. was supported by the Italian Regional Grant (Regione Piemonte) "Studi sui fattori che favoriscono le epidemie di flavescenza dorata in Piemonte e loro superamento," subproject; B.M.S. was supported by a grant from Iranian Ministry of Science, Research and Technology.

References

1. Hogenhout SA et al (2008) Phytoplasmas: bacteria that manipulate plants and insects. Mol Plant Pathol 9:403–423
2. Kakizawa S, Oshima K, Namba S (2006) Diversity and functional importance of phytoplasma membrane proteins. Trends Microbiol 14:254–256
3. Suzuki S et al (2006) Interaction between the membrane protein of a pathogen and insect microfilament complex determines insect-vector specificity. Proc Natl Acad Sci U S A 103:4252–4257
4. Galetto L et al (2011) The major antigenic membrane protein of '*Candidatus* Phytoplasma asteris' selectively interacts with ATP synthase and actin of leafhopper vectors. PLoS One 6:e22571
5. Fabre A, Danet JL, Foissac X (2011) The stolbur phytoplasma antigenic membrane protein gene *stamp* is submitted to diversifying positive selection. Gene 472:37–41
6. Kakizawa S et al (2009) Cloning of immunodominant membrane protein genes of phytoplasmas and their *in planta* expression. FEMS Microbiol Lett 293:92–101
7. Kakizawa S et al (2006) Positive selection acting on a surface membrane protein of the plant-pathogenic phytoplasmas. J Bacteriol 188:3424–3428
8. Bai X et al (2006) Living with genome instability: the adaptation of phytoplasmas to diverse environments of their insect and plant hosts. J Bacteriol 188:3682–3696
9. Kube M et al (2008) The linear chromosome of the plant-pathogenic mycoplasma '*Candidatus* Phytoplasma mali'. BMC Genomics 9:306
10. Oshima K et al (2004) Reductive evolution suggested from the complete genome sequence of a plant-pathogenic phytoplasma. Nat Genet 36:27–29
11. Tran-Nguyen LTT et al (2008) Comparative genome analysis of '*Candidatus* Phytoplasma australiense' (subgroup *tuf*-Australia I; *rp*-A) and '*Ca.* Phytoplasma asteris' strains OY-M and AY-WB. J Bacteriol 190: 3979–3991
12. Galetto L et al (2008) Characterization of putative membrane protein genes of the '*Candidatus* Phytoplasma asteris', chrysanthemum yellows isolate. Can J Microbiol 54:341–351
13. Sambrook J, Russell DW (2001) Molecular cloning: a laboratory manual, 3rd edn. Cold Spring Harbor Laboratory Press, New York
14. Hirokawa T, Boon-Chieng S, Mitaku S (1998) SOSUI: classification and secondary structure prediction system for membrane proteins. Bioinformatics 14:378–379
15. Singh SM, Panda AK (2005) Solubilization and refolding of bacterial inclusion body proteins. J Biosci Bioeng 99:303–310
16. de Marco A (2011) Molecular and chemical chaperones for improving the yields of soluble recombinant proteins. Methods Mol Biol 705:31–51
17. Bernard A, Payton M (2001) Selection of Escherichia coli expression systems. In: Coligan JE, Dunn BM, Ploegh HL, Speicher DW, Wingfield PT (eds). Curr Protoc Protein Sci vol.1. Wiley and Sons Inc, New York, Unit 5.2

Chapter 31

Phytoplasma Plasmid DNA Extraction

Mark T. Andersen and Lia W. Liefting

Abstract

Phytoplasma plasmids have generally been detected from DNA extracted from plants and insects using methods designed for the purification of total phytoplasma DNA. Methods include extraction from tissues that are high in phytoplasma titre, such as the phloem of plants, with the use of CsCl–bisbenzimide gradients that exploit the low G+C content of phytoplasma DNA. Many of the methods employed for phytoplasma purification have been described elsewhere in this book. Here we describe in detail two methods that are specifically aimed at isolating plasmid DNA.

Key words: Alkaline lysis, CsCl gradients, Plasmid

1. Introduction

1.1. Plasmids

Plasmids are self-replicating extrachromosomal DNA molecules found in both prokaryotes and eukaryotes. Although generally covalently closed circular molecules, linear plasmids have been reported (1). Plasmids were first reported in phytoplasmas in 1988 in maize bushy stunt phytoplasma (2), and since then extrachromosomal DNA has been found in more than 20 different phytoplasmas (3, 4). Plasmids are of interest because of their potential involvement with pathogenicity and virulence (5, 6), as well as a potential role in exchange of genetic material with the phytoplasma chromosome (7) and increasing the genetic diversity of phytoplasmas. A possible evolutionary link between phytoplasma plasmids and geminiviruses is also a source of ongoing investigation (8).

1.2. Overview of Methods

Phytoplasma plasmids have been isolated and detected in DNA extracted by a variety of methods. Some methods have involved isolation of intact phytoplasma cells using differential centrifugation (9) or sieve element isolation (10). Others have adapted

Matt Dickinson and Jennifer Hodgetts (eds.), *Phytoplasma: Methods and Protocols*, Methods in Molecular Biology, vol. 938, DOI 10.1007/978-1-62703-089-2_31,

commonly used "hot CTAB" methods of Doyle and Doyle (11) and Dellaporta et al. (12) for both large- and small-scale plant DNA extractions (3, 13). Commercial kits for plant DNA extraction have also been used successfully (3, 7). Phytoplasma plasmid DNA has been successfully separated, along with phytoplasma chromosomal DNA, from plant DNA by the use of CsCl–bisbenzimide gradients using modifications of the method of Kollar et al. (13–15). Phytoplasma plasmids have even been isolated using methods where they should have been excluded. Four plasmids were detected in aster yellows witches' broom DNA isolated using pulsed-field gel electrophoresis, where in principle the smaller plasmids should have separated away from the phytoplasma chromosomal DNA (16).

Two methods that specifically targeted the isolation of phytoplasma plasmids were those of Denes and Sinha (17) and Kuske and Kirkpatrick (18). The former employed a modification of the alkaline lysis method of Birnboim and Doly (19) in which covalently closed circular DNA is resistant to denaturation in alkaline conditions. Kuske and Kirkpatrick (18) used the CsCl–ethidium bromide gradient method of separating covalently closed circular plasmid DNA from linear chromosomal DNA—a standard procedure in molecular biology (20, 21) (see Note 1).

2. Materials

Unless otherwise stated, all solutions should be made in distilled or deionized water with an electrical conductivity of 18 MΩ or less at 25°C.

2.1. Alkaline Lysis Method

1. Mortar and pestle.
2. Centrifuge—low speed with suitable tubes to match.
3. Liquid nitrogen.
4. Lysing buffer: 100 mM Tris–HCl (pH 8.0), 50 mM EDTA (pH 8.0), 1% sodium dodecyl sulfate (SDS) (see Note 2).
5. NaCl: 5 M.
6. NaOH: 3 M (see Note 3).
7. Phenol/$CHCl_3$ (see Note 4).
8. $CHCl_3$ (see Note 4).
9. PEG 8000: 50% w/v.
10. Tris–HCl: 1 M pH 8.0 (see Note 5).
11. Tris–HCl: 2 M pH 7.0 (see Note 5).
12. EDTA: 0.5 M pH 8.0 (see Note 6).
13. SDS: 10% w/v (see Note 7).

2.2. CsCl–Ethidium Bromide Gradients

1. Ultracentrifuge, rotors and tubes.
2. Centrifuge, low speed.
3. Syringe, 3 mL with luer-lock connection.
4. Needles, 18 gauge to fit syringe.
5. Needles, 23 gauge or similar.
6. UV light source.
7. Personal protective equipment (PPE): nitrile gloves, UV-resistant face shield, UV-resistant glasses.
8. TE buffer (pH 8.0): 10 mM Tris (pH 8.0), 1 mM EDTA (pH 8.0).
9. CsCl (see Note 8).
10. CsCl solution: 1.55 g/mL dissolved in TE (pH 8.0).
11. Ethidium bromide (EtBr): 10 mg/mL (see Note 9).
12. 1-Butanol saturated with H_2O (see Note 10).

3. Methods

3.1. Alkaline Lysis Method

This is an adaptation of the alkaline lysis plasmid purification method that uses the characteristic of covalently closed circular DNA molecules to resist alkaline denaturation (17).

1. Grind plant material into a powder using a mortar and pestle with liquid nitrogen.
2. Add 10 mL of lysing buffer per gram of tissue and allow the cells to lyse for 1 h at room temperature (RT).
3. Adjust the viscous lysate to pH 12.4 by addition of 3 M NaOH with gentle but thorough mixing with a plastic pipette.
4. Incubate at RT for 30 min, then adjust lysate to pH 8.4 with 2 M Tris–HCl (pH 7.0).
5. Adjust solution to 1 M with respect to NaCl, and mix by gentle inversion.
6. Incubate on ice for 6 h.
7. Precipitate SDS/NaCl complex by centrifugation at 17,000 × g for 30 min at 4°C.
8. Extract supernatant with an equal volume of phenol/$CHCl_3$.
9. Extract upper aqueous layer with an equal volume of $CHCl_3$.
10. Transfer upper aqueous layer to a fresh tube, add 50% (w/v) PEG 8000 to give a final concentration of 10%.
11. Mix gently and leave on ice overnight (see Note 11).

12. Centrifuge samples at 12,000 × *g* for 30 min at 4°C to pellet the DNA.
13. Resuspend the pellet in 50 mM Tris–HCl (pH 8.0) and 20 mM EDTA (pH 8.0).

The resulting DNA can then be used for further purification procedures such as CsCl–bisbenzimide gradient separation that will separate the phytoplasma and host DNA.

3.2. CsCl–Ethidium Bromide Gradients

CsCl–ethidium bromide gradients are a reliable method of plasmid purification. However, they are lengthy, expensive and complicated procedures, which involve the use of ethidium bromide—a toxic and potentially mutagenic dye. For these reasons, researchers have tended to use quicker, cheaper, and less complicated methods; however these are only appropriate where lower yields and co-isolation with phytoplasma chromosomal DNA is not a problem.

Note of caution: compounds containing ethidium bromide should be treated with the utmost care. Nitrile gloves should be worn when handling this chemical, and waste containing this compound should be disposed of in accordance with accepted safe laboratory practice.

1. Add 1.01 g of solid CsCl per gram of DNA solution. Mix gently (warm to 30°C if necessary) (see Note 12).
2. Add 100 μL of 10 mg/mL EtBr for each 5 g of original DNA solution.
3. Check density by using a refractometer, or weigh 1 mL of solution. Adjust until refractometer measures 1.3860 or density is 1.55 g/mL.
4. A dark red precipitate may appear and this can be removed by a low speed 1,000–2,000 × *g* centrifugation for 5 min at RT (see Note 13).
5. Transfer solution to an ultracentrifuge tube using a Pasteur pipette. The tube chosen will depend on the rotor selected for ultracentrifugation (see Note 14).
6. Ensure centrifuge tubes are in balanced pairs and seal tube according to manufacturer's instructions.
7. Run gradients at 15–20°C at an appropriate speed and time. (This is dependent on the rotor selected).
8. After centrifugation, the tubes should be examined for the presence of bands (see Note 15).
9. After successful centrifugation there should be two bands. The top band is nicked circular or linear DNA, and the lower band is closed circular plasmid DNA (see Note 16).
10. Remove the lower (plasmid) band (see Note 17): using a small gauge needle, puncture the top of the tube leaving it in place,

to enable air to enter to tube and the lower band to be drawn out.

11. Attach an 18 gauge needle to a 3 mL syringe, and insert carefully into centrifuge tube with beveled side up (see Note 18).
12. Draw the lower band into the syringe, moving the needle around to capture as much of the DNA as possible. Draw the needle out, leaving the rest of the gradient to flow into a waste container, or seal needle hole with tape or vacuum grease.
13. Remove the needle and transfer DNA to a sterile glass or polypropylene test tube.
14. Remove EtBr from the DNA by extracting with an equal volume of 1-butanol saturated with water (see Note 19).
15. Discard (in an appropriate waste container) the top aqueous layer and continue until top aqueous layer remains clear.
16. Add 3 volumes of sterile distilled water (see Note 20).
17. Add 2 volumes of EtOH and leave at +4°C overnight or until required (see Note 21).
18. Pellet precipitated DNA by centrifugation at 20,000 × *g* for 15 min at +4°C.

3.3. Restriction Enzyme Digestion

Phytoplasma plasmids can be linearized or fragmented (for further analysis such as cloning and sequencing) by the use of restriction enzymes. Restriction enzymes are endonucleases that recognize and cut specific double-stranded DNA sequences. There is a vast array of restriction enzymes recognizing different DNA sequences, and they can be broadly categorized by the length in base pairs (bp) of the recognition sequence as 4, 6, or 8 cutters. The probability of the number of recognition sequences in any given stretch of DNA is inversely proportional to the length of the recognition sequence. A specific 4 base sequence is likely to appear once in every 256 (4^4) bp, 6 base sequence once in every 4,096 (4^6) bp, and 8 base sequence once in every 65,536 (4^8) bp. Restriction enzymes recognition sequences can also be AT or GC dominant, which could impact on the probability of a specific restriction site being present if, as for phytoplasmas, the DNA is A+T rich. For restriction enzyme digestions 1–5 μg of DNA is cut in accordance with the supplier's instructions.

4. Notes

1. Most protocols for plasmid purification are based on circular molecules, but not all plasmids are circular. It is possible that linear plasmids may be excluded by using such extraction methods. Likewise the CsCl–ethidium bromide purification

technique uses the low G+C content of phytoplasma DNA and the differential that exists between this and the DNA of the plant or insect host. Plasmids are generally considered to have a low G+C content, and therefore such extraction methods should be sufficient. If a plasmid did not have a low G+C content then it may be not be purified using such methods.

2. Lysing buffer can be made up from stock solutions.
3. Sodium hydroxide is harmful and very corrosive, and should be handled with care.
4. Phenol and chloroform ($CHCl_3$) are harmful chemicals and should not be used without proper PPE. Latex gloves or equivalent should be worn when handling these chemicals, and where possible, work should be done in a fume hood. Phenol should be water- or buffer-saturated. Phenol should not be used if it is pink or yellow unless 8-hydroxyquiniline has been added to a final concentration of 0.1%. Isoamyl alcohol (IAA) can be added to chloroform as an antifoaming agent in a ratio of 24 $CHCl_3$: 1 IAA. The ratio of phenol to $CHCl_3$ is 1:1. If IAA is added to $CHCl_3$ then the ratio of phenol:$CHCl_3$:IAA will be 25:24:1.
5. Tris–HCl can be made as a 1 or 2 M stock solution. Adjust to the correct pH by the addition of HCl. The stock solution should be autoclaved.
6. The disodium salt of EDTA will not go into solution until it is close to pH 8.0. This is achieved by the addition of NaOH pellets. Approximately 20 g of NaOH pellets are required for 1 L solution. The stock solution should be autoclaved.
7. SDS is also called sodium lauryl sulfate. Stock solutions of 10 or 20% can be made. Care should be taken with this chemical, and powder should be weighed out in a cytoxic hood or fume hood. Heating may be required to get powder into solution, and a precipitate can form in cold temperatures. To redissolve, simply reheat the solution. SDS should not be autoclaved.
8. A good quality grade (AnalaR or molecular biology grade) of CsCl should be used.
9. Great care must be taken with handling ethidium bromide. Nitrile gloves should be worn, and to prevent exposure to the dust ethidium bromide powder must be weighed out in a cytotoxic or fume hood. Alternatively ethidium bromide can be purchased as either preweighed tablets, or as a ready-made stock solution.
10. This is to avoid a flow of water from the DNA fraction into the 1-butanol fraction resulting in reduction in fraction size, and potential precipitation of CsCl when saturation has been achieved.

11. Keep ice in cold room to avoid ice melting.
12. To avoid precipitation of CsCl avoid temperatures below 15°C.
13. After low-speed centrifugation, the dark red flocculent material will either float on top (protein–EtBr complex) or form a pellet (cell debris). Using a plastic disposable Pasteur pipette, transfer the clear red solution to ultracentrifuge tube.
14. Rotor selection will depend on volumes to be centrifuged and what may be available. Vertical and near-vertical rotors will enable shorter run times. Some examples (20) are:

Beckman NVT 65 rotor	Spun at 366,000 × *g* (62,000 rpm)	For 6 h
Beckman VTi65 rotor	Spun at 194,000 × *g* (45,000 rpm)	For 16 h
Beckman Type 65Ti rotor	Spun at 314,000 × *g* (60,000 rpm)	For 24 h

15. Treat the gradients with great care. Keep them upright and with as little disturbance as possible. UV illumination (low-intensity shortwave) may be required if the DNA concentration is low. Ensure that proper protection is used to prevent damage to eyes and skin from exposure to UV irradiation. UV-resistant safety glasses and a full face shield should be worn while UV light is in use.
16. There may be a protein layer near the top and an RNA pellet at the bottom. With vertical rotors the shift of the gradient from the horizontal to the vertical can mean some of the protein layer may appear on the side of the tube. This should be avoided when extracting the lower band.
17. The most effective method of removing the band of interest is to puncture the centrifuge tube with a large-bore (18 gauge) hypodermic needle. This is a destructive process and care must be taken to contain the ethidium bromide solution that will flow from the tube. Ensure that everything is assembled and ready to be used prior to puncturing the tube.
18. There is more control of the process if the needle is attached to a syringe. (Prior to insertion: ensure plunger of syringe moves freely and draw the plunger out slightly). Penetration of the tube can be difficult and some force may be required. Ensure that fingers holding the tube are not directly opposite the needle entry point, as the needle under pressure may inadvertently penetrate the far side of the tube. A slight twisting or rotation of the needle can assist in getting penetration through the tube wall; however once the tip of the needle has broken the tube surface it will need to be fully inserted into the tube to prevent

leakage. A small piece of clear adhesive tape should be put over the proposed entry point prior to prevent leaks. If the needle becomes blocked as a result of the insertion process, leave it in place and insert another needle.

19. EtBr is extracted into the top aqueous layer making it pink. To extract, mix samples together in a suitable tube such as a Corex glass tube. The tube opening should be covered by a piece of clingfilm and then Parafilm and held firmly while the tube is inverted two to three times. The samples will settle into two phases, but a short low-speed centrifugation should facilitate a quicker separation.
20. The purpose here is to dilute the CsCl to a concentration at which it remains in solution after the ethanol precipitation step. An alternative is to remove the CsCl by dialysis.
21. CsCl can precipitate out at temperatures lower than 4°C.

References

1. Hinnebusch J, Tilly K (1993) Linear plasmids and chromosomes in bacteria. Mol Microbiol 10:917–922
2. Davis MJ et al (1988) Cloning of chromosomal and extrachromosomal DNA of the mycoplasmalike organism that causes maize bushy stunt disease. Mol Plant Microbe Interact 1: 295–302
3. Tran-Nguyen LTT, Gibb KS (2006) Extrachromosomal DNA isolated from tomato big bud and *Candidatus* Phytoplasma australiense phytoplasma strains. Plasmid 56:153–166
4. Namba S, Oshima K, Gibb K (2005) Phytoplasma genomics. In: Blanchard A, Browning GS (eds) Mycoplasmas: molecular biology, pathogenicity and strategies for control. Horizon Bioscience, Norfolk, pp 97–133
5. Panopoulos NJ, Peet RC (1985) The molecular genetics of plant pathogenic bacteria and their plasmids. Annu Rev Phytopathol 23: 381–419
6. Vivian A, Murillo J, Jackson RW (2001) The roles of plasmids in phytopathogenic bacteria: mobile arsenals? Microbiology 147:763–780
7. Toruño TY et al (2010) Phytoplasma PMU1 exists as linear chromosomal and circular extrachromosomal elements and has enhanced expression in insect vectors compared with plant hosts. Mol Microbiol 77:1406–1415
8. Nishigawa H et al (2002) Evidence of intermolecular recombination between extrachromosomal DNAs in phytoplasma: a trigger for the biological diversity of phytoplasma? Microbiology 148:1389–1396
9. Niemark H, Kirkpatrick BC (1993) Isolation and characterization of full-length chromosomes from non-culturable plant-pathogenic *Mycoplasma-like* organisms. Mol Microbiol 7:21–28
10. Lee I-M, Davis RE (1983) Phloem-limited prokaryotes in sieve elements isolated by enzyme treatment of diseased plant tissues. Phytopathology 73:1540–1543
11. Doyle JJ, Doyle JL (1990) Isolation of plant DNA from fresh tissue. Focus 12:13–15
12. Dellaporta SL, Wood J, Hicks JB (1983) A plant DNA minipreparation: version II. Plant Mol Biol Rep 1:19–21
13. Liefting LW et al (2006) Comparative analysis of the plasmids from two isolates of "*Candidatus* Phytoplasma australiense". Plasmid 56: 138–144
14. Harrison NA et al (1991) Molecular cloning and detection of chromosomal and extrachromosomal DNA of mycoplasmalike organisms associated with witches'-broom disease of pigeon pea in Florida. Mol Plant Microbe Interact 4:300–307
15. Kollar A et al (1990) Isolation of the DNA of various plant pathogenic mycoplasmalike organisms from infected plants. Phytopathology 80:233–237
16. Bai X et al (2006) Living with genome instability: the adaptation of phytoplasmas to diverse environments of their insect and plant hosts. J Bacteriol 188:3682–3696

17. Denes AS, Sinha RC (1991) Extrachromosomal DNA elements of plant pathogenic mycoplasmalike organisms. Can J Plant Pathol 13: 26–32

18. Kuske CR, Kirkpatrick BC (1990) Identification and characterization of plasmids from the western aster yellows mycoplasmalike organism. J Bacteriol 172:1628–1633

19. Birnboim HC, Doly J (1979) A rapid alkaline extraction procedure for screening recombinant plasmid DNA. Nucleic Acids Res 7:1513–1523

20. Sambrook J, Russell DW (2001) Molecular cloning: a laboratory manual, 3rd edn. Cold Spring Harbour Press, New York

21. Ausubel FM et al (eds) (1992) Current protocols in molecular biology. Wiley, New York

Chapter 32

Cesium Chloride-Bisbenzimide Gradients for Separation of Phytoplasma and Plant DNA

Lucy T.T. Tran-Nguyen and Bernd Schneider

Abstract

CsCl–bisbezimide gradient centrifugation is a method to separate phytoplasma from host plant DNA. Bisbenzimide forms a complex with A + T-rich DNA thereby lowering its relative density. During centrifugation the A + T-rich phytoplasma DNA is spatially separated from the less A + T-rich host plant DNA. The difference in buoyant density between phytoplasma DNA and plant DNA varies according to the host–pathogen combination. The phytoplasma DNA forms a distinct band above the host plant DNA and can be collected. Depending on the phytoplasma titer and the scale of extraction, highly purified DNA is obtained in sufficient quantities for the construction of a genomic library, a sequencing project or hybridization studies.

Key words: Bisbenzimide, Buoyant density gradient centrifugation, Cesium chloride, DNA extraction, Hoechst 33258

1. Introduction

Cesium chloride (CsCl) density gradient centrifugation is used to purify and separate DNA in an analytical and preparative scale. The purified DNA can be used for downstream applications such as library construction, genome sequencing, or hybridization studies. CsCl solutions form a linear density gradient after appropriate time and centrifugal force. The DNA accumulates in a CsCl gradient at a position equal to its relative density (~1.7 g/cm^3), which is related to its G + C content (1). Based on this, DNA with different base compositions can be spatially separated. However, the separation distance is low and individual fractions are difficult to collect. This situation is improved by complexing DNA with bisbenzimide (Hoechst 33258) (2, 3). The fluorescent dye interacts with A + T-rich sequences in the minor groove of DNA strands (4, 5)

Matt Dickinson and Jennifer Hodgetts (eds.), *Phytoplasma: Methods and Protocols*, Methods in Molecular Biology, vol. 938, DOI 10.1007/978-1-62703-089-2_32, © Springer Science+Business Media, LLC 2013

proportional to the content of A+T bases, thereby lowering its relative density compared to G+C-rich DNA. The DNA–bisbenzimide complex fluoresces blue, with excitation and emission wavelengths of 365 and 465 nm, respectively, allowing easy visualization of DNA under UV light.

The A+T content of phytoplasma DNA ranges from 78.6 to 72.6 mol% for '*Candidatus* Phytoplasma mali' and '*Candidatus* Phytoplasma australiense', respectively (6–8). Phytoplasma DNA is, in general, A+T rich compared with plant DNA and can be separated according to the principles outlined earlier. The separation distance of phytoplasma and host plant DNA depends on the combination of pathogen and host. Whereas aster yellows phytoplasma DNA (27–29 mol% G+C) and periwinkle (32.9 mol% G+C) DNA are closely spaced, DNA from the apple proliferation phytoplasma (21.4 mol% G+C) is well separated from periwinkle, tobacco (35 mol% G+C), or apple (35.1 mol% G+C) DNA (6, 9).

Prior to commencing phytoplasma DNA purification for the first time from a host plant, it is advisable to examine the healthy plant DNA in CsCl–bisbenzimide gradients. Plants often contain satellite DNAs which can be either A+T- or G+C-rich and produce distinct DNA bands apart from the main chromosomal DNA (10, 11). A+T-rich satellites can superimpose on phytoplasma DNA or be mistakenly interpreted as phytoplasma DNA. Periwinkle (*Catharanthus roseus* (L.) G. Don.), the preferred experimental host plant for many phytoplasmas, shows a distinct band (26.5 mol% G+C) above the main chromosomal DNA after CsCl–bisbenzimide centrifugation (Fig. 1) (6). G+C-rich satellites are located below the chromosomal DNA band. The G+C content of DNA from most phytoplasma host plants is undetermined and it is advisable to run a reference DNA (i.e., *Spiroplasma citri*) with known G+C content in parallel. *S. citri* is easy to grow and its G+C content (26 mol%) is similar to phytoplasma DNA.

Preparations for CsCl–bisbenzimide centrifugation depend on the scale of phytoplasma DNA purification. The staining procedure in particular is critical. Analytical amounts of DNA (<50 μg/mL) can be stained by direct addition of diluted dye solutions (<0.5 mg/mL) to DNA using gentle mixing. Concentrated DNA solutions (>50 μg/mL) have to be stained by dialysis to avoid precipitation of DNA caused by the stain. After staining, the refractive index (*n*) of the CsCl–DNA solution is adjusted to an experimentally calculated value which is close to the buoyant density of stained phytoplasma DNA (6). Following equilibrium centrifugation, the DNA bands are visualized under UV light and the uppermost band, representing the phytoplasma DNA, is collected. In general, the phytoplasma DNA is subjected to further rounds of centrifugation until no remnants of plant DNA are visible. After the final centrifugation, the phytoplasma DNA is destained and desalted prior to downstream applications.

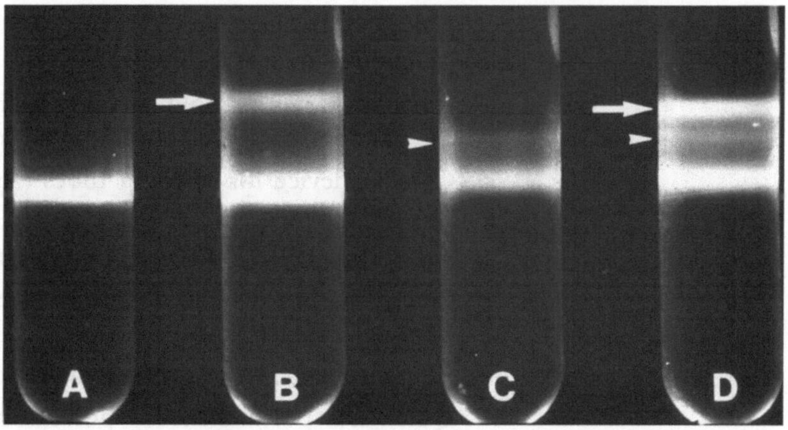

Fig. 1. Purification of '*Candidatus* Phytoplasma mali' DNA by CsCl–bisbenzimide centrifugation from its natural host *Malus domestica* (**b**) and its experimental host *Catharanthus roseus* (**d**). (**a**, **c**) Healthy apple and periwinkle, respectively. *Arrows* indicate phytoplasma DNA. *Arrowheads* indicate A+T-rich periwinkle satellite DNA. Reprinted from Kollar A, Seemüller E, Bonnet F, Saillard C, Bove JM (1990) Isolation of the DNA of various plant pathogenic mycoplasmalike organisms from infected plants. Phytopathology 80:234.

2. Materials

2.1. Chemicals

1. DNA extracted using the phytoplasma enrichment protocol (see Notes 1 and 2) or cetyltrimethylammonium bromide (CTAB) extraction method.
2. Cesium chloride, analytical grade: 1.4 g/mL in 10 mM Tris–HCl, pH 8.0.
3. Bisbenzimide (Hoechst 33528): 0.5 mg/mL in ddH_2O.
4. Tris(hydroxymethyl)-aminomethane: 10 mM in ddH_2O, pH 8.0.
5. TE buffer: 10 mM Tris–HCl, pH 8.0; 1 mM EDTA, pH 8.0.
6. Isopropanol.
7. Ethanol: 70 and 80%.
8. TAE buffer (50× stock): 242 g Tris base, 57.1 mL glacial acetic acid, 100 mL 0.5 M EDTA (pH 8.0) in 1 L.
9. Bacteriophage lambda DNA.
10. Agarose, molecular biology grade: 1% in 1× TAE buffer.
11. Ethidium bromide: 10 mg/mL in ddH_2O.
12. DNA loading dye (6× stock): 25 mg bromophenol blue, 25 mg xylene cyanol FF, 3.3 mL glycerol, and 6.71 mL ddH_2O to make 10 mL.

2.2. Equipment

1. Dialysis tubing MWCO 12,000–14,000: diameter depends on volume of staining solution, prewet tubing in ddH_2O in a beaker for at least 15 min.

2. Dialysis clips (optional, or simply knot the tube).
3. Magnetic stirrer/stirring bar.
4. Rack or device for holding tubes (suitable for visualization of the gradient on a UV transilluminator).
5. Cutting device for opening tubes (secateurs or sharp scissors recommended).
6. Teflon tube: 5–8 cm, 1 mm inner diameter.
7. Pipette (Gilson P1000 or P5000 or equivalent) or disposable syringe (5 or 10 mL).
8. Refractometer: Kruess refractometer or comparable device.
9. Quick-seal polyallomer or polycarbonate ultracentrifugation tubes: 13.5 mL or larger (Beckman and Coulter).
10. Quick-seal device (Beckman and Coulter).
11. Ultracentrifuge rotor(s): Beckmann Ti, NVT or VT series.
12. Ultracentrifuge: Beckmann, L8-70 M or comparable device.
13. UV transilluminator.
14. Power supply.
15. Electrophoresis chamber.
16. UV-protective clothing (face shield, gloves).
17. Spectrophotometer.

3. Methods

1. Dissolve solid CsCl in 10 mM Tris–HCl, pH 8.0 at a concentration of 1.4 g/mL (see Note 3). The solution should be filtered if insoluble material is present. Prepare bisbenzimide (Hoechst 33528) solution (see Note 4).
2. The nucleic acid pellet is carefully dissolved in 5–10 mL of CsCl solution. The volume needed for complete solubilization depends on the amount and purity of DNA (see Note 5).
3. Estimate the amount of DNA by agarose gel electrophoresis or spectrophotometry (see Note 6).
4. Calculate the total amount of stain needed. The ratio of DNA: stain should be 2:1 (e.g. 50 μg of DNA: 25 μg bisbenzimide). The stain is either added directly in small quantities with gentle mixing or filled in a prewetted dialysis tube (see Notes 7 and 8). Place a stirring bar inside the dialysis tube and seal by either clipping or knotting. Add the dialysis tube to the DNA–CsCl solution. The DNA–CsCl solution should be in a small beaker or Erlenmeyer flask. The depth of the DNA–CsCl solution in the beaker/flask should be the same level as the dialysis tube. The flask is placed on a magnetic stirrer and adjusted to a speed

that the dialysis tube is just moving. *High speed causes shearing of DNA strands.* Dialyze at room temperature overnight in a dark place such as a cupboard or simply cover the flask with aluminum foil. Remove the dialysis tube the next day.

5. Adjust the refractive index of the solution with a precision refractometer to a value between 1.3942 and 1.3945 (see Note 9). Use the DNA–CsCl solution sparingly as multiple reads might be necessary for accurate adjustment.
6. Measure the volume of the DNA–CsCl solution. Take the appropriate number of polyallomer tubes and fill the tubes with identical volumes. If the tubes are not completely filled, top up with CsCl in 10 mM Tris–HCl solution of identical refractive index. A small space should be left between the solution surface and the neck of the tube (see Note 10).
7. Weigh tubes before sealing. Seal the tubes with the quick-seal device.
8. Select an appropriate rotor for ultracentrifugation (see Note 11). Place balanced tubes opposite each other, cover with a spacer, and close the lid tightly. Run the gradient to equilibrium at 71,830 × *g* [r_{max}] for 72 h (e.g., Ti70.1 rotor 99,746 × g) at 20°C without deceleration.
9. Remove tubes from rotor carefully and place on the UV transilluminator.
10. Open each centrifuge tube carefully (see Notes 12 and 13). Switch on UV light. Insert the Teflon tube to the position of the phytoplasma DNA band (Fig. 2). Collect the phytoplasma

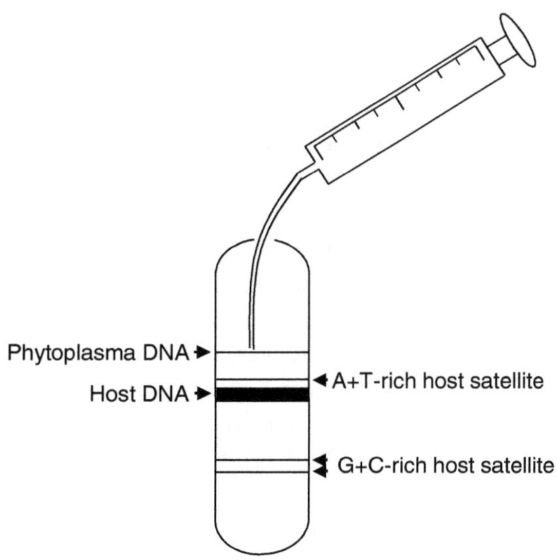

Fig. 2. Graphical display of the DNA banding pattern after the first round of CsCl–bisbenzimide gradient centrifugation of *C. roseus* infected by '*Ca.* Phytoplasma mali' and method of removing the phytoplasma DNA.

DNA fraction keeping the volume as small as possible. Pool the collected phytoplasma DNA if multiple tubes were used for centrifugation.

11. Measure the amount of recovered DNA by either of the methods mentioned in step 3. Calculate the amount of stain as described in step 4. Staining is performed by direct addition of bisbenzimide under slight agitation. Adjust the refractive index as described before and run the gradient with the same parameters. In cases where a vertical rotor is used the running time can be reduced.
12. Repeat steps 9–11 (see Notes 14 and 15).
13. The collected pure phytoplasma DNA is destained by adding an equal volume of isopropanol and gently inverting the tube several times (see Note 16). Remove the upper isopropanol layer, then add an equal amount of isopropanol again, repeating twice more. To control the destaining procedure, the CsCl solution is briefly exposed to UV light: if no fluorescence is observed, destaining is complete.
14. The phytoplasma DNA is precipitated by adding 2.5 volumes of room temperature 80% ethanol. The DNA is recovered by centrifuging in a microcentrifuge at 16,060× g (15,000×*g*) for 20 min at room temperature. The pellet is washed twice with 70% ethanol (see Note 17).
15. Dissolve the DNA in a small amount of TE buffer and measure the quantity of recovered DNA. The DNA is ready for downstream applications.

4. Notes

1. Due to the narrow spacing or superposition of the main chromosomal DNA band or A+T-rich satellite from *C. roseus*, respectively, with DNA of some phytoplasmas (i.e., '*Ca.* Phytoplasma asteris'), dodder transmission of the phytoplasma to an alternative host plant such as tobacco, or use of pulsed-field gel electrophoresis (PFGE), should be considered. However, the A+T-rich satellite of *C. roseus* is often present in lower quantities compared to the phytoplasma DNA. Depending on the experimental purpose, some small plant DNA contamination might be tolerable.
2. The DNA can be extracted according to any protocol yielding high quality DNA. CTAB-based protocols are the most common methods for large-scale plant DNA extraction (12, 13). Regardless of the extraction procedure, the amount of plant DNA always exceeds phytoplasma DNA. The balance, however,

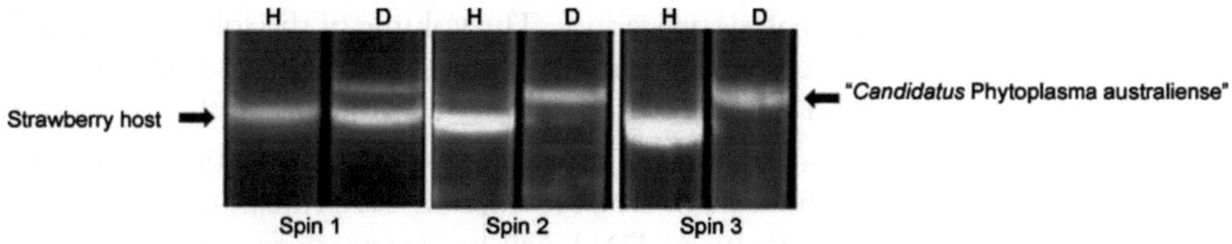

Fig. 3. CsCl–bisbenzimide gradients of '*Ca.* Phytoplasma australiense' DNA from *Fragaria* spp. showing the host and phytoplasma band positions after each centrifugation step.

can be shifted to some extent towards phytoplasma DNA when plant material with a high phytoplasma titer is used. Phloem tissue is the richest source of phytoplasma DNA. In some cases the phloem tissue of woody plants or from lignified parts of herbaceous plants can be prepared by peeling off the phloem from a ring-barked specimen. With most plants, however, this procedure is not feasible or it is too cumbersome. In such cases, plant parts with a high phytoplasma but low plant DNA content, i.e. stems, roots, midribs, peduncles, or flowers should be used. Excessive plant DNA causes broad host DNA bands in CsCl–bisbenzimide gradients, interfering or even masking the phytoplasma DNA. If only DNA-rich material (i.e. leaves) is available, a phytoplasma enrichment procedure according to Ahrens and Seemüller (13) should be performed. For plant hosts that contain high concentrations of polysaccharides and polyphenols, such as strawberry (*Fragaria* spp.), it is recommended that the DNA extraction method be modified as described by Porebski et al. (14). The '*Ca.* Phytoplasma australiense' DNA band is positioned very close to the strawberry host band (Fig. 3); this may lead to a small amount of plant DNA carry over which can be tolerated depending on the applications post purification.

Use disposable sterile plasticware whenever handling the purified phytoplasma DNA. The DNA pellet can be dissolved in several ways but degradation or shearing should always be kept to a minimum. Large DNA pellets tend to be recalcitrant to dissolve. Resuspend the DNA by spreading the alcohol-washed wet pellet to a thin layer, air-dry the pellet and add a few mL of CsCl solution. The larger the surface, the faster the DNA dissolves. The process can be aided by slight warming of the soaked DNA or by carefully chopping the DNA in finer chunks. Time is important; leaving the DNA–CsCl solution on the bench overnight often gives good results. CsCl solution is not the best solvent but protects the DNA from the action of degrading enzymes. Other methods of DNA solubilization can be applied provided the integrity of the DNA is guaranteed. Smaller remaining chunks will dissolve in the solution during the dialysis procedure. The volume of the solvent should always be kept to a minimum at the start, otherwise large amounts of solid CsCl need to be added to adjust the refractive index.

3. Good quality CsCl is expensive. The quantities needed depend on the scale of DNA extraction. As a rule of thumb, 1 mL CsCl per 150 μg of DNA will be necessary for the first centrifugation. If large quantities of CsCl solution are prepared, the crystals should be added slowly because the solution cools while CsCl dissolves. To speed up the process, the solution can be warmed to 60°C in a microwave oven or water bath.

4. Although the toxicology of bisbenzimide is not fully investigated, it is recommended that the operator wear a lab coat, protective gloves, and safety glasses at any time when handling the stock solution or solutions containing the stain.

5. The DNA is stable in buffered CsCl solutions at room temperature. No cooling is required between individual steps if the complete procedure is done reasonably quickly. However, solutions containing bisbenzimide should be kept in a dark place.

6. The amount of DNA can be estimated by spectrophotometry at 260 nm. One optical density unit (OD) corresponds to 50 μg of DNA. The nucleic acid pellet usually contains high amounts of RNA after CTAB extraction which results in inaccurate calculations (usually too high). A simple method for accurate estimation of quantity and quality is by conducting gel electrophoresis. An aliquot of the dissolved DNA is electrophoresed in comparison with a known DNA standard, e.g. lambda DNA. Due to the high CsCl concentration, the solution has to be diluted fivefold with TE buffer prior to electrophoresis. Serial dilutions of extracted and lambda DNA should be prepared and examined in a 1% agarose gel stained with ethidium bromide. There are several alternative methods for estimation of DNA concentration, although most of them will not determine the quality and so are not recommended.

   If care is taken from the grinding of plant material to the resuspension of the precipitated DNA, an average DNA fragment size of 20–35 kb (estimated by PFGE, Schneider unpublished) can be obtained. Large amounts of degraded DNA result in unfocused fuzzy bands and A+T-rich plant DNA sequences contaminate phytoplasma DNA. The amount of RNA does not affect the quality of the gradient, because RNA ($\rho = 1.85$ g/cm^3) sediments and proteins ($\rho = 1.3$ g/cm^3) float under the conditions applied.

7. Bisbenzimide is a yellow to yellow-green powder with a water solubility of 10 mg/mL, but stock solutions are usually prepared as 0.5 mg/mL or less. The dissolved stain can be stored

at 4°C in the dark. The staining procedure depends on the concentration of the DNA in solution. If the DNA concentration is high (<0.1 mg/mL), direct addition of the staining solution results in precipitation (yellow strings formed). This can, to a certain degree, be avoided by using a less concentrated bisbenzimide stock and by swirling the DNA solution while adding the stain in small quantities (dropwise). Addition of stain should stop immediately if precipitation is apparent. Concentrated DNA solutions have to be stained by dialysis. The stirring bar should be larger than the diameter of the tube so it does not rotate inside. During the staining procedure a large amount of water will pass through the dialysis tube and dilute the CsCl solution. Slow overnight agitation on a magnetic stirrer gives excellent staining results without DNA precipitation. Other methods of agitation without using a stirring bar have been trialed but did not provide acceptable results (Schneider unpublished). The efficiency of staining can be verified by exposing the DNA solution to UV light after removing the dialysis bag. The solution should fluoresce bright blue. A yellowish color indicates poor staining or little DNA in the solution. The stirring bar can be recovered and cleaned with isopropanol.

8. The amount of DNA per mL should not exceed 100–150 μg. Most of the DNA will be from the host plant DNA and higher concentrations result in a broad host DNA band. This can be a problem when the spatial separation between phytoplasma and host DNA is low. As a result, the phytoplasma DNA cannot be removed without heavy contamination by host DNA. In the worst case, the host band covers the phytoplasma DNA band.

9. For details on how to operate the refractometer, refer to the manufacturer's instructions. Adjust the refractive index to a value of 1.3942–1.3945 by adding solid CsCl. If the refractive index is too high, 10 mM Tris–HCl can be added. Solid CsCl or Tris–HCl should be added carefully and in small quantities to assess the degree of change. The DNA–CsCl solution should be gently mixed before the next measurement. Avoid adding too much solid CsCl as it cools down the solution, preventing solubilization. The refractive index was estimated experimentally (6) and is slightly higher than the refractive index of the CsCl solution at which DNA from '*Ca.* Phytoplasma mali' migrates ($n = 1.3940$ [$\rho = 1.630$ g/cm^3]).

10. The neck of the tube must be free of CsCl drops or encrusted material. The difference in weight of the tubes (in opposite positions in the rotor) should not exceed 10 mg. It is recommended to balance the tubes using a digital balance. If necessary, adjust by adding or removing CsCl solution. This is best done with a clean needle attached to a syringe. After sealing,

the tubes should be checked for leaks by moderate pressing between thumb and forefinger. To avoid any risk of contaminating yourself if the tube is not sealed correctly, place them in a plastic bag before applying pressure.

11. There is a range of rotors available holding tubes of different volumes suited for gradient centrifugation at analytical or preparative scale (e.g. Beckman Ti70.1, VTi50, and NVTi70). For a preparative scale experiment, start with rotor type VTi50, holding 8 tubes each with a maximum content of 42 mL. Vertical rotors reduce centrifugation times markedly. For analytical scale or recentrifuging, use Ti70.1 or NVTi70 with 72 or 42 h running time, respectively.

12. The gradient is stable for some time but it is better to proceed immediately after the centrifuge has stopped. Place the tube in a rack or tube holder that allows a clear view of the entire tube from top to bottom. Commercial plastic racks can be modified to allow the UV light to penetrate freely. The DNA banding pattern depends on the phytoplasma–plant combination. Examples provided include '*Ca.* Phytoplasma mali'-infected *Malus domestica* and *C. roseus* (Fig. 1) and '*Ca.* Phytoplasma australiense'-infected strawberry (Fig. 3). The most prominent DNA band is from the host and is located in the lower half of the centrifuge tube. The brightness and breadth of this band depends on the amount and quality of DNA and the efficiency of the staining. In the case of '*Ca.* Phytoplasma mali', the distance to the plant host DNA band is about 1 cm (Fig. 1). The brightness of this band varies and depends on the phytoplasma titer in your plant and the extraction procedure. Often the band is much fainter (Fig. 3), sometimes barely visible compared to the situation depicted in Fig. 1. An A+T-rich plant satellite can also be visible between the phytoplasma and host DNA band (Fig. 1d). The fluorescence of this band resembles the phytoplasma DNA in brightness but the separation distance from the main band is smaller. As mentioned earlier, the distance of the phytoplasma DNA band to the *C. roseus* chromosomal DNA band depends on the A+T content and is usually less than with '*Ca.* Phytoplasma mali' DNA. Record the gradient by taking a photograph. Mark the upper and lower border of all DNA bands with a permanent marker. Handling time under UV light should be kept to a minimum to avoid DNA damage and self-exposure. DNA quantities of 5 μg and less are visible as a distinct band under UV light in 12.5 mL polyallomer tubes.

13. Secateurs or very sharp scissors are an ideal tool for opening quick-seal tubes. If a regular pipette is used, ensure the piston is pressed down and the air is removed before inserting the Teflon tube. Air bubbles will disturb the gradient. Place the

tip of the Teflon tube right over the DNA band of interest (Fig. 2). Very slowly and with extreme care, suck the CsCl solution in the pipette or syringe, performing a rotary movement with the tip of the Teflon tube. While doing so, slowly move the Teflon tip from the upper margin to the lower margin of the marked band. Failure to do this step will result in the collection of CsCl from above, where the solution is less dense than the solution underneath. If the phytoplasma and host band are very close together, the permanent marker will be a helpful guide. Avoid touching or disturbing the layer of chromosomal DNA. This step might need some practice. DNA is very sticky, and if the chromosomal DNA layer is touched it is difficult to stop it entering the tube. In this instance, remove the pipette and start again. Puncturing the centrifuge tube with a syringe needle from the side to collect the phytoplasma DNA was trialed but, due to difficulties in controlling the flow and the sticky nature of DNA, this method is not recommended.

In cases where the phytoplasma titer in the host plant is low or the DNA is too dilute, it will be difficult to see a distinct phytoplasma DNA band after the first centrifugation. This does not mean there is no phytoplasma DNA. The band might just be too faint to see or the fluorescence of the host DNA is too strong. In this case, the CsCl fraction from above the host DNA band can be taken. The dots from the permanent marker will be a good guide for what fraction to take. In most cases a phytoplasma DNA band will appear after recentrifugation.

It is always advisable to recover a small amount of the plant DNA. An aliquot can be taken from the healthy plant reference DNA spun in parallel.

14. The DNA is now in a pure state and a spectrophotometer reading is reliable. Use a microcuvette to reduce loss of precious phytoplasma DNA. If very high readings are obtained the fraction is probably contaminated by plant DNA. Staining must be done carefully to avoid precipitation. If the pooled volume of the phytoplasma fractions is very low (1–4 mL), CsCl solution with a refractive index of 1.3942–1.3945 should be added, up to 12 mL, before adding the stain. This reduces the risk of DNA precipitation.

15. The recovered fraction of the plant DNA should be treated the same as the phytoplasma DNA and run as a comparison. Depending on the spatial separation of phytoplasma and host DNA and the skills of the operator, recentrifugation should be repeated until no visible host DNA band is present.

16. Destaining is performed by adding the same volume of room temperature isopropanol to the DNA–CsCl solution. The solution is gently mixed two to three times and left until phase separation

is apparent. A quick spin in a microcentrifuge speeds up the process. The upper isopropanol phase is discarded. Isopropanol withdraws water from the CsCl solution, thereby increasing the CsCl concentration and causing salt precipitation. If the solution becomes turbid, the addition of a few microliters of 10 mM Tris–HCl reverses the process. If too much Tris–HCl is added, the phase separation might disappear. Separation of phases can be restored by adding small amounts of solid CsCl until the phases reappear. After the final destaining step any remaining isopropanol is removed. The success of the procedure can be visualized on the UV transilluminator. The DNA–CsCl solution should be nonfluorescent.

17. An alternative method to remove the CsCl is by column chromatography through Sepharose 4B equilibrated with TE. The DNA–CsCl solution is loaded on top of the column and, due to physical properties of the matrix, the migration of the CsCl is retarded relative to the phytoplasma DNA. The eluted DNA fraction (in TE buffer) can be identified by an absorbance monitor. The shape and size of the peak is an indicator for quality and quantity. After column chromatography the volume increases and the DNA can be recovered by sedimentation in an ultracentrifuge (277,072 × *g* for 24 h at 4°C with maximum deceleration) using polycarbonate tubes. Immediately after the run, discard the supernatant and place the inverted tube on absorbent paper to remove any residual buffer. The translucent DNA pellet can be dissolved in a small amount of TE buffer. After column chromatography the DNA can also be concentrated with various commercial devices (e.g. Millipore Centricon Ultracell YM100).

Acknowledgements

The authors would like to thank Erich Seemüller (Julius Kühn-Institute, Federal Research Centre for Cultivated Plants, Institute for Plant Protection in Fruit Crops and Viticulture) for critically reading the chapter.

References

1. Schildkraut CL, Marmur J, Doty P (1962) Determination of the base composition of deoxyribonucleic acid from its buoyant density in CsCl. J Mol Biol 4:430–443
2. Bontemps H, Houssier C, Fredericq E (1975) Physio-chemical study of the complexes of "33258 Hoechst" with DNA and nucleohistone. Nucleic Acids Res 2:971–984
3. Manuelidis L (1977) A simplified method for preparation of mouse satellite DNA. Anal Biochem 78:561–568
4. Burckhardt G et al (1985) Two binding modes of netropsin are involved in the complex formation with poly(dA-dT).poly(dA-dT) and other alternating DNA duplex polymers. J Biomol Struct Dyn 2:721–736

5. Kopka ML et al (1985) Binding of an antitumor drug to DNA, Netropsin and C-G-C-G-A-A-T-T-BrC-G-C-G. J Mol Biol 183:553–563
6. Kollar A, Seemüller E (1989) Base composition of the DNA of mycoplasma-like organisms associated with various plant diseases. J Phytopathol 127:177–186
7. Kube M et al (2008) The linear chromosome of the plant-pathogenic mycoplasma '*Candidatus* Phytoplasma mali'. BMC Genomics 9:306
8. Tran-Nguyen LTT et al (2008) Comparative genome analysis of '*Candidatus* Phytoplasma australiense' (subgroup tuf Australia I; rp-A) and 'Ca. Phytoplasma asteris' strains OY-M and AY-WB. J Bacteriol 190:3979–3991
9. Berg M, Seemüller E (1999) Chromosomal organization and nucleotide sequence of the genes coding for the elongation factors G and Tu of the apple proliferation phytoplasma. Gene 226:103–109
10. Capesius I et al (1975) An A + T-rich satellite DNA in a monocotyledonous plant, Cymbidium. Biochim Biophys Acta 395: 67–73
11. Sacristán MD, Dobrigkeit I (1973) Variable content of a (GC) rich satellite DNA in tumorous and normal cultures of *Crepis capillaris.* Z Naturforschung C 28:564–567
12. Doyle JJ, Doyle JL (1990) Isolation of plant DNA from fresh tissue. Focus 12:13–15
13. Ahrens U, Seemüller E (1992) Detection of DNA of plant pathogenic mycoplasma-like organisms by a polymerase chain reaction that amplifies a sequence of the 16S rRNA gene. Phytopathology 82:828–832
14. Porebski S, Bailey LG, Baum BR (1997) Modification of a CTAB DNA extraction protocol for plants containing high polysaccharide and polyphenol components. Plant Mol Biol Rep 15:8–15

Chapter 33

Pulsed-Field Gel Electrophoresis for Isolation of Full-Length Phytoplasma Chromosomes from Plants

Carmine Marcone

Abstract

Pulsed-field gel electrophoresis (PFGE) is a powerful technique for genomic studies of unculturable plant-pathogenic phytoplasmas, which enables separation of full-length phytoplasma chromosomes from contaminating host plant nucleic acids. The PFGE method described here involves isolation of phytoplasmal DNA from high-titer phytoplasma-infected herbaceous plants using a phytoplasma enrichment procedure, embedding of phytoplasma chromosomes in agarose blocks, and separation of entire phytoplasma chromosomes from contaminating host plant nucleic acids by electrophoresis. Full-length phytoplasma chromosomes are resolved as single, discrete bands in the gel. The identity of these bands can be confirmed by Southern blot hybridization using a ribosomal DNA fragment as a probe. The method does not utilize gamma-irradiation to linearize phytoplasma chromosomes prior to electrophoresis.

Key words: Agarose blocks, *Catharanthus roseus*, CHEF system, Full-length phytoplasma chromosome, Phloem preparation, Pulsed-field gel electrophoresis

1. Introduction

Pulsed-field gel electrophoresis (PFGE) is a powerful technique in bacterial genomics which enables separation of high-molecular-weight DNA ranging in size from 10 kb to more than 10 Mb, well beyond the upper resolution limit of standard agarose gel electrophoresis (see ref. (1) for review). Unlike standard agarose gel electrophoresis, PFGE separates DNA molecules by applying an electric field that periodically changes direction in the gel matrix during the run. Separation is achieved because different-sized DNA molecules differ in their abilities to reorientate and align themselves to the new field direction. The larger a molecule is, the longer it will take to change to the new direction, and this time difference can be used to separate molecules. PFGE has had a major impact on genomic studies of unculturable plant-pathogenic phytoplasmas, mainly due to its

Matt Dickinson and Jennifer Hodgetts (eds.), *Phytoplasma: Methods and Protocols*, Methods in Molecular Biology, vol. 938, DOI 10.1007/978-1-62703-089-2_33,

ability to physically separate full-length phytoplasma chromosomes from contaminating host plant nucleic acids. In various studies chromosome sizes of more than 100 phytoplasmas, belonging to major phylogenetic groups and subgroups, have been estimated by comparing electrophoretic mobilities of full-length chromosomes to those of DNA size markers in PFGE gels (2–6). Physical maps of phytoplasmas have been constructed using rare-cutting restriction endonucleases on PFGE-purified full-length chromosomes, resolving the fragments with PFGE (7–11). Phytoplasma genomic libraries have also been generated from PFGE-purified high quality DNA. The availability of such libraries has made possible sequencing projects targeting either selected chromosomal fragments or entire chromosomes (12–16).

Previous work has shown that the yield of phytoplasma full-length chromosomal DNA isolated by PFGE largely depends on the host plant. Substantial amounts of chromosomal DNA, sufficient for restriction enzyme mapping, could only be obtained from plants in which phytoplasmas occur in very high titer or which have at least a fairly high titer and are, in addition, suitable for phloem preparation. As phytoplasmas occur in plants only in the phloem, preparation of phloem leads to a considerable phytoplasma enrichment. For some phytoplasmas, sufficient amounts of PFGE-purified DNA can be obtained from *Catharanthus roseus* (Madagascar periwinkle), the most important experimental phytoplasma host in which many phytoplasmas are routinely maintained by periodic grafting. In cases where periwinkle has proved to be an unsuitable host, other host plants such as celery, *Nicotiana tabacum* (tobacco) cv. Samsun, *Nicotiana glutinosa*, tomato, or faba bean have been much better as a DNA source (7, 8, 10, 11). *N. tabacum* cv. Samsun and *N. glutinosa* can support not only a high phytoplasma titer, but have also proved suitable for phloem preparations (8, 10). There are no reports of successful isolation of full-length phytoplasma chromosomes by PFGE from phloem preparations of woody plants, e.g., extraction of '*Candidatus* Phytoplasma mali' chromosomes from highly colonized phloem root of apple trees has been reported as unsatisfactory (17).

The PFGE method described in this chapter is based on three major steps:

Subheading 3.1 isolation of phytoplasmal DNA from high-titer phytoplasma-infected herbaceous plants using a phytoplasma enrichment procedure.

Subheading 3.2 embedding of phytoplasma chromosomes in agarose blocks.

Subheading 3.3 separation of entire phytoplasma chromosomes from contaminating host plant nucleic acids by electrophoresis.

To confirm the identity of resolved bands, an additional step (Subheading 3.4) which involves Southern blot hybridization with a phytoplasma ribosomal DNA fragment as probe may be carried out. The method described does not use γ-irradiation to linearize phytoplasma chromosomes prior to electrophoresis.

2. Materials

Use sterile, distilled deionized water to prepare all reagents.

2.1. Chromosomal DNA Isolation

1. High-titer phytoplasma-infected herbaceous plants.
2. Phytoplasma grinding buffer: 125 mM potassium phosphate, 10% sucrose, 0.15% bovine serum albumin, 2% polyvinylpyrrolidone, 30 mM ascorbic acid, pH 7.6. To prepare 1 L of this buffer, dissolve 21.7 g $K_2HPO_4 \cdot 3H_2O$ (or 16.7 g K_2HPO_4), 4.1 g KH_2PO_4, 100 g sucrose, 1.5 g bovine serum albumin fraction V, 20 g polyvinylpyrrolidone (PVP-40), and 5.3 g ascorbic acid in water to 1 L. Add ascorbic acid just before use. After adding ascorbic acid, adjust pH to 7.6 with 1 M NaOH. The stock buffer without ascorbic acid can be stored at 4°C or −20°C.
3. Tris-sucrose (TS) buffer: 20 mM Tris–HCl, 10% sucrose, 50 mM EDTA, pH 8.
4. Refrigerated centrifuge.
5. 50 mL centrifuge tubes.
6. Cheesecloth.
7. Mortar and pestle.
8. Ice bucket.

2.2. Preparation of PFGE Blocks

1. 2× TES buffer: 0.2 M Tris–HCl, 0.2 M NaCl, 20 mM EDTA, pH 8.
2. Lysis buffer: 0.5 M EDTA, 1% sodium dodecyl sulfate (SDS), pH 8.
3. Proteinase K solution: 20 mg/mL in water.
4. Storage buffer: 0.5 M EDTA, 1% *N*-lauroylsarcosine, pH 8.
5. Agarose certified for preparing chromosomal DNA blocks prior to PFGE (e.g., InCert agarose, FMC Bioproducts): 1% in 2× TES buffer.
6. Plastic moulds for casting PFGE blocks with a small plastic tab for extruding blocks from moulds. These can be purchased commercially from PFGE apparatus manufacturers (see Note 1).
7. Screw cap glass vials.
8. Water baths.

2.3. Pulsed-Field Gel Electrophoresis

1. TE buffer: 10 mM Tris–HCl, 1 mM EDTA, pH 8.
2. Agarose for the preparation of PFGE gel (e.g., SeaPlaque GTG agarose, FMC Bioproducts): 1% in 0.5× TBE buffer.
3. 5× Tris-borate EDTA (TBE) electrophoresis buffer: 450 mM Tris base, 450 mM boric acid, 10 mM EDTA, pH 8.
4. PFGE apparatus, e.g., CHEF-DR III Variable Angle System or CHEF Mapper XA System (Bio-Rad Laboratories). These systems are equipped with essentially identical electrophoresis chambers, chiller systems, variable speed pumps, combs, casting stands, etc.
5. DNA size markers: *Hansenula wingei* chromosomes (Bio-Rad Laboratories), *Saccharomyces cerevisiae* chromosomes (Bio-Rad Laboratories or Pharmacia Biotech), and concatemeric lambda DNA ladder (Bio-Rad Laboratories).
6. Ethidium bromide (EtBr) stock solution: 10 mg/mL in water.
7. UV light box equipped with camera or image-capture system.

3. Methods

3.1. Chromosomal DNA Isolation

1. Select approximately 5.0 g of leaf midribs, fresh shoots with leaves, shoot tips, cortical stem tissue including the phloem, or stem phloem preparations from herbaceous host plants with symptoms and high phytoplasma titer (see Note 2). For plants suitable for phloem preparations, phloem is prepared by peeling off 2- to 3-mm-wide strips of cortical stem tissue and removing the fine layer of vascular phloem with a scalpel.
2. Cut specimens into small pieces with a scissor and incubate for 10 min in 15 mL of ice-cold phytoplasma grinding buffer in a mortar on ice.
3. Grind thoroughly with a cold pestle (see Note 3), add another 60 mL of fresh buffer and repeat grinding until the tissue is completely broken up.
4. Transfer the homogenate to two cold 50 mL centrifuge tubes and keep tubes on ice until all samples are ground.
5. Place the tubes in a cold rotor and centrifuge at 1,500 × *g* for 5 min at 4°C. Carefully transfer the supernatant to clean, cold 50 mL centrifuge tubes by filtering through several layers of cheesecloth.
6. Centrifuge the supernatant at 18,000 × *g* for 25 min at 4°C. Carefully discard the supernatant and drain tubes thoroughly.
7. Gently resuspend the pellet in 10 mL of cold TS buffer (see Note 4). Pool the resulting suspensions together in one 50 mL

centrifuge tube, and add additional TS buffer to bring the volume near the top of the centrifuge tube.

8. Repeat centrifugation at 1,500 × *g* for 5 min at 4°C, transfer the supernatant to a clean 50 mL centrifuge tube, and centrifuge at 18,000 × *g* for 25 min at 4°C. Discard the supernatant, briefly drain the tube, and then place it in an ice bucket, being careful that the phytoplasma-enriched pellet is slightly covered with a film of TS buffer.
9. Add 50 μL of cold TS buffer and gently but thoroughly resuspend the phytoplasma-enriched pellet (see Note 5). Transfer the contents to a cold 1.5 mL microcentrifuge tube, and keep the tube on ice until just before preparing the PFGE blocks.

3.2. Preparation of PFGE Blocks

1. Prepare a solution containing 1% InCert Agarose in 2× TES buffer. Heat to approximately 90°C to dissolve agarose, then transfer it to a water bath and keep at 40°C.
2. Warm the phytoplasma-enriched suspension to 37°C for 1–2 min, then quickly add an equal volume of previously melted agarose maintained at 40°C and mix thoroughly using cutoff pipette tips.
3. Rapidly dispense aliquots of the above mix into 50-μL wells of an ice-cold plastic block mould. Leave the mould on ice for 30 min to allow agarose blocks to solidify completely. The resulting rectangular blocks are 2 × 4 × 6 mm in size.
4. Carefully extrude the hardened blocks from the moulds (see Note 6) into a glass vial containing prewarmed lysis buffer. For every 50-μL block, a volume of 500 μL of lysis buffer is used.
5. Add freshly prepared proteinase K solution to the lysis buffer to a final concentration of 2 mg/mL, mix thoroughly, and incubate at 52°C for 3 days. Remove lysis buffer containing proteinase K and replace with fresh lysis buffer and proteinase K solution, two times a day at least. Gently swirl the vial occasionally to keep the blocks from sticking to each other. If blocks are still opaque in the center after 3 days of incubation, incubate for additional 24 h by changing lysis buffer and proteinase K solution (see Note 7). Completely digested blocks should appear clear.
6. Once the blocks are clear, pour off lysis buffer containing proteinase K and replace with storage buffer. Incubate at 52°C for approximately 1 h to equilibrate the blocks with the buffer. The vial containing the blocks can now be stored at 4°C until needed (see Note 8).

3.3. Pulsed-Field Gel Electrophoresis

1. Prior to electrophoresis, wash agarose blocks three times with TE buffer, for at least 30 min per wash.

2. Prepare a solution containing 1% SeaPlaque GTG agarose in 0.5× TBE buffer. Heat to approximately 90°C to dissolve agarose, then transfer it to a water bath and keep at 40°C.
3. Set up the PFGE gel casting stand on a level surface following the manufacturer's instructions. Place agarose blocks, including those containing DNA size markers, on the platform on which the gel is cast using a clean spatula. Press the blocks gently against each tooth of the comb, and pour the SeaPlaque GTG agarose solution to a thickness of approximately 6–7 mm. Allow the gel to solidify at room temperature for at least 30 min. Carefully remove the comb from the gel. The blocks will remain in place when the comb is removed. Seal the wells by adding melted agarose and allow it to solidify (see Note 9).
4. Remove the gel with the underlying platform from the casting stand and place them into the electrophoresis chamber of the PFGE apparatus. The gel should be completely submerged in the buffer.
5. Run the gel employing the following protocol: temperature 14°C, voltage 6 V/cm, pulse parameters 20–120 s, running time 21 h, angle 120°, and buffer 0.5× TBE.
6. Remove the gel with the underlying platform from the electrophoresis chamber. Carefully transfer the gel to a developing tray which contains enough distilled water that the gel can move freely. Add ethidium bromide solution to a final concentration of 1 μg/mL, and shake gently for 1 h.
7. Destain the gel by shaking gently in distilled water for 1–2 h, changing the water several times. Photograph the gel under UV light. The full-length phytoplasma chromosome is resolved as a single, discrete band in the gel (Fig. 1) (see Note 10).

3.4. Southern Blot Hybridization

1. Full-length phytoplasma chromosomes resolved by PFGE are transferred from the agarose gel to a nylon membrane following standard methods (18). To facilitate transfer, DNA is depurinated by treating the gel twice in 0.2 M HCl for 10 min with gentle shaking and then rinsing briefly in distilled water before continuing with denaturation and neutralization steps.
2. Transfer is carried out in 10× SSC buffer (18) using upward blotting for at least 12 h. The transferred DNA can be bound to the membrane by UV cross-linking.
3. Hybridization is carried out according to the protocol of Hoheisel et al. (19) using a radiolabeled ribosomal DNA probe which encompasses the entire 16S rRNA gene, the 16S/23S rDNA spacer region, and the 5′ region of the 23S rRNA gene of phytoplasmas (3, 10). An example of Southern blot hybridization of PFGE gel is shown in Fig. 2 (see Note 11).

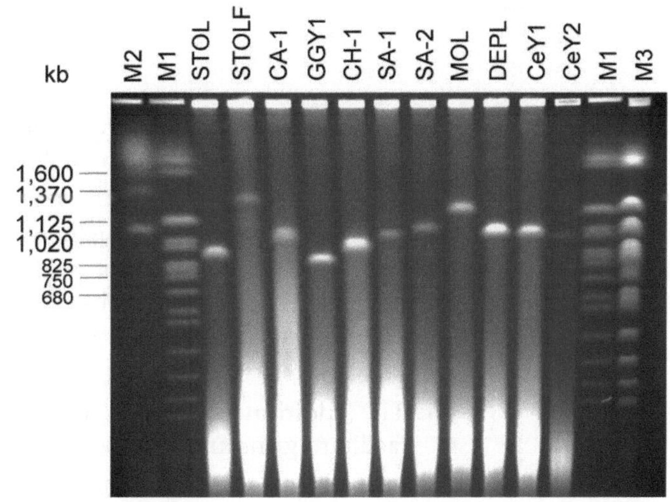

Fig. 1. Full-length chromosomes of phytoplasmas belonging to stolbur or 16SrXII group isolated by pulsed-field gel electrophoresis from the experimental host *Catharanthus roseus* (periwinkle) (STOL through DEPL) and naturally infected field-collected plants of celery (CeY1 and CeY2). DNA size markers were: *M2 Hansenula wingei* chromosomes (Bio-Rad Laboratories); *M1 Saccharomyces cerevisiae* chromosomes (Bio-Rad Laboratories); *M3 S. cerevisiae* chromosomes (Pharmacia Biotech). Smears suggestive of fragmented DNA in the lower portion of the lanes are visible.

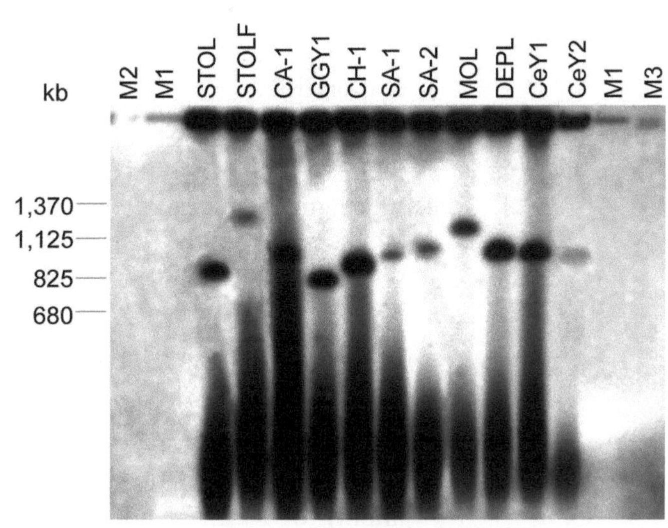

Fig. 2. Southern blot hybridization of phytoplasma chromosomes shown in Fig. 1, probed with a ribosomal DNA fragment which encompasses the entire 16S rRNA gene, the 16S/23S rDNA spacer region, and the 5′ region of the 23S rRNA gene of phytoplasmas.

4. Notes

1. Block moulds can also be made in the laboratory. Any plastic or glass item into which melted agarose can be poured can serve as a block mould.
2. The best results are usually obtained when plants are fully symptomatic but do not show signs of decline. Periwinkle flowers, instead of midribs, turned out to be the best phytoplasma source for isolation of '*Candidatus* Phytoplasma australiense' chromosomes (15).
3. The addition of a small amount of glass beads or sand will facilitate grinding.
4. This can be accomplished using a loose-fitting homogenizer such as a plastic transfer pipette bulb.
5. To minimize damage to phytoplasma chromosomes by mechanical shearing, use cutoff or large-bore pipette tips. Also, avoid repeatedly pipetting up and down when dissolving pellet and do not mix vigorously. Gentle disruptions of clumps can be accomplished using a plastic transfer pipette bulb as homogenizer.
6. Plastic block moulds purchased commercially come with a small plastic tab designed for pushing blocks out of the moulds. However, a clean spatula will suffice.
7. It is suggested to seal glass vials with Parafilm to prevent evaporation.
8. Blocks can be stored for many months or even years in storage buffer at 4°C in a glass vial sealed with Parafilm.
9. Alternatively, loading of samples can be done by inserting blocks into wells after the comb is removed, and then sealing the wells containing blocks with melted agarose. For ease of loading, the size of wells should be slightly larger than that of blocks.
10. Full-length phytoplasma chromosome bands are obtained without using γ-irradiation presumably by using physical shearing, endogenous nuclease activity, or both during sample preparation to obtain linearized chromosomes. In contrast to the finding of Neimark and Kirkpatrick (2), recent studies have shown that γ-irradiation of the agarose blocks did not improve the yield of linearized chromosomes but increased markedly the nonspecific DNA background in the gel (3, 8, 9). Also, the occurrence of linear phytoplasma chromosomes has recently been reported (16).
11. Previous studies showed that for some phytoplasma isolates, more than one band was observed in the PFGE gel. The simplest

explanation for the occurrence of multiple banding patterns is that the plants from which the samples were prepared were infected with distinct phytoplasmas containing different-sized chromosomes (2, 3). In this case, Southern blot hybridization using suitable phytoplasma probes would help to elucidate the nature of additional bands.

References

1. Nassonova ES (2008) Pulsed field gel electrophoresis: theory, instruments and application. Cell Tissue Biol 2:557–565
2. Neimark H, Kirkpatrick BC (1993) Isolation and characterization of full-length chromosomes from non-culturable plant-pathogenic *Mycoplasma-like* organisms. Mol Microbiol 7:21–28
3. Marcone C et al (1999) Chromosome sizes of phytoplasmas composing major phylogenetic groups and subgroups. Phytopathology 89: 805–810
4. Marcone C et al (2001) Updating and extending genetic characterization and classification of phytoplasmas from wild and cultivated plants in southern Italy. J Plant Pathol 83: 133–138
5. Griffiths HM et al (1999) The phytoplasma associated with ash yellows and lilac witches'-broom: '*Candidatus* Phytoplasma fraxini'. Int J Syst Bacteriol 49:1605–1614
6. Seemüller E, Schneider B (2007) Differences in virulence and genomic features of strains of '*Candidatus* Phytoplasma mali', the apple proliferation agent. Phytopathology 97: 964–970
7. Firrao G, Smart CD, Kirkpatrick BC (1996) Physical map of the Western X-disease phytoplasma chromosome. J Bacteriol 178: 3985–3988
8. Lauer U, Seemüller E (2000) Physical map of the chromosome of the apple proliferation phytoplasma. J Bacteriol 182:1415–1418
9. Padovan AC et al (2000) Chromosome mapping of the sweet potato little leaf phytoplasma reveals genome heterogeneity within the phytoplasmas. Microbiology 146:893–902
10. Marcone C, Seemüller E (2001) A chromosome map of the European stone fruit yellows phytoplasma. Microbiology 147:1213–1221
11. Malembic-Maher S et al (2008) A chromosome map of the Flavescence dorée phytoplasma. Microbiology 154:1454–1463
12. Liefting LW, Kirkpatrick BC (2003) Cosmid cloning and sample sequencing of the genome of the uncultivable mollicute, Western X-disease phytoplasma, using DNA purified by pulsed-field gel electrophoresis. FEMS Microbiol Lett 221:203–211
13. Oshima K et al (2004) Reductive evolution suggested from the complete genome sequence of a plant-pathogenic phytoplasma. Nat Genet 36:27–29
14. Bai X et al (2006) Living with genome instability: the adaptation of phytoplasmas to diverse environments of their insect and plant hosts. J Bacteriol 188:3682–3696
15. Tran-Nguyen LTT et al (2008) Comparative genome analysis of "*Candidatus* Phytoplasma australiense" (subgroup *tuf*-Australia I; *rp*-A) and "*Ca.* Phytoplasma asteris" strains OY-M and AY-WB. J Bacteriol 190:3979–3991
16. Kube M et al (2008) The linear chromosome of the plant-pathogenic mycoplasma '*Candidatus* Phytoplasma mali'. BMC Genomics 9:306
17. Lauer U, Seemüller E (1998) Physical map of the apple proliferation phytoplasma genome. 12th International Organisation of Mycoplasmology Conference, 22–28 July, 1998, Sydney, Australia, p 155–156
18. Sambrook J, Fritsch EF, Maniatis T (1989) Molecular cloning: a laboratory manual, 2nd edn. Cold Spring Harbor Laboratory Press, New York
19. Hoheisel JD et al (1991) Use of high coverage reference libraries of *Drosophila melanogaster* for relational data analysis. A step towards mapping and sequencing of the genome. J Mol Biol 220:903–914

Chapter 34

Mapping the Phytoplasma Chromosome

Sylvie Malembic-Maher and Patricia Carle

Abstract

Physical and genetic mapping of the phytoplasma chromosome can be a useful tool in a genome sequencing project in order to assemble the *in silico*-predicted contigs robustly. Mapping consists of four distinct steps: preparation of phytoplasma chromosomes from infected plants, single- and double-digestion of chromosomes with rare-cutting restriction enzymes, separation of large DNA fragments by pulsed-field gel electrophoresis, and hybridization with various genetic markers. Materials and methods needed for each step are described and the technique is illustrated using the flavescence dorée phytoplasma genome map as an example.

Key words: Flavescence dorée phytoplasma, Genetic map, *Mollicutes*, Physical map, Pulsed-field gel electrophoresis, Restriction endonucleases, Southern blot

1. Introduction

The analysis and comparison of phytoplasma genomes has enabled important progress to be made in the knowledge of phytoplasma biology, as for example the identification of molecular determinants of phytopathogenicity (see ref. (1, 2) for reviews). During a genome sequencing project, physical and genetic mapping of the phytoplasma chromosome can be helpful in order to align and orientate the genome sections predicted *in silico*. Physical and genetic maps of phytoplasma chromosomes have been determined for '*Ca. Phytoplasma pruni*' (3), sweet potato little leaf phytoplasma (4), '*Ca. Phytoplasma mali*' (5), '*Ca. Phytoplasma prunorum*' (6), and the flavescence dorée (FD) phytoplasma (7). The methods used in these studies share similarities that can be divided into four different stages: (1) the enrichment of phytoplasma chromosomes from infected plant material and their inclusion in agarose blocks (4); (2) the single- and double-digestion of DNA in the blocks by rare-cutting restriction enzymes (8); (3) the separation of the large

Matt Dickinson and Jennifer Hodgetts (eds.), *Phytoplasma: Methods and Protocols*, Methods in Molecular Biology, vol. 938, DOI 10.1007/978-1-62703-089-2_34, © Springer Science+Business Media, LLC 2013

DNA molecules by pulsed-field gel electrophoresis (PFGE) (9, 10); and (4) Southern blotting followed by hybridization with selected genetic markers (11).

Other techniques, called optical mapping (12), have been developed for the mapping of bacterial chromosomes (13). These avoid the PFGE step, need only small amounts of DNA, and preserve the order of DNA fragments. Nevertheless, optical mapping is unsuitable for use on uncultivated phytoplasma because of the need for highly purified chromosomes. In this chapter, we describe the detailed materials and methods of the four steps we performed in order to obtain a physical and genetic map of the flavescence dorée phytoplasma chromosome (7).

2. Materials

2.1. Preparation of Chromosomes

Prepare all solutions with ultrapure water, filter with 0.45 μm filters, keep sterile, and store at 4°C.

1. 0.5 M Na_2HPO_4.
2. 0.3 M NaH_2PO_4.
3. 1 M Tris–HCl pH 8.0.
4. 1 M NaCl.
5. 20 % (w/v) polyvinylpyrrolidone (PVP) (MW 40,000), dissolved at 55°C.
6. 0.5 M EDTA pH 8.0.
7. Bovine serum albumin (BSA), 5 %w/v.
8. 10 % (w/v) *N*-lauroylsarcosine.
9. Isolation buffer: 0.1 M Na_2HPO_4, 0.03 M NaH_2PO_4, 10 % (w/v) sucrose, 2 % (w/v) PVP, 10 mM EDTA, pH 7.6, with 0.15 % (w/v) BSA and 1 mM *iso*-ascorbic acid added fresh.
10. Suspension buffer or TSE: 20 mM Tris–HCl, 10 % (w/v) sucrose, 50 mM EDTA, pH 8.0.
11. 2× TES buffer: 0.2 M Tris–HCl, 0.2 M NaCl, 20 mM EDTA, pH 8.0.
12. Lysis buffer: 1 % (w/v) *N*-lauroylsarcosine, 0.5 M EDTA, pH 8.0, 1 mg/mL proteinase K.
13. Agarose low melting point (e.g., from Bio-Rad).
14. Ball-crushing apparatus (BIOREBA AG) with polyethylene bags 12 × 12 cm with nylon mesh C010.
15. 30 mL Corex® II tubes.
16. High performance centrifuge (e.g., Avanti® J-E with JA20 rotor).

17. Sterile cheesecloth.
18. Sterile glass micro-pestles.
19. Reusable plug moulds for 100 μL agarose plugs (e.g., from Bio-Rad).

2.2. Restriction Endonuclease Digestion

Prepare all solutions with ultrapure water, sterilize through a 0.45 μm filter.

1. 0.1 M phenylmethylsulphonyl fluoride (PMSF) in isopropanol, store at –20°C.
2. 10 mM dithiothreitol (DTT), store at –20°C.
3. 1 M spermidine, store at –20°C.
4. 10 mg/mL BSA, store at –20°C.
5. 1× TE buffer: 10 mM Tris–HCl, 1 mM EDTA, pH 8.0.
6. Digestion buffer: 4 mM DTT, 0.1 mg/mL BSA, 4 mM spermidine, and 1× restriction enzyme buffer supplied by the manufacturer, stored at –20°C.
7. 0.5 M EDTA, pH 8.0.
8. Rare-cutting restriction endonucleases, e.g., *Apa*I, *Bss*HII, *Eag*I, I-*Ceu*I, *Kpn*I, *Mlu*I, *Pvu*II, *Sal*I, *Sma*I, and *Xho*I with their respective 10× digestion buffers.
9. Small spatulas.
10. Sterile razor blades.

2.3 Pulsed-Field Gel Electrophoresis

1. 10× TBE (Bio-Rad).
2. Pulsed-field certified agarose (Bio-Rad).
3. Yeast chromosome PFG (*Saccharomyces cerevisiae* YPH80) and low range PFG size markers (mixture of lambda DNA-*Hin*dIII fragments and concatemers, New England Biolabs).
4. CHEF-DR® III Pulsed-Field Electrophoresis System with cooling module and variable speed pump (Bio-Rad).
5. Standard casting stand (14 × 13 cm frame and platform, Bio-Rad).
6. 15-well comb, 14 cm wide, 1.5 mm thick, adjustable height (Bio-Rad).
7. Quantity One® basic software (Bio-Rad).
8. Sterile razor blades.
9. Small spatulas.
10. Ethidium bromide solution, 2 μg/mL.
11. UV transilluminator with camera.

2.4. Southern Blot and Hybridization

Prepare all solutions with deionized water, sterilize in an autoclave, and store at room temperature.

2.4.1. Transfer to Solid Support

1. Depurination solution: 0.25 M HCl.
2. Denaturation buffer: 0.5 M NaOH, 1.5 M NaCl.
3. Neutralization buffer: 1 M Tris–HCl, 1.5 M NaCl, pH 7.6.
4. Transfer buffer 10× SSC: 1.5 M NaCl, 0.15 M tri-sodium citrate.
5. Nylon N+ membrane (Nytran® Super Charge, Schleicher and Schuell).
6. Plastic or glass tray.
7. Two glass plates.
8. Flat-topped shaker.
9. Plastic wrap.
10. Whatman 3MM paper.
11. Dry paper towels.
12. Blunt pincer.
13. Oven at 80°C.

2.4.2. Probe Labeling

1. 100 ng/μL DNA from a phytoplasma-infected plant, store at −20°C.
2. 100 μM PCR primers, store at −20°C.
3. 25 mM $MgCl_2$, store at −20°C.
4. 5 mM dNTPs, store at −20°C.
5. 5 units/μL Taq DNA polymerase, store at −20°C.
6. 1 mM digoxigenin-dUTP (DIG-11-dUTP alkali-labile, Roche), store at −20°C.
7. Sterile deionized water.
8. Microfuge tubes (0.5 mL volume).
9. Thermal cycler.

2.4.3. Hybridization

1. Purified DNA probe labeled with DIG-11-dUTP, alkali-labile by PCR amplification, store at −20°C.
2. Salmon sperm DNA, 10 mg/mL, store at −20°C.
3. 20× SSC buffer: 3 M NaCl, 0.3 M tri-sodium citrate.
4. Sodium dodecyl sulphate (SDS) solution, 10 % (w/v) in deionized water. Store at room temperature because SDS precipitates at 4°C.
5. Hybridization buffer: 5× SSC, 0.1 % (w/v) *N*-lauroylsarcosine, 0.02 % SDS (v/v), 50 % (v/v) formamide, 2 % (w/v) blocking reagent (Roche). Store at room temperature.

6. Sealable plastic bags.
7. Bag sealer.
8. Water bath at 100°C and then at 50°C.
9. Incubator at 37°C.

2.4.4. Chemiluminescence Detection

1. Maleic acid buffer: 0.1 M maleic acid, 0.15 M NaCl, pH 7.5.
2. Washing buffer: 0.1 M maleic acid, 0.15 M NaCl, pH 7.5, 0.3 % (v/v) Tween 20.
3. Detection buffer: 0.1 M Tris–HCl, 0.1 M NaCl, pH 9.5.
4. 10× blocking stock solution: blocking reagent 10 % (w/v) in maleic acid buffer, stirred constantly on a heating block at 65°C. The solution remains opaque. Store at −20°C. Dilute the blocking solution in maleic acid buffer.
5. Antidigoxigenin-AP FAb fragment labeled with alkaline phosphatase (Roche). Dilute antidigoxigenin-AP 1:10,000 in blocking solution.
6. CDP-*Star*® substrate (Roche). The CDP-*Star*® working solution is prepared by diluting CDP-*Star*® 1:100 in detection buffer.
7. Plastic or glass tray.
8. Flat-topped rocking shaker.
9. Incubator at 37°C.
10. X-ray film (Amersham Hyperfilm ECL, GE Healthcare).
11. Film cassette for X-ray film (Kodak).
12. Film developer (Kodak GBX, Sigma-Aldrich).
13. Film fixer (Kodak GBX, Sigma-Aldrich).

3. Methods

3.1. Preparation of Chromosomes

All steps must be performed at 4°C.

1. Collect the top 15–20 cm of infected plants and dissect midribs, petioles, and stems.
2. Grind 2 g of fresh material in 10 mL isolation buffer (see Note 1).
3. Homogenize the extract, centrifuge at 1,500 × *g* for 5 min, and filter the supernatant through multiple layers of cheesecloth in order to remove the debris.
4. Centrifuge at 18,000 × *g* for 30 min and gently resuspend the pellet in 10 mL TSE (see Note 2).
5. Repeat the low and high speed centrifugation and filtration steps and resuspend the final pellet in TSE in order to obtain 50 μL of final phytoplasma suspension (see Note 3).

6. Heat the suspension at 55°C and gently mix it with an equal volume of molten 2 % (w/v) pulsed-field certified agarose dissolved in 2× TES (and maintained at 55°C).
7. Pour 100 μL in the plastic moulds, avoiding air bubbles, and let the agarose solidify for 15 min at 4°C.
8. Transfer the blocks into lysis buffer (1 mL per block) and incubate for 60 h at 50°C with 4 buffer changes, then for 12 h in lysis buffer without proteinase K. Store at 4°C in 0.5 M EDTA pH 8.0. Perform all the same procedures with healthy control plants.

3.2. Restriction Endonuclease Digestion

Endonuclease digestion is carried out in agarose blocks prepared from infected plants, with healthy plants as a negative control. Half a block provides enough material for electrophoresis.

1. After two rinses of 2 h with 10 volumes of 1× TE, incubate the blocks overnight at 4°C in 10 volumes of 1× TE plus 1 mM PMSF in order to inactivate the proteinase K, then rinse again with 10 volumes of 1× TE, four times for 2 h at 4°C.
2. Incubate each block separately for 1 h at 4°C in 5 volumes of digestion buffer then transfer to 2 volumes of digestion buffer plus 30 units of the restriction enzyme and leave overnight at the temperature recommended by the manufacturer.
3. For double-digestions, inactivate the first enzyme by incubating the blocks for 2 h at 4°C in 10 volumes of 0.5 M EDTA pH 8.0, then rinse three times and repeat the digestion procedure with the second enzyme. Perform single- and double-digestions separately.

3.3. Pulsed-Field Gel Electrophoresis

PFGE is performed using the clamped homogenous electrical field (CHEF) technique with the CHEF-DR ®-III system (Bio-Rad).

1. Place the 0.5× TBE in a cold chamber in order to cool the buffer to between 10 and 15°C.
2. Melt 150 mL of 0.8 % (w/v) pulsed-field certified agarose in 0.5× TBE with a microwave.
3. Cast approximately 100 mL of agarose solution in the casting stand with the comb (see Note 4).
4. Let it solidify for 1 h; keep the rest of the agarose at 65°C.
5. Gently remove the comb and insert the agarose blocks into the wells using small spatulas.
6. Place the blocks containing the size markers in the first and the last wells. Avoid making bubbles between the gel and the blocks.
7. Fill the wells with molten agarose.
8. Release the gel from the mould by removing the frame, and let it lie on the lower black plate. Remove any surplus of agarose which

may be under the plate and place the gel in the adapted frame in the center of the electrophoresis chamber (see Note 5).

9. Slowly fill the chamber with 2.2 L of cold 0.5× TBE buffer.
10. Turn on the power, adjust the cooling system to 14°C, and activate the pump for circulation of the buffer. Electrophoresis is performed at 6 V/cm, with an angle of 120° for 18 h and various ramped pulse times: 0.7–13 s for low size range resolution, 2–40 s and 2–60 s for medium size range resolution, and 5–90 s for high size range resolution (see Note 6).
11. After the electrophoresis, clean the electrophoresis chamber with deionized water three times.
12. Incubate the gel for 15 min in ethidium bromide solution, and destain by washing for 1 h in distilled water before visualization under UV light. Take a picture of every gel with a ruler positioned in the same way. Figure 1a shows an example of PFGE performed on digested chromosomes of the flavescence dorée phytoplasma.

3.4. Southern Blot Hybridization

All the procedures are carried out at room temperature.

3.4.1. DNA Transfer

1. Place the gel in a tray on the rocking shaker. The gel should be completely covered by approximately 100 mL of the successive buffers and under gentle rocking throughout the different procedures. Rinse the gel in sterile deionized water between each step.

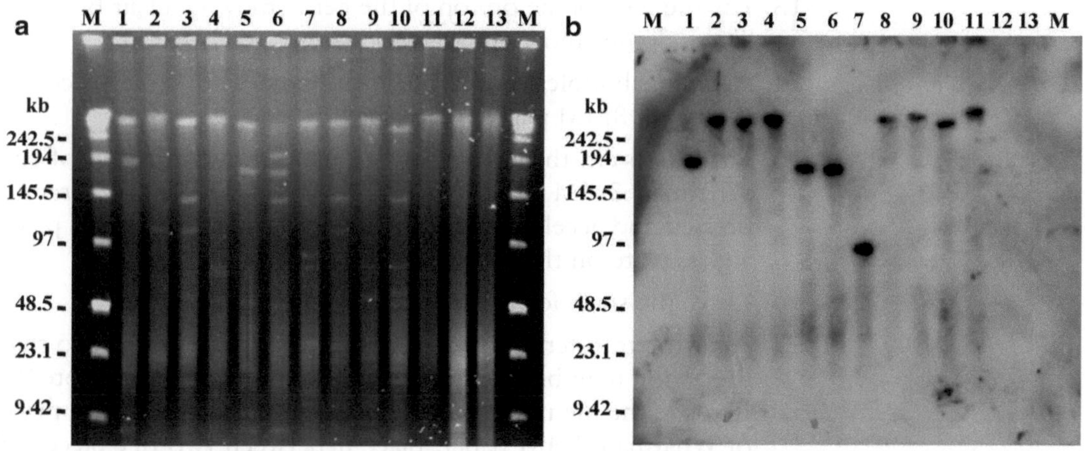

Fig. 1. (**a**) PFGE of DNA prepared from flavescence dorée phytoplasma-infected (lanes 1–11) and healthy (lanes 12, 13) broad beans, single- or double-digested with various rare-cutting restriction enzymes. Lane 1, *Sal*I; 2, *Bss*HII; 3, *Eag*I; 4, *Mlu*I; 5, *Bss*HII + *Sal*I; 6, *Eag*I + *Sal*I; 7, *Mlu*I + *Sal*I; 8, *Eag*I + *Bss*HII; 9, *Mlu*I + *Bss*HII; 10, *Mlu*I + *Eag*I; 11, undigested; 12, *Sal*I; 13, *Bss*HII; M, low range PFG marker (kb). (**b**) Southern blot hybridization profile of the same gel probed with the *uvrB-degV* marker.

2. Perform the depurination by incubating the gel twice for 10 min with the depurination solution.
3. Perform denaturation with denaturation buffer twice for 20 min.
4. Neutralization is then carried out first for 30 min and then for 15 min.
5. After neutralization, discard the solution and rinse the gel in 10× SSC.
6. Place the gel on a piece of plastic wrap and accurately measure the length and the width of the gel. Put the gel back into 10× SSC buffer.
7. Cut one piece of Nylon N+ membrane filter to the dimensions of the gel. Place the membrane in a tray and wet it with 10× SSC buffer for 10 min.
8. Cut three sheets of Whatman 3MM paper into pieces of the same dimensions of the gel.
9. Prepare a wick by cutting one piece of Whatman 3MM paper about 2 cm wider than the width of the gel and 30–40 cm long.
10. Place 500 mL of 10× SSC in a large tray.
11. Put a glass plate over the tray and place the wick on the plate with both ends hanging over into 10× SSC. Wet the wick thoroughly with 10× SSC. Remove air bubbles trapped between the wick and the glass plate.
12. Lay the gel on top of the wick and remove air bubbles between the gel and the paper.
13. Lay the membrane on top of the gel, removing all air bubbles. Then, do not move the transfer system (see Note 7).
14. Avoiding bubbles, cover the membrane with three pieces of Whatman 3MM paper (see Note 8).
15. Put a 5–6 cm-thick stack of dry paper towels on top of the Whatman 3MM paper. In order to keep the entire pyramid pressed and accelerate the capillary movement of buffer, place a glass plate on the top with a weight (~300 g) on top.
16. Leave the stack for the transfer overnight.
17. After the transfer is completed, take the pyramid apart so that the Nylon membrane filter is still lying on the gel (see Note 9). Remove the membrane from the gel and dry it on a fresh piece of Whatman 3MM paper, place it between two dry pieces of Whatman 3MM paper and bake for 2 h at 80°C.
18. Store the membranes wrapped in plastic film at 4°C until needed.

3.4.2. Probe Labeling

Probe labeling is performed using the DIG DNA Labeling and Detection Kit (Roche). The probe is labeled with DIG-11-dUTP, alkali-labile by PCR amplification. Nested PCR should be performed in order to obtain efficient labeling, adding digoxigenin solely in the second PCR. We describe here only the second PCR:

1. Set up a 50 μL reaction mixture in a 0.5 mL microfuge tube. The mixture consists of 1 μL of 100 ng/μL DNA of phytoplasma-infected plant with 1 μM of each PCR primer, 2 mM $MgCl_2$, 200 μM dNTPs, 20 μM DIG-11-dUTP, and 0.04 unit/μL *Taq* polymerase in a final volume of 50 μL. A reaction without DIG-11-dUTP must be performed as a negative control.
2. The PCR conditions employed for the synthesis of the probes depend on the size and the G+C content of the amplified fragment. In our case, after 5 min denaturation at 95°C, we perform 35 cycles using the following temperature profile: denaturation 94°C for 30 s; primer annealing 52°C for 30 s; and primer extension 66°C for 60 s. Cycling was concluded with a final extension at 66°C for 5 min.
3. Confirm the successful DNA amplification and incorporation of DIG-11-dUTP by electrophoresis on an agarose gel. The size of the DIG-11-dUTP-amplified fragment must be higher than the one without DIG-11-dUTP.

3.4.3. Hybridization

1. Place the baked membrane filter in a sealable bag. Measure the surface of the membrane. Add 6–10 mL of hybridization solution (depending on the size of the filter, 1 mL per 10 cm^2) and 200 μg/mL denatured salmon sperm boiled 5 min in water (see Note 10).
2. Seal the bag, avoiding bubbles, and then prehybridize for 1 h 30 min at 37°C. During that time place the hybridization buffer at 37°C.
3. At the end of prehybridization, cut a corner of the plastic bag and remove the solution.
4. Add fresh hybridization solution (1 mL per 15 cm^2).
5. Denature 200 μg/mL salmon sperm DNA (in one tube) and about 20 ng/μL of purified labeled probes (1 μL/mL of hybridization buffer used in step 2, in a second tube) by boiling for 5 min in water, and then place the two tubes immediately on ice.
6. Add the denatured salmon sperm and the labeled probe solutions into the sealable bag. Spread the solution over the filter as above and seal the bag without bubbles.
7. Incubate at 37°C overnight.

8. The next day, wash the filter in a plastic bag four times for 15 min in 200 mL of 2× SSC + 0.5 % SDS at room temperature with shaking.
9. Wash the filter in a plastic bag twice for 30 min in 200 mL 0.1× SSC + 0.1 % SDS at 50°C with shaking.

3.4.4 Chemiluminescence Detection

All the procedures are carried out at room temperature with gentle rocking.

1. At the end of the washes, place the filter in a tray and rinse for 2 min in 100 mL of washing buffer.
2. Incubate the membrane for 30 min in 100 mL of freshly prepared 1× blocking solution.
3. Incubate the membrane for 30 min in 20 mL of antidigoxigenin-AP (diluted 1:10,000).
4. Wash the membrane twice for 15 min in 100 mL of washing buffer.
5. Equilibrate the membrane for 5 min in 20 mL of detection buffer.
6. Place the membrane filter with the DNA side facing up, on a development folder or hybridization bag and apply 1–2 mL CDP-*Star*® working solution (see Note 11). Immediately cover the membrane with the second sheet of the folder to distribute the substrate evenly and without air bubbles over the membrane.
7. Incubate the membrane for 5 min at 37°C.
8. Squeeze out excess liquid using absorbent paper and seal the edges of the development folder (see Note 12).
9. For the autoradiography, the membrane is exposed to X-ray film in a film cassette. In a dark room, place the film on the membrane. Multiple time exposures (from few seconds to 1 h) can be performed to achieve the desired signal strength.
10. The film is developed for 30 s in the developer solution, rinsed for 2 min in water, and then placed for 30 s in the fixer solution. At the end, the film is rinsed in water for 30 s and then air-dried. Figure 1b shows an example of a Southern blot hybridization performed with a *uvrB-degV* probe after PFGE of the digested flavescence dorée phytoplasma chromosome.

3.5. Data Analysis

The molecular weight of the restriction fragments is estimated by comparison to the appropriate marker using the Quantity One® software. The mean size of each fragment is calculated from at least three different PFGEs. Each fragment must be confirmed to be phytoplasma DNA by comparing it with healthy plant DNA prepared under the same conditions and by Southern blot hybridization with the appropriate probe. Perform gels with undigested DNA in order to estimate the size of the chromosome. Calculate the sum of the fragment sizes for each digestion and compare it with the size of the chromosome.

The construction of the physical and genetic map of the phytoplasma chromosome is then possible by combining the results of the single- and double-digestions with different enzymes and those of the Southern blot hybridizations.

4. Notes

1. One floresence dorée-infected broad bean gives approximately 2 g of material which is enough to make one agarose block.
2. Resuspension of pellets must be performed very carefully with truncated cones or pipettes; a micro-pestle can be used to gently crush the aggregates.
3. Take into account the volume of the pellet for the final volume of resuspension.
4. Be sure that the casting stand is on a flat surface, eliminate bubbles which can be trapped under the plate by gently pressing it with a tip.
5. Make a joint with some molten agarose between the frame and the gel in order to avoid the gel floating in the buffer.
6. Only start the electrophoresis when the temperature is 14°C and the current is between 110 and 150 mA. Regularly eliminate bubbles in the tubing during electrophoresis to ensure good circulation of the buffer.
7. With a pencil, trace a line on the membrane to mark the lower position of the wells of the gel in order to know the exact position of the wells after hybridization.
8. Make sure that the pieces of Whatman 3MM paper have the same dimensions as the membrane.
9. With scissors, make a little notch on the filter in the upper right part to indicate the correct orientation of the gel.
10. The hybridization solution should be added to the bottom of the bag. The filter is wetted by moving the bag back and forth across the edge of a bench.
11. Be careful: CDP-*Star*® is very sensitive to light.
12. Be careful: never touch the membrane with fingers because fingerprints will make marks.

Acknowledgements

This research was funded by the French National Institute of Agronomical Research (INRA) and by the Interprofessional Council of Bordeaux Wine (CIVB). We thank our colleagues Pascal Salar and Géraldine Gourgues for elaborating the protocols for phytoplasma chromosome isolation and for PFGE.

References

1. Hogenhout SA, Šeruga MM (2010) Phytoplasma genomics, from sequencing to comparative and functional genomics—what have we learnt? In: Weintraub PG, Jones P (eds) Phytoplasmas: genomes, plant hosts and vectors. CABI Publishing, Oxfordshire, pp 19–36
2. Sugio A et al (2011) Diverse targets of phytoplasma effectors: from plant development to defense against insects. Annu Rev Phytopathol 49:175–195
3. Firrao G, Smart CD, Kirkpatrick BC (1996) Physical map of the Western X-disease phytoplasma chromosome. J Bacteriol 178:3985–3988
4. Padovan AC et al (2000) Chromosome mapping of the sweet potato little leaf phytoplasma reveals genome heterogeneity within the phytoplasmas. Microbiology 146:893–902
5. Lauer U, Seemüller E (2000) Physical map of the chromosome of the apple proliferation phytoplasma. J Bacteriol 182:1415–1418
6. Marcone C, Seemüller E (2001) A chromosome map of the European stone fruit yellows phytoplasma. Microbiology 147:1213–1221
7. Malembic-Maher S et al (2008) A chromosome map of the Flavescence dorée phytoplasma. Microbiology 154:1454–1463
8. Bautsch W (1988) Rapid physical mapping of the *Mycoplasma mobile* genome by two-dimensional field inversion gel electrophoresis techniques. Nucl Acids Res 16:11461–11467
9. Herschleb J, Ananiev G, Schwartz DC (2007) Pulsed-field gel electrophoresis. Nat Protoc 2:677–684
10. Schwartz DC, Cantor CR (1984) Separation of yeast chromosome-sized DNAs by pulsed field gradient gel electrophoresis. Cell 37:67–75
11. Sambrook J, Fritsch EF, Maniatis T (1989) Molecular cloning: a laboratory manual, 2nd edn. Cold Spring Harbor Laboratory Press, New York
12. Schwartz DC et al (1993) Ordered restriction maps of *Saccharomyces cerevisiae* chromosomes constructed by optical mapping. Science 262:110–114
13. Latreille P et al (2007) Optical mapping as a routine tool for bacterial genome sequence finishing. BMC Genomics 8:321

INDEX

Matt Dickinson and Jennifer Hodgetts (eds.), *Phytoplasma: Methods and Protocols*, Methods in Molecular Biology, vol. 938, DOI 10.1007/978-1-62703-089-2, © Springer Science+Business Media, LLC 2013

Q

R

S

T

V

W

Z

MIX
Papier aus verantwortungsvollen Quellen
Paper from responsible sources
FSC® C105338

If you have any concerns about our products,
you can contact us on
ProductSafety@springernature.com

In case Publisher is established outside the EU,
the EU authorized representative is:
Springer Nature Customer Service Center GmbH
Europaplatz 3, 69115 Heidelberg, Germany

Printed by Libri Plureos GmbH
in Hamburg, Germany